MULTISTAGE SELECTION AND RANKING PROCEDURES

STATISTICS: Textbooks and Monographs

A Series Edited by

D. B. Owen, Founding Editor, 1972–1991

1. The Generalized Jackknife Statistic, *H. L. Gray and W. R. Schucany*
2. Multivariate Analysis, *Anant M. Kshirsagar*
3. Statistics and Society, *Walter T. Federer*
4. Multivariate Analysis: A Selected and Abstracted Bibliography, 1957–1972, *Kocherlakota Subrahmaniam and Kathleen Subrahmaniam*
5. Design of Experiments: A Realistic Approach, *Virgil L. Anderson and Robert A. McLean*
6. Statistical and Mathematical Aspects of Pollution Problems, *John W. Pratt*
7. Introduction to Probability and Statistics (in two parts), Part I: Probability; Part II: Statistics, *Narayan C. Giri*
8. Statistical Theory of the Analysis of Experimental Designs, *J. Ogawa*
9. Statistical Techniques in Simulation (in two parts), *Jack P. C. Kleijnen*
10. Data Quality Control and Editing, *Joseph I. Naus*
11. Cost of Living Index Numbers: Practice, Precision, and Theory, *Kali S. Banerjee*
12. Weighing Designs: For Chemistry, Medicine, Economics, Operations Research, Statistics, *Kali S. Banerjee*
13. The Search for Oil: Some Statistical Methods and Techniques, *edited by D. B. Owen*
14. Sample Size Choice: Charts for Experiments with Linear Models, *Robert E. Odeh and Martin Fox*
15. Statistical Methods for Engineers and Scientists, *Robert M. Bethea, Benjamin S. Duran, and Thomas L. Boullion*
16. Statistical Quality Control Methods, *Irving W. Burr*
17. On the History of Statistics and Probability, *edited by D. B. Owen*
18. Econometrics, *Peter Schmidt*

19. Sufficient Statistics: Selected Contributions, *Vasant S. Huzurbazar (edited by Anant M. Kshirsagar)*
20. Handbook of Statistical Distributions, *Jagdish K. Patel, C. H. Kapadia, and D. B. Owen*
21. Case Studies in Sample Design, *A. C. Rosander*
22. Pocket Book of Statistical Tables, *compiled by R. E. Odeh, D. B. Owen, Z. W. Birnbaum, and L. Fisher*
23. The Information in Contingency Tables, *D. V. Gokhale and Solomon Kullback*
24. Statistical Analysis of Reliability and Life-Testing Models: Theory and Methods, *Lee J. Bain*
25. Elementary Statistical Quality Control, *Irving W. Burr*
26. An Introduction to Probability and Statistics Using BASIC, *Richard A. Groeneveld*
27. Basic Applied Statistics, *B. L. Raktoe and J. J. Hubert*
28. A Primer in Probability, *Kathleen Subrahmaniam*
29. Random Processes: A First Look, *R. Syski*
30. Regression Methods: A Tool for Data Analysis, *Rudolf J. Freund and Paul D. Minton*
31. Randomization Tests, *Eugene S. Edgington*
32. Tables for Normal Tolerance Limits, Sampling Plans and Screening, *Robert E. Odeh and D. B. Owen*
33. Statistical Computing, *William J. Kennedy, Jr., and James E. Gentle*
34. Regression Analysis and Its Application: A Data-Oriented Approach, *Richard F. Gunst and Robert L. Mason*
35. Scientific Strategies to Save Your Life, *I. D. J. Bross*
36. Statistics in the Pharmaceutical Industry, *edited by C. Ralph Buncher and Jia-Yeong Tsay*
37. Sampling from a Finite Population, *J. Hajek*
38. Statistical Modeling Techniques, *S. S. Shapiro*
39. Statistical Theory and Inference in Research, *T. A. Bancroft and C.-P. Han*
40. Handbook of the Normal Distribution, *Jagdish K. Patel and Campbell B. Read*
41. Recent Advances in Regression Methods, *Hrishikesh D. Vinod and Aman Ullah*
42. Acceptance Sampling in Quality Control, *Edward G. Schilling*
43. The Randomized Clinical Trial and Therapeutic Decisions, *edited by Niels Tygstrup, John M Lachin, and Erik Juhl*
44. Regression Analysis of Survival Data in Cancer Chemotherapy, *Walter H. Carter, Jr., Galen L. Wampler, and Donald M. Stablein*
45. A Course in Linear Models, *Anant M. Kshirsagar*
46. Clinical Trials: Issues and Approaches, *edited by Stanley H. Shapiro and Thomas H. Louis*
47. Statistical Analysis of DNA Sequence Data, *edited by B. S. Weir*
48. Nonlinear Regression Modeling: A Unified Practical Approach, *David A. Ratkowsky*
49. Attribute Sampling Plans, Tables of Tests and Confidence Limits for Proportions, *Robert E. Odeh and D. B. Owen*
50. Experimental Design, Statistical Models, and Genetic Statistics, *edited by Klaus Hinkelmann*
51. Statistical Methods for Cancer Studies, *edited by Richard G. Cornell*
52. Practical Statistical Sampling for Auditors, *Arthur J. Wilburn*
53. Statistical Methods for Cancer Studies, *edited by Edward J. Wegman and James G. Smith*

54. Self-Organizing Methods in Modeling: GMDH Type Algorithms, *edited by Stanley J. Farlow*
55. Applied Factorial and Fractional Designs, *Robert A. McLean and Virgil L. Anderson*
56. Design of Experiments: Ranking and Selection, *edited by Thomas J. Santner and Ajit C. Tamhane*
57. Statistical Methods for Engineers and Scientists: Second Edition, Revised and Expanded, *Robert M. Bethea, Benjamin S. Duran, and Thomas L. Boullion*
58. Ensemble Modeling: Inference from Small-Scale Properties to Large-Scale Systems, *Alan E. Gelfand and Crayton C. Walker*
59. Computer Modeling for Business and Industry, *Bruce L. Bowerman and Richard T. O'Connell*
60. Bayesian Analysis of Linear Models, *Lyle D. Broemeling*
61. Methodological Issues for Health Care Surveys, *Brenda Cox and Steven Cohen*
62. Applied Regression Analysis and Experimental Design, *Richard J. Brook and Gregory C. Arnold*
63. Statpal: A Statistical Package for Microcomputers—PC-DOS Version for the IBM PC and Compatibles, *Bruce J. Chalmer and David G. Whitmore*
64. Statpal: A Statistical Package for Microcomputers—Apple Version for the II, II+, and IIe, *David G. Whitmore and Bruce J. Chalmer*
65. Nonparametric Statistical Inference: Second Edition, Revised and Expanded, *Jean Dickinson Gibbons*
66. Design and Analysis of Experiments, *Roger G. Petersen*
67. Statistical Methods for Pharmaceutical Research Planning, *Sten W. Bergman and John C. Gittins*
68. Goodness-of-Fit Techniques, *edited by Ralph B. D'Agostino and Michael A. Stephens*
69. Statistical Methods in Discrimination Litigation, *edited by D. H. Kaye and Mikel Aickin*
70. Truncated and Censored Samples from Normal Populations, *Helmut Schneider*
71. Robust Inference, *M. L. Tiku, W. Y. Tan, and N. Balakrishnan*
72. Statistical Image Processing and Graphics, *edited by Edward J. Wegman and Douglas J. DePriest*
73. Assignment Methods in Combinatorial Data Analysis, *Lawrence J. Hubert*
74. Econometrics and Structural Change, *Lyle D. Broemeling and Hiroki Tsurumi*
75. Multivariate Interpretation of Clinical Laboratory Data, *Adelin Albert and Eugene K. Harris*
76. Statistical Tools for Simulation Practitioners, *Jack P. C. Kleijnen*
77. Randomization Tests: Second Edition, *Eugene S. Edgington*
78. A Folio of Distributions: A Collection of Theoretical Quantile-Quantile Plots, *Edward B. Fowlkes*
79. Applied Categorical Data Analysis, *Daniel H. Freeman, Jr.*
80. Seemingly Unrelated Regression Equations Models: Estimation and Inference, *Virendra K. Srivastava and David E. A. Giles*
81. Response Surfaces: Designs and Analyses, *Andre I. Khuri and John A. Cornell*
82. Nonlinear Parameter Estimation: An Integrated System in BASIC, *John C. Nash and Mary Walker-Smith*
83. Cancer Modeling, *edited by James R. Thompson and Barry W. Brown*
84. Mixture Models: Inference and Applications to Clustering, *Geoffrey J. McLachlan and Kaye E. Basford*

85. Randomized Response: Theory and Techniques, *Arijit Chaudhuri and Rahul Mukerjee*
86. Biopharmaceutical Statistics for Drug Development, *edited by Karl E. Peace*
87. Parts per Million Values for Estimating Quality Levels, *Robert E. Odeh and D. B. Owen*
88. Lognormal Distributions: Theory and Applications, *edited by Edwin L. Crow and Kunio Shimizu*
89. Properties of Estimators for the Gamma Distribution, *K. O. Bowman and L. R. Shenton*
90. Spline Smoothing and Nonparametric Regression, *Randall L. Eubank*
91. Linear Least Squares Computations, *R. W. Farebrother*
92. Exploring Statistics, *Damaraju Raghavarao*
93. Applied Time Series Analysis for Business and Economic Forecasting, *Sufi M. Nazem*
94. Bayesian Analysis of Time Series and Dynamic Models, *edited by James C. Spall*
95. The Inverse Gaussian Distribution: Theory, Methodology, and Applications, *Raj S. Chhikara and J. Leroy Folks*
96. Parameter Estimation in Reliability and Life Span Models, *A. Clifford Cohen and Betty Jones Whitten*
97. Pooled Cross-Sectional and Time Series Data Analysis, *Terry E. Dielman*
98. Random Processes: A First Look, Second Edition, Revised and Expanded, *R. Syski*
99. Generalized Poisson Distributions: Properties and Applications, *P. C. Consul*
100. Nonlinear L_p-Norm Estimation, *Rene Gonin and Arthur H. Money*
101. Model Discrimination for Nonlinear Regression Models, *Dale S. Borowiak*
102. Applied Regression Analysis in Econometrics, *Howard E. Doran*
103. Continued Fractions in Statistical Applications, *K. O. Bowman and L. R. Shenton*
104. Statistical Methodology in the Pharmaceutical Sciences, *Donald A. Berry*
105. Experimental Design in Biotechnology, *Perry D. Haaland*
106. Statistical Issues in Drug Research and Development, *edited by Karl E. Peace*
107. Handbook of Nonlinear Regression Models, *David A. Ratkowsky*
108. Robust Regression: Analysis and Applications, *edited by Kenneth D. Lawrence and Jeffrey L. Arthur*
109. Statistical Design and Analysis of Industrial Experiments, *edited by Subir Ghosh*
110. *U*-Statistics: Theory and Practice, *A. J. Lee*
111. A Primer in Probability: Second Edition, Revised and Expanded, *Kathleen Subrahmaniam*
112. Data Quality Control: Theory and Pragmatics, *edited by Gunar E. Liepins and V. R. R. Uppuluri*
113. Engineering Quality by Design: Interpreting the Taguchi Approach, *Thomas B. Barker*
114. Survivorship Analysis for Clinical Studies, *Eugene K. Harris and Adelin Albert*
115. Statistical Analysis of Reliability and Life-Testing Models: Second Edition, *Lee J. Bain and Max Engelhardt*
116. Stochastic Models of Carcinogenesis, *Wai-Yuan Tan*
117. Statistics and Society: Data Collection and Interpretation: Second Edition, Revised and Expanded, *Walter T. Federer*
118. Handbook of Sequential Analysis, *B. K. Ghosh and P. K. Sen*
119. Truncated and Censored Samples: Theory and Applications, *A. Clifford Cohen*
120. Survey Sampling Principles, *E. K. Foreman*

121. Applied Engineering Statistics, *Robert M. Bethea and R. Russell Rhinehart*
122. Sample Size Choice: Charts for Experiments with Linear Models: Second Edition, *Robert E. Odeh and Martin Fox*
123. Handbook of the Logistic Distribution, *edited by N. Balakrishnan*
124. Fundamentals of Biostatistical Inference, *Chap T. Le*
125. Correspondence Analysis Handbook, *J.-P. Benzécri*
126. Quadratic Forms in Random Variables: Theory and Applications, *A. M. Mathai and Serge B. Provost*
127. Confidence Intervals on Variance Components, *Richard K. Burdick and Franklin A. Graybill*
128. Biopharmaceutical Sequential Statistical Applications, *edited by Karl E. Peace*
129. Item Response Theory: Parameter Estimation Techniques, *Frank B. Baker*
130. Survey Sampling: Theory and Methods, *Arijit Chaudhuri and Horst Stenger*
131. Nonparametric Statistical Inference: Third Edition, Revised and Expanded, *Jean Dickinson Gibbons and Subhabrata Chakraborti*
132. Bivariate Discrete Distribution, *Subrahmaniam Kocherlakota and Kathleen Kocherlakota*
133. Design and Analysis of Bioavailability and Bioequivalence Studies, *Shein-Chung Chow and Jen-pei Liu*
134. Multiple Comparisons, Selection, and Applications in Biometry, *edited by Fred M. Hoppe*
135. Cross-Over Experiments: Design, Analysis, and Application, *David A. Ratkowsky, Marc A. Evans, and J. Richard Alldredge*
136. Introduction to Probability and Statistics: Second Edition, Revised and Expanded, *Narayan C. Giri*
137. Applied Analysis of Variance in Behavioral Science, *edited by Lynne K. Edwards*
138. Drug Safety Assessment in Clinical Trials, *edited by Gene S. Gilbert*
139. Design of Experiments: A No-Name Approach, *Thomas J. Lorenzen and Virgil L. Anderson*
140. Statistics in the Pharmaceutical Industry: Second Edition, Revised and Expanded, *edited by C. Ralph Buncher and Jia-Yeong Tsay*
141. Advanced Linear Models: Theory and Applications, *Song-Gui Wang and Shein-Chung Chow*
142. Multistage Selection and Ranking Procedures: Second-Order Asymptotics, *Nitis Mukhopadhyay and Tumulesh K. S. Solanky*

Additional Volumes in Preparation

MULTISTAGE SELECTION AND RANKING PROCEDURES

Second-Order Asymptotics

NITIS MUKHOPADHYAY
University of Connecticut
Storrs, Connecticut

TUMULESH K. S. SOLANKY
University of New Orleans
New Orleans, Louisiana

Marcel Dekker, Inc. New York • Basel • Hong Kong

Library of Congress Cataloging-in-Publication Data

Mukhopadhyay, Nitis
Multistage selection and ranking procedures: second-order asymptotics / Nitis Mukhopadhyay, Tumulesh K. S. Solanky.
p. cm. -- (Statistics, textbooks and monographs: v. 142)
Includes bibliographical references and indexes.
ISBN 0-8247-9078-2 (acid-free)
1. Ranking and selection (Statistics). 2. Asymptotic distribution (Probability theory). I. Solanky, Tumulesh K. S. II. Title. III. Series.
QA278.75.M85 1994
519.5'4--dc20 93-47524
CIP

The publisher offers discounts on this book when ordered in bulk quantities. For more information, write to Special Sales/Professional Marketing at the address below.

This book is printed on acid-free paper.

MARCEL DEKKER, INC.
270 Madison Avenue, New York, New York 10016

Current printing (last digit):
10 9 8 7 6 5 4 3 2 1

PRINTED IN THE UNITED STATES OF AMERICA

To my wife, Mahua

N. Mukhopadhyay

To my inspiring parents

T.K.S. Solanky

PREFACE

The area of selection and ranking within multiple decision theory gained much recognition through the pioneering efforts of Bechhofer (1954) and Gupta (1956). The three major monographs, namely the ones by Bechhofer, Kiefer and Sobel (1968), Gibbons, Olkin and Sobel (1977), and Gupta and Panchapakesan (1979), essentially charted and guided in many ways the course of research in this area at various phases of development. One notes on the other hand that nonlinear renewal theory and its applications to sequential analysis started to make its mark in the late seventies, through important contributions of Lai and Siegmund (1977, 1979) and Woodroofe (1977), among others. During the eighties and more recently, one finds significant progress made in the areas of triple sampling [Hall (1981) and Mukhopadhyay (1990)] as well as accelerated sequential sampling techniques [Hall (1983) and Mukhopadhyay and Solanky (1991)]. During this period there has been a steady flow of journal articles addressing interesting problems of selection and ranking and showing sophisticated uses of these otherwise difficult theories of multistage sampling methodologies. Customarily though, most researchers in modern sequential analysis (i.e. after the advent of nonlinear renewal theory) have shied away from selection and ranking problems, while many researchers in selection and ranking have stayed away from second–order asymptotic analysis. Actually, the connections are not very obvious, and hence this book was written to consolidate research showing the interplay of these two important areas, each known to be very rich in its own right, and to present second–order asymptotic analysis and its uses for selection and ranking problems, largely under the indifference zone framework of Bechhofer (1954).

We address the theoretical and practical aspects of two–stage, three–stage, accelerated sequential and purely sequential methodologies in the light of certain selection and ranking problems. We do not aim to provide an overview of the whole area of selection and ranking or sequential analysis. We focus only on a few selection and ranking problems where modern sequential analysis has thus far played key roles, and this book intends to show how much one area has indeed enriched the other.

The book contains a thorough synthesis (Chapter 2) of second–order asymptotics in the area of multistage sampling methodologies for selection and ranking. The emphasis lies in the second–order theory and how such tools can be fruitfully applied in certain interesting problems of selection and ranking. It starts with the necessary unification of the available second–order results in general situations, and then these developments are fully exploited in a whole array of problems including those in selection and ranking, and estimation of ordered parameters. The readers will benefit from the depth of our coverage and the breadth and up–to–date nature of the literature citations. In Chapters 3 and 4, we address the problems of selecting the best among normal and negative exponential populations, and also summarize moderate sample size performances of the competing methodologies under consideration, after first providing all the necessary tables that will facilitate the evaluation of second–order "correction terms". This particular aspect will enhance the usefulness of the second–order theory in the eyes of both practitioners and theoreticians alike. These chapters include results obtained from fresh computer simulations so that the "numbers" and "performances" become comparable and then attempts are made to formulate various "guidelines" for practical implementation of the methodologies whenever appropriate. There is a chapter each on multivariate selection problems and estimation of ordered parameters. The chapter on estimation after selection and ranking opens up possible new directions for further investigation. Also, the last chapter includes a host of selection and ranking problems and some associated approaches where the thrust of second–order asymptotics has not yet been felt with as much vigor, but the included problem areas are quite possibly some of the prime targets for such development. Each chapter provides a section on Notes and Overviews that lists interesting topics for further study and research, and ends with Complements and Problems, supplying hints in many of these.

The book is written for graduate students, researchers and

practitioners. Researchers will discover many challenging problems here to work on and enhance their programs. For a Ph.D. level course based on these materials, a background in mathematical statistics and probability at the level of T.S. Ferguson's (1967) or J.O. Berger's (1985) book would essentially suffice. Chapters 3 onward primarily use the tools from Chapter 2. Even without knowing all the sophisticated details of limit theorems, readers are expected to grasp the important ideas and values of the results provided in subsequent chapters. For a thorough study of Chapter 2, a background in limit theorems at the level of R.J. Serfling's (1980) book is essential. On the other hand, an appropriate Masters level course can also be tailored out of this book. We envision this as a possible textbook at various levels. There is certainly enough material to cover a full–year graduate course. The emphasis one will give on certain topics will clearly depend on the interests and mix of a particular audience. Suggestions have been added at the end of the introductory chapter.

We would like to thank the students who took courses based on these materials or parts thereof at the Department of Theoretical Statistics at the University of Minnesota at Minneapolis, the Indian Statistical Institute at Calcutta, and the Department of Statistics at the University of Missouri at Columbia, Oklahoma State University at Stillwater, and the University of Connecticut at Storrs. We would also like to thank Ms. Cathy Randazzo for her patience in typing of the manuscript. It has been a real pleasure to work with the editorial and production staff members at Marcel Dekker, Inc., in planning and completing this project. Finally, we express the deepest sense of gratitude to our families, for their constant encouragement and patience during the preparation of this book.

Nitis Mukhopadhyay
Tumulesh K.S. Solanky

CONTENTS

MULTISTAGE SELECTION AND RANKING PROCEDURES

1

INTRODUCTION

1.1 The background

In everyday life, one selects or decides on the *best* product, *best* treatment, *best* candidate for a position or the *best* route for a destination etc. among many other decisions. Customarily, the various possible "treatment" effects are measured by means of an appropriately designed statistical experiment that is followed up by means of *analysis of variance* (ANOVA) and standard multiple comparisons between "treatments" of interest. The null hypothesis that is tested by the ANOVA typically describes that the possible treatment effects or various products under consideration are similar, the alternative hypothesis being that these are not. Quite often the ANOVA F–test rejects the null hypothesis of equal treatment effects at a reasonable level of type I error α. At this stage, the experimenter often turns around and wishes to acquire more in–depth information and detect those treatments which primarily contribute towards eventual rejection of the null hypothesis about equal treatment effects via ANOVA. One may also quite possibly wish to select or identify the treatment that is most responsible, in some sense, for the rejection of the customary null hypothesis. On the other hand, the experimenter may simply like to order all possible choices of treatments from the "worst" to the "best". The area of *multiple comparisons* is devoted to such and other related problems. The recently published book of Hochberg and Tamhane (1987) gives an excellent account of this area. One should get interesting views from Tukey (1991, 1993).

After carrying out the established methodologies for testing the null hypothesis via ANOVA, when the experimenter selects the best treatment or orders the treatments, one of the primary concerns then lies in evaluating the probability that the experimenter is indeed correct in making a particular final decision. In the multiple comparisons scenario, this step often turns out to be a big hurdle. Also, even if one can evaluate such probabilities in some situations, one may however find that the probability of the final decision being correct is possibly unacceptably low for that final decision to be of any practical use because of the poor choice of some design parameters, such as the sample sizes. One may also end up with "large" confidence regions for the parameters of interest. These are practical drawbacks. On the other hand, from a foundational point of view, one may argue that the treatments or products under consideration are quite possibly different anyway and hence one does not have to design an experiment merely to go through the ANOVA exercise and find out that the treatments are indeed different. Usually, one does eventually reject the null hypothesis anyway! In other words, the experimenter should not perhaps gear the sampling design merely to see how the ANOVA fares in a particular situation; instead the experimenter should formulate the ultimate problem of selecting the best treatment or ordering all the treatments in an appropriate fashion and obtain solutions via suitable sampling designs and methodologies such that the probability of making a correct decision is at a predetermined level. That is, the experimenter should keep the eventual *goal* in focus and try to get there, observing, however, that the ANOVA route is quite possibly not the only road to heaven.

1.2 SELECTION AND RANKING – TWO BREAKTHROUGHS

In the fifties, some researchers such as Bahadur (1950), Bahadur and Robbins (1950), Bechhofer (1954) and Gupta (1956) deviated from the so called ANOVA tradition and brought in remarkable new ideas and tools in this area. Two pioneers in this area, Bechhofer (1954) and Gupta (1956), came up with two fundamentally different formulations for the selection and ranking problem. Both these approaches considered only single–stage sampling strategies, that is, the necessary sampling was carried out in one single batch. Bechhofer (1954) formulated the problem from the *indifference zone* point of view, where the experimenter wishes to select the best treatment, say, when the best and the second best treatment are separated by a certain distance; the part of the parameter space that is called the *preference zone*. For various types of problems, the Bechhofer (1954) approach leads to a selection procedure such that the probability of a correct decision can be preset whenever the unknown parameters come from the preference zone. In his fundamental approach, Gupta (1956), on the other hand, controlled the probability of correct decision in the whole parameter space. In the Gupta (1956) approach, one finally comes up with a subset of a random number of treatments and claims that the best treatment is included in the selected subset with a preset probability. The latter approach is known in the literature as the *subset selection* approach. There are several other types of formulations as well, however Bechhofer's (1954) *indifference zone approach* and Gupta's (1956) *subset selection approach* are undeniably the two most fundamental ones.

For a review in this area, one should consider Bechhofer et al. (1968), Gibbons et al. (1977), Gupta and Panchapakesan (1979), Gupta and Huang (1981). One should also look at the categorized bibliography of Dudewicz and Koo (1981). The recent review articles of Gupta and Panchapakesan (1988, 1991), Laan and Verdooren (1989) and Mukhopadhyay (1993a) are very relevant as well.

1.2.1 SPECIFIC FORMULATIONS

In this Section, some specific details are provided with reference to the two most fundamental approaches in the area of selection and ranking. Suppose we have π_i, $i = 1,\ldots,k(\geq 2)$, independent populations and we have available a sequence of independent and identically distributed (i.i.d.) random variables $X_{i1}, X_{i2},\ldots,X_{in},\ldots$ having the probability density function (p.d.f.) $f_i(x;\ \theta_i,\ \xi)$ with respect to Lebesgue measure, $i = 1,\ldots,k$. We assume that the support of these p.d.f.'s is given by $\mathscr{X} \subseteq R$ and $\theta_1,\ldots,\theta_k$ and ξ are unknown parameters where $\theta_i \in R$ and $\xi \in R^+$, $i = 1,\ldots,k$. The parameters $\theta_1,\ldots,\theta_k$ are assumed to be of primary importance while ξ is assumed to be a nuisance parameter. Let us write $\theta_{[1]} \leq \theta_{[2]} \leq \cdots \leq \theta_{[k]}$ for the ordered θ–parameter values and the goal is to identify the population associated with the largest

θ–parameter, namely $\theta_{[k]}$, along with high probability of correct selection or identification. We assume that there is only one population associated with $\theta_{[k]}$.

Having recorded $X_{i1},...,X_{in}$ from π_i, let T_{in} be a natural estimator of θ_i, such as the maximum likelihood estimator, $i = 1,...,k$ and suppose that one uses a natural selection rule as follows.

$$\tilde{\mathscr{P}}_n: \text{ Select } \pi_j \text{ as the population associates with } \theta_{[k]} \text{ if}$$
$$T_{jn} = \max_{1 \le i \le k} T_{in}. \tag{1.2.1}$$

1.2.1.1 Indifference Zone Approach

Bechhofer's (1954) indifference zone formulation and the motivation behind it can be explained as follows. For notational convenience, let us write $\pi_{(i)}$ as the population associated with $\theta_{[i]}$ and suppose $T_{(in)}$ corresponds to the estimator coming out of $\pi_{(i)}$. Then, under the selection rule $\tilde{\mathscr{P}}_n$, one can write the probability of *correct selection* (CS) as

$$P_n(CS) = P\{T_{(in)} \le T_{(kn)},\ i = 1,...,k-1\}. \tag{1.2.2}$$

When θ's are location parameters and ξ is a scale parameter, for "reasonable" estimators T_{in}'s, $P_n(CS)$ given by (1.2.2) often turns out to be a function of n, $\{\theta_{[k]} - \theta_{[i]}\}$ for $i = 1,...,k-1$, and ξ. Write $\delta_i = \theta_{[k]} - \theta_{[i]}$ and then $P_n(CS)$ can be written as $f^*(n, \delta_1,...,\delta_{k-1}, \xi)$ for some suitably defined function $f^*(\cdot)$. We denote $\underset{\sim}{\theta} = (\theta_1,...,\theta_k)$ and let

$$\Omega = \{\underset{\sim}{\theta} = (\theta_1,...,\theta_k) : \theta_i \in R,\ i = 1,...,k\}. \tag{1.2.3}$$

In other words, Ω is the whole parameter space for $\underset{\sim}{\theta}$. The selection problems we address in Chapters 3 and 4 are such that

$$\operatorname*{Inf}_{\underset{\sim}{\theta}\in\Omega} P_n(CS) = 1/k, \tag{1.2.4}$$

in other words, under these setups, the selection rule may as well be equivalent to picking any one of the k populations $\pi_1,\ldots,\pi_k$ randomly and with equal probability.

Bechhofer (1954) introduced the following approach. Suppose one has a preassigned number $P^*\epsilon(1/k, 1)$ and one wishes to implement the selection rule $\tilde{\mathscr{P}}_n$ given by (1.2.1) in such a way that $P_n(CS) \geq P^*$ for all $\underset{\sim}{\theta}$ such that $\theta_{[k]}$ and $\theta_{[k-1]}$ are "separated" enough. The motivation is that since one cannot achieve a preassigned high target probability of correct selection via $\tilde{\mathscr{P}}_n$ in Ω, perhaps one should attempt to achieve such a goal when θ's are "sufficiently" separated from one another. Bechhofer (1954) defined

$$\Omega(\delta^*) = \{\underset{\sim}{\theta}\epsilon\Omega : \theta_{[k]} - \theta_{[k-1]} \geq \delta^*\}, \tag{1.2.5}$$

where $\delta^*(>0)$ is a preassigned number. $\Omega(\delta^*)$ is called the *preference zone* and $\Omega^c(\delta^*)$ $(= \Omega - \Omega(\delta^*))$ is known as the *indifference zone.* In the space $\Omega^c(\delta^*)$, the experimenter is indifferent about the θ–values and does not wish to select the population associated with $\theta_{[k]}$, because in $\Omega^c(\delta^*)$ the θ's are not separated enough from one another, However, in the *preference zone* $\Omega(\delta^*)$, we wish to have $P_n(CS) \geq P^*$, and Bechhofer (1954) formulated the goal as follows: We require

$$\operatorname*{Inf}_{\underset{\sim}{\theta}\in\Omega(\delta^*)} f^*(n, \delta_1,\ldots,\delta_{k-1}, \xi) = P^*. \tag{1.2.6}$$

In the parameter subset $\Omega(\delta^*)$, one notes that $\delta_i \geq \delta^*$, for $i = 1,\ldots,k$ and in several specific instances one can evaluate the infimum given in (1.2.6). One will find details in Chapters 3 and 4. Now, one will solve (1.2.6) for n and often come up with a solution that the fixed sample size n required from each π will have to be of at least certain magnitude; that is n will be the smallest integer no smaller than something like $\tilde{n}^* = p\xi^q/\delta^{*r}$ and $p = p(k, P^*, f^*)$, where q and r are known positive real numbers.

Hence, if ξ were known, one will take $n = [\tilde{n}^*]^* + 1$ random samples from each π and implement the selection procedure $\tilde{\mathscr{P}}_n$ given by (1.2.1). Here and elsewhere, $[x]^*$ stands for the largest integer $< x$. In the ξ known case, one can achieve (1.2.6) in several situations via suitable single stage procedures by taking appropriate number of samples from each π. Bechhofer (1954), in his pioneering work, put forth such methodologies in the normal distribution framework. Desu and Sobel (1968) provided directions for the analogous selection problems for two parameter negative exponential populations. Details are included in Chapters 3 and 4.

But, then, what happens when ξ is also unknown and hence $\tilde{n}^*$ in turn is unknown? Bechhofer et al. (1954) had proposed a *two–stage selection procedure* in the normal case. In a two–stage procedure, one first starts with some pilot samples from each π, and then estimates $\tilde{n}^*$ by means of a suitable positive integer valued random variable, say, N that is constructed based on pilot samples alone. Usually N will be considerably larger than the pilot sample size and one then samples the difference in the second stage from each π. Then, one uses the corresponding selection rule $\tilde{\mathscr{P}}_N$ based on all combined samples obtained from the two stages of sampling. How good such a procedure is will depend on many considerations. For example, one should at least look at E(N), Var(N), and certain analogs of (1.2.6) when the sample size is random and evaluate such expressions up to various *orders of accuracies*, and for this one arguably needs some general theoretical foundations. Section 2.2 of Chapter 2 provides such tools. In such problems, one has available other types of *multistage sampling* techniques and associated selection procedures, namely the modified two–stage, three–stage, purely sequential and accelerated sequential methodologies.

Chapter 2 unifies theories of various *multistage sampling* procedures in general. Chapter 2 is useful for selection and ranking problems and we should add that the usefulness of these tools are by no means limited to the specific problems considered in this book. In Chapter 2, we cite examples primarily from the theory of estimation areas in order to present a rather broad view. This chapter puts together in one place most of the major useful tools available for the theory of multistage sampling as such. In all the chapters that follow, we apply various theoretical results from Chapter 2 to show how rich the *theory of second-order asymptotics* in the areas of selection and ranking really is. The specifics of "selection and ranking" basically starts in Chapter 3 and these developments continue and expand in various directions subsequently.

1.2.1.2 Subset Selection Approach

Gupta (1956) approached the problem from a different angle in his pioneering contribution. The methodology is known as the *subset selection* approach where one will select a nonempty subset S of populations from $\pi_1,\ldots,\pi_k$ and declare that the population associated with $\theta_{[k]}$ is included in it. The advantage is that one can claim that for given $P^*\epsilon(1/k,\, 1)$

$$\operatorname*{Inf}_{\underset{\sim}{\theta}\in\Omega} P\{\pi_{(k)} \in S\} = P^*, \tag{1.2.7}$$

that is, the infimum now is taken over the whole parameter space Ω. Of course, what we mean is that such a subset can be appropriately constructed in many situations, for example, we may cite that Gupta (1956) provided such a methodology in the normal case. But one will note that the subset size in this case becomes a positive integer valued random variable taking values in the set $\{1, 2,\ldots,k\}$.

Let us explain the methodology briefly when the populations are normal, that is, $f_i(x;\, \theta_i,\, \xi) = (2\pi)^{-\frac{1}{2}}\xi^{-1}\exp\{-(x-\theta_i)^2/2\xi^2\}$. Here, one takes $T_{in} = \overline{X}_{in}(= n^{-1}\sum_{j=1}^{n} X_{ij})$. In the known ξ case, Gupta (1956) had proposed the following selection procedure.

$$\tilde{\mathscr{P}}_n^*\text{: Include } \pi_j \text{ in } S \text{ if } \overline{X}_{jn} \geq \max_{1\leq i\leq k} \overline{X}_{in} - n^{-\frac{1}{2}} p_0^* \xi, \tag{1.2.8}$$

where $p_0^* = p_0^*(k, P^*)$ is an appropriate positive number.

If ξ is unknown, then Gupta (1956) proposed the following selection procedure for $n \geq 2$:

$$\tilde{\mathscr{P}}_n^{**}\text{: Include } \pi_j \text{ in } S \text{ if } \overline{X}_{jn} \geq \max_{1\leq i\leq k} \overline{X}_{in} - n^{-\frac{1}{2}} p_1^* U^{\frac{1}{2}}(n), \tag{1.2.9}$$

where $U(n) = (kn-k)^{-1} \sum_{i=1}^{k} \sum_{j=1}^{n} (X_{ij} - \overline{X}_{in})^2$, the pooled sample variance, and $p_1^* = p_1^*(n, k, P^*)$ is an appropriate positive number, tabulated in Gupta and Sobel (1957), among others.

In both scenarios, for $n(\geq 2)$ fixed, it is possible to construct a *subset* S of *random size such that (1.2.7) holds.* We emphasize that as such one does not need a multistage subset selection approach in the normal case when ξ is unknown, that is a single stage procedure suffices. In the Bechhofer approach, we did not have that feature.

Most of this book is devoted to studies of asymptotic second–order characteristics of selection and ranking problems essentially under the indifference zone formulation. In the last Chapter, we include a variation of Gupta's (1956) subset selection formulation where one does need to implement suitable multistage sampling techniques and here, naturally second–order asymptotics are emphasized.

1.3 Sequential and Multistage Procedures

From this point onwards, we consider only those selection problems where the experimenter has to make decisions under the presence of nuisance parameters. For example, one may wish to select the population associated with the largest location parameter when the common variance is also unknown where the populations are, say, normal or two–parameter negative exponentials. The books of Bechhofer et al. (1968) and Gupta and Panchapakesan (1979) contain various types of sequential and two–stage procedures for a long array of problems. At times, several of these procedures are validated through likelihood comparisons, while many others have often been proposed from ad hoc principles focusing on practical considerations and common sense.

The area of sequential decision theory is well established, starting with the truly classic work of Wald (1947). Since then, several other books and monographs have appeared successively to fill voids and synthesize later advancements in this area of research. See Ghosh (1970), Wetherill (1975), Govindarajulu (1981), Woodroofe (1982) and Siegmund (1985). The area of sequential nonparametrics and associated invariance principles is also very rich. See, for example, the two monographs of Sen (1981, 1985). A recently published handbook of sequential analysis edited by Ghosh and Sen (1991) will also prove to be an invaluable resource in both parametric as well as nonparametric statistics.

In order to handle the topics discussed in this book (with the exception of certain materials in Chapter 8), we assume that one does not have available appropriate single–stage methodologies because of the presence of unknown nuisance parameters. What that means is that the experimenter must sample in at least two stages. In all the problems considered in this book, one will find algebraic expressions of quantities like $\tilde{n}^*$, say, which are called the *optimal fixed sample sizes* required from each treatment or population, followed up by an appropriate selection or estimation rule, had the nuisance parameters been known. Such optimal fixed sample size selection procedures will also control the *associated risk* at a certain preset level in some sense. But, then, $\tilde{n}^*$ would be unknown in practice, even though their algebraic forms are known. As we have added earlier, a *two–stage* procedure is the one where one starts with a few samples from each population and estimates $\tilde{n}^*$ from the pilot, and depending on this estimate, samples the difference from each population, all in one batch, at the second stage. In a *three–stage* procedure, one starts out like a two–stage procedure but then estimates only a fraction of $\tilde{n}^*$ at the second stage, followed by a third stage where one samples the difference of estimate of $\tilde{n}^*$ obtained from second stage samples and the estimate of the fraction of $\tilde{n}^*$ obtained from the first stage samples, that is the pilot. In other words, a three–stage procedure consists of batch sampling in three steps. A *purely sequential* procedure is the one where one keeps taking one sample at a time from each population, after the start, till some boundary condition is met involving the sample size and the corresponding estimate of $\tilde{n}^*$. An *accelerated sequential* methodology typically sits right between the corresponding three–stage and purely sequential procedures; in other words, here one starts off sequentially and terminates the sampling process early, obtains the estimate of $\tilde{n}^*$ at this stage, which is then followed by batch sampling in order to take the remaining samples.

1.4 The following chapters

The book consists of eight chapters. In Chapter 2, we discuss general formulations of two–stage, modified two–stage, three–stage, purely sequential and accelerated sequential sampling. The first–order asymptotic properties are presented in detail for the two–stage and modified two–stage procedures in general. Hall (1981, 1983) introduced three–stage and accelerated sequential fixed–width confidence interval problems for the mean of a normal population when the variance is unknown. Mukhopadhyay (1976) however gave a different type of three–stage procedure for the same problem. The unified theories of three–stage and accelerated sequential sampling have been available only recently. The associated second–order asymptotics are developed in Mukhopadhyay (1990) and Mukhopadhyay and Solanky (1991) and these are synthesized here fully for immediate use. On the other hand, the nonlinear renewal theory for purely sequential sampling has been in existence for some time now. See, for example, Lai and Siegmund (1977, 1979), Woodroofe (1977, 1982, 1991) and Siegmund (1985). Hence, in the purely sequential set up, we have summarized the available important results, giving only partial proofs of some of the important results and generalizations. The synthesis given in Chapter 2 forms the very basis for all the following chapters. To show the breadth of applications of such results, here we cite examples mostly from the areas of theory of estimation, primarily for the ease of understanding the developments.

In Chapters 3 – 7, we discuss various selection and ranking as well as estimation problems in the contexts of normal and two–parameter negative exponential populations having unknown means, and unknown and common variance. The *emphasis lies* in the derivations of *asymptotic second–order properties* of the associated methodologies. We consider *only* the *indifference zone formulation* of Bechhofer (1954) throughout Chapters 3 and 4. Since the populations have equal variance, we consider sequential and multistage procedures by taking an equal number of samples from each population. Of course, there can be and there are other variations of the proposed sampling strategies discussed here. But, these are essentially left out because second–order theories of such methodologies have not yet been fully developed.

The area of considerable practical interest consists of selection procedures where some type of "check" is built in, thereby eliminating the "inferior" populations as sampling proceeds. The original elimination–type selection procedure of Paulson (1964) obviously comes to mind. The results of Swanepoel and Geertsema (1973) and the more recent improvement over Paulson's (1964) methodology by Hartmann (1988, 1991) also raise much hope. But, we add that the types of second–order results which we present in Chapters 3 and 4 do not immediately carry over in the elimination–type scenarios. Such expected transitions are indeed very nontrivial. It is not obvious how the results of sections 2.3 – 2.5 have to be modified to suit such realistic formulations. Similar comments also hold in the case of likelihood based selection procedures which are available in Bechhofer et al. (1968). Again, one should note that there has not been much success in obtaining second–order characteristics of such likelihood based methodologies. Hence, *all the methodologies considered in Chapters 3 – 7 are* of *nonelimination type.* Chapters 3 – 7 however describe a rich body of literature associated with the area of selection and ranking where asymptotic second–order theory has been successfully applied.

Chapter 3 considers the selection of the normal population associated with the largest mean when each population has a common and unknown variance. Here, various notions of *first–order* and *second–order efficiencies* are introduced, and the procedures are compared theoretically among themeselves. Chapter 4 is also full of similar ideas and results in the context of two–parameter negative exponential populations. There is, however, a big difference between these two chapters. In exponential populations, one has to work with spacings between order statistics, and hence to apply the second–order theories of sections 2.3 – 2.5, an embedding technique of Lombard and Swanepoel (1978) turns out to be most essential so that the stopping variables can be indirectly written as functions of sample sums of i.i.d. random variables. In Chapters 3 – 4, we also provide guidelines with reference to moderate sample performances of the proposed selection methodologies. In Chapter 5, we construct fixed–width confidence intervals as well as bounded maximal risk and minimax risk point estimators for some parameters of interest for both normal and the negative exponential populations. Chapter 6 consists of a few analogous multivariate normal problems. Chapter 7 consists of some new ideas and results; we consider a few estimation problems and their associated properties when the estimators are constructed after one of the selection procedures in Chapter 3 or 4 has stopped sampling. This chapter is called, quite rightfully, estimation after selection and ranking, even though our interests do not lie in the traditional aspects of estimating certain parameters in the "selected" population.

Chapter 8 includes some "additional topics", such as (i) elimination type selection procedures, (ii) likelihood based selection procedures, (iii) selection procedures when variances are unknown and unequal, (iv) a variation of Gupta's (1956) subset selection formulation, and (v) some nonparametric selection procedures. In this last chapter, we first summarize some of the basic ideas of "elimination" of inferior populations and associated results. Next we view the selection problem under the Bechhofer (1954) approach as a multiple hypotheses testing problem and introduce some of the available techniques. Some of the relevant problems associated with those considered in Chapters 3 and 4 in the case when the populations have unknown and unequal variances are also included here. Then, we examine how the theory of second–order asymptotics can be fruitfully applied in a subset selection problem of picking a high proportion of "good" populations. This chapter ends with the discussion of a few simple nonparametric selection problems.

The last section in each chapter consists of "Notes and Overviews" where a few general comments and some research problems are included. It will be of interest to note that each chapter ends with "Complements and Problems" which lists a number of exercises. A word of caution is in order here. The bibliography given at the end of the book does not attempt to review the whole literature of sequential analysis or sequential selection and ranking problems. It will indeed be too ambitious to attempt to do that and it is not our objective either. It is hoped that the cited books, monographs and review articles will serve that need more successfully. We only refer to those works which have most direct bearing on the topics considered.

1.5 To The Readers

It is our belief that readers with various levels and types of backgrounds will be able to learn different aspects of multistage selection and ranking problems from this textbook. Guidelines given below differ as the audience and objectives change. First, we give some suggestions for a one semester course.

Masters Level Course: Students are assumed to have knowledge of basic mathematical statistics and probability theory at the level of Hogg and Craig (1978). After going through Chapter 1, the instructor will move to Chapter 3 and introduce the basic "selection problem" and discuss all the multistage procedures one by one. Some of the "elementary" properties of such procedures can be proved in class. One should be very detailed in the coverage of Section 3.8 and, if possible, assign appropriate simulation projects in class. Deal with Chapter 4 in the same manner. Some of the selected topics from Chapter 8 should be emphasized. Here, the goal is to achieve the basic understanding of the proposed methodologies and their implementations.

Ph.D. Level Regular Course: Students are assumed to have background at the Ferguson (1967) or Berger (1985) level mathematical statistics. Knowledge of "limit theorems" at the level of Serfling (1980) would help. One should first follow the guidelines given at the Masters Level, but at a significantly faster pace. Here, one will also discuss "proofs" from Chapters 3 and 4, *assuming* the results given in Chapter 2. Here, the theoretical depth will be considerably more demanding. The coverage will be expanded to include selected topics from Chapters 5 – 7.

Ph.D. Level Seminar Course: Here, the audience is mixed. Probably, most students will be well–versed with Lehmann (1983, 1986). Start with Chapter 2 and give the foundations, then follow up with an appropriate mix of topics, always emphasizing sections 3.8 and 4.8.

These are some brief guidelines for a one semester course. The readers will, however, quickly discover that this textbook contains more than enough materials for a solid one year course as well. The textbook can certainly be used for independent study and as a guide for those advanced students searching for thesis topics. The textbook is also useful for practitioners in the area of selection and ranking; particularly discussions in Chapters 3 and 4 and certain topics of Chapter 8 are noted to have considerable practical use.

REFERENCES

Bahadur (1950), Bahadur and Robbins (1950), Bechhofer (1954), Bechhofer, Dunnett and Sobel (1954), Bechhofer, Kiefer and Sobel (1968), Berger (1985), Desu and Sobel (1968), Dudewicz and Koo (1981), Ferguson (1967), Ghosh (1970), Ghosh and Sen (1991), Gibbons, Olkin and Sobel (1977), Govindarajulu (1981), Gupta (1956), Gupta and Huang (1981), Gupta and Panchapakesan (1979, 1988, 1991), Gupta and Sobel (1957), Hall (1981, 1983), Hartmann (1988, 1991), Hochberg and Tamhane (1987), Hogg and Craig (1978), Laan and Verdooren (1989), Lai and Siegmund (1977, 1979), Lehmann (1983, 1986), Lombard and Swanepoel (1978), Mukhopadhyay (1976, 1990, 1993a), Mukhopadhyay and Solanky (1991), Paulson (1964), Sen (1981, 1985), Serfling (1980), Siegmund (1985), Swanepoel and Geertsema (1973), Tukey (1991, 1993), Wald (1947), Wetherill (1975), Woodroofe (1977, 1982, 1991).

2

THEORY OF SEQUENTIAL AND MULTISTAGE PROCEDURES

2.1 INTRODUCTION

In this chapter we consider general types of purely sequential, accelerated sequential and three–stage procedures. In each set–up, an abstract formulation is discussed first and then this is followed by appropriate second–order expansions of various moments of the associated stopping variables. One will note that the corresponding two–stage and modified two–stage procedures do not quite enjoy such second–order properties, and yet for historical reasons, completeness and proper perspectives, we include such developments here as well. All these procedures included here are easily understood in the contexts of classical theory of estimation. In fact, these procedures were originally developed for various specific estimation problems and then the general unified theory eventually followed. After each general formulation and the discussions of associated characteristics, one will find a few appropriate simple examples from the theory of estimation which will clarify the concepts. The relevant details on applications of such procedures in the area of selection and ranking are developed in their fullest in the subsequent chapters. Throughout, we write $\phi(y) = (2\pi)^{-\frac{1}{2}} \exp(-y^2/2)$ and $\Phi(y) = \int_{-\infty}^{y} \phi(x)dx$ for all $y \in R$.

2.2 Two–stage procedure

Mahalanobis (1940) pioneered the idea of utilizing pilot samples in order to gain efficiency in large scale sample surveys. According to Wald (1947, p. 2), this idea may be regarded as a forerunner of sequential analysis. Later, Stein (1945, 1949) proposed an ingenious technique of sampling in two–stages in order to construct fixed–width confidence intervals for a normal mean (variance unknown), by achieving the exact preassigned confidence level in the sense of (2.2.2). Chow and Robbins' (1965) purely sequential procedure fails to achieve this property. Theorem 2.2.1 is quite general, and it shows that a fixed–width confidence interval can be constructed in a reasonable way to achieve exact confidence level even without normality. One observes that in Stein's construction, normality is not really crucial and we have replaced this by (i) independence of the estimators, and (ii) pivotal nature of the estimators in some sense. Utilizing the ideas of Mukhopadhyay (1980), a modified version of Stein's two–stage procedure is proposed in Section 2.2.2, and we study and compare various asymptotic properties. Some of these points are highlighted through normal and negative exponential distributions. The general discussions have been adapted from Mukhopadhyay (1982).

Let $X_1, X_2, \ldots$ be i.i.d. with a continuous p.d.f. (with respect to Lebesgue measure) on $\mathscr{X}$ depending on two parameters θ and ξ where $(\theta,\xi) \in \mathrm{R} \times \mathrm{R}^+$, while the domain space $\mathscr{X}$ may depend on θ alone. Let $T_n = T_n(X_1,\ldots,X_n)$, $U_n = U_n(X_1,\ldots,X_n)$ be estimators of θ and ξ respectively based on samples $X_1,\ldots,X_n$, and suppose they satisfy the following conditions:

(a) For any fixed n (possibly ≥ 2), T_n is independent of $(U_2,\ldots,U_n)$.

(b) (i) For some $\omega > 0$, for a measurable function $g\colon \mathrm{R}^+ \to \mathrm{R}^+$, the distribution of $n^{\omega}(T_n - \theta)/g(\xi)$ does not depend on n, θ and ξ, and we write

$$F(a) = P\left[\frac{n^{\omega}|T_n - \theta|}{g(\xi)} \leq a\right], \quad a > 0.$$

(ii) The distribution of $(T_n - \theta)/g(U_n)$ does not depend on θ and ξ, and $P\left[\frac{n^{\omega}|T_n - \theta|}{g(U_n)} \le b_n\right] = 1 - \alpha$, say, where $\alpha \in (0, 1)$.

Given two preassigned numbers $d(>0)$ and $0 < \alpha < 1$, we wish to construct a confidence interval I for θ having width 2d, and confidence coefficient at least $(1 - \alpha)$. Having recorded n observations, we will propose the interval $I_n = [T_n - d, T_n + d]$ for θ, keeping in mind that the lower or upper confidence limit will be modified suitably if $\mathscr{X}$ depends on θ. Suppose $F(a) = 1 - \alpha$. Now

$$P(\theta \in I_n) = P(|T_n - \theta| \le d)$$

$$= P\left[\frac{n^{\omega}|T_n - \theta|}{g(\xi)} \le \frac{dn^{\omega}}{g(\xi)}\right]$$

$$= F\left[\frac{dn^{\omega}}{g(\xi)}\right] \ge 1 - \alpha,$$

if and only if $dn^{\omega}/g(\xi) \ge a$, using condition b(i); that is $n \ge [ag(\xi)/d]^{1/\omega} = n^*$, say. Here and throughout, $[x]^*$ will stand for the largest integer $< x$ for $x \in R$.

Case 1. If ξ is known, we take a sample of size $[n^*]^* + 1$, and propose the corresponding interval I for θ.

Case 2. Suppose ξ is unknown. In this case we propose a two–stage procedure along the lines of Stein (1945). Start the experiment with a sample size m (possibly ≥ 2). Define

$$N = N(d) = \max\left\{m, \left[\left\{\frac{b_m g(U_m)}{d}\right\}^{1/\omega}\right]^* + 1\right\}. \tag{2.2.1}$$

If $N = m$, we do not take any more samples at the second stage, however, if $N > m$, then we sample the difference $(N - m)$ at the second stage. Based on the totality of all samples $X_1, \ldots, X_N$, we propose the fixed–width confidence interval $I_N = [T_N - d, T_N + d]$ for θ. Observe that I_N is well defined.

Theorem 2.2.1. *For the two-stage procedure (2.2.1), under the assumptions (a) and (b), for all* $(\theta, \xi) \in \mathbb{R} \times \mathbb{R}^+$, *we have:*

$$P\{\theta \in I_N\} \geq 1 - \alpha. \tag{2.2.2}$$

Proof. First we notice the basic inequality:

$$\left\{\frac{b_m g(U_m)}{d}\right\}^{1/\omega} \leq N \leq \left\{\frac{b_m g(U_m)}{d}\right\}^{1/\omega} + m. \tag{2.2.3}$$

Now,

$$\begin{aligned} P\{\theta \in I_N\} &= \Sigma_{n=m}^{\infty} P\{|T_N - \theta| \leq d \cap N = n\} \\ &= \Sigma_{n=m}^{\infty} P\{|T_n - \theta| \leq d \cap N = n\} \\ &= \Sigma_{n=m}^{\infty} P\{|T_n - \theta| \leq d \,|\, N = n\} P(N = n) \\ &= \Sigma_{n=m}^{\infty} P\{|T_n - \theta| \leq d\} P(N = n). \end{aligned} \tag{2.2.4}$$

The last step follows from condition (a), since the random variable $I(N = n)$ depends only on U_m for any fixed n. Thus from (2.2.4) we can write

$$\begin{aligned} P\{\theta \in I_N\} &= E\left\{F\left[\frac{dN^{\omega}}{g(\xi)}\right]\right\} \geq E\left\{F\left[\frac{b_m g(U_m)}{g(\xi)}\right]\right\} \\ &= E\left[P\left\{|V| \leq \frac{b_m g(U_m)}{g(\xi)} \,\middle|\, U_m\right\}\right] \end{aligned}$$

where V is independent of U_m and has the same distribution as that of $n^{\omega}(T_n - \theta)/g(\xi)$. Hence, the proof is complete since

$$P\{\theta \in I_N\} \geq P\left\{\frac{|V| g(\xi)}{g(U_m)} \leq b_m\right\} = 1 - \alpha. \qquad \square$$

Example 2.2.1. Suppose X's are from $N(\mu, \sigma^2)$ where $(\mu, \sigma) \in \mathbb{R} \times \mathbb{R}^+$. Choose $\theta = \mu$, $\xi = \sigma$, $T_n = n^{-1}\Sigma_{i=1}^n X_i = \overline{X}_n$, $S_n^2 = (n-1)^{-1} \Sigma_{i=1}^n (X_i - \overline{X}_n)^2$, $U_n = S_n$ for $n \geq 2$. In that case $\omega = \frac{1}{2}$, and $g(x) = x(>0)$, and the distributions in b(i) and b(ii) are respectively N(0, 1) and Student's t with (n–1) degrees of freedom. Consider $I_n = [\overline{X}_n - d, \overline{X}_n + d]$. Then from the procedure in (2.2.1) we get Stein's (1945) procedure as a special case since condition (a) is naturally satisfied, with b_m = upper $100(\alpha/2)\%$ point of Student's t–distribution with (m–1) degrees of freedom. See Problems 2.1 – 2.2.

Example 2.2.2. We consider an example where the $\mathscr{X}$–space depends on the parameter to be estimated. Suppose $X_1, X_2, \ldots$ are independent random variables with the p.d.f.

$$f(x; \mu, \sigma) = \sigma^{-1} \exp(-\tfrac{x-\mu}{\sigma})\, I(x > \mu), \tag{2.2.5}$$

where $I(\cdot)$ stands for the indicator function of $(\cdot)$ throughout the book. Note that $\theta = \mu$, $\xi = \sigma$, $T_n = \tilde{X}_n = \min\{X_1, \ldots, X_n\}$, $U_n = \sigma_n = (n-1)^{-1} \Sigma_{i=1}^n (X_i - \tilde{X}_n)$ for $n \geq 2$. In this case $\omega = 1$, and $g(x) = x(>0)$. We propose the fixed–width confidence interval $I_n = [\tilde{X}_n - 2d, \tilde{X}_n]$ for μ. Using joint distributions of order statistics, one can show that

(i) $\tilde{X}_n$ is dependent of $(\sigma_2, \ldots, \sigma_n)$ for all fixed $n(\geq 2)$ which is (a),

(ii) $n(\tilde{X}_n - \mu)/\sigma$ has a p.d.f. $f(x; 0, 1)$, and $n(\tilde{X}_n - \mu)/\sigma_n$ has the F distribution with degrees of freedom 2 and 2(n–1), which is (b). See Ghurye (1958) and Problems 2.3 – 2.4.

For this problem, (2.2.1) will look like

$$N = \max\left\{m, \left[\frac{b_m \sigma_m}{2d}\right]^* + 1\right\} \tag{2.2.6}$$

where $m(\geq 2)$ is the starting sample size, b_m is upper $100\alpha\%$ point of F–distribution with 2, 2(m–1) degrees of freedom and we propose the interval $I_N = [\tilde{X}_N - 2d, \tilde{X}_N]$ for μ. For further discussions on sequential estimation of μ, one is referred to Ghurye (1958), Basu (1971, 1991) and Mukhopadhyay (1974, 1988, 1993b).

Example 2.2.3. Suppose we have the usual intra–class model, and that $(X_1,\ldots,X_n)'$ is n–dimensional $(n = 1, 2,\ldots)$ normal with mean vector $\underset{\sim}{\mu} = \mu\underset{\sim}{1}$, and dispersion matrix $\Sigma = \sigma^2(\varrho_{ij})$ with $\varrho_{ii} = 1$, $\varrho_{ij} = \varrho^*(i\neq j=1,\ldots,n)$. Suppose $\underset{\sim}{v}' = (\mu,\sigma,\varrho^*)\in \mathbb{R}\times\mathbb{R}^+\times(-1, 0)$. Given $d(>0)$ and $\alpha(0 < \alpha < 1)$, we wish to construct a confidence interval I for μ having width 2d, and confidence coefficient at least $(1-\alpha)$. Let us consider $I_n = [\overline{X}_n - d, \overline{X}_n + d]$ for μ. Then, one obtains

$$P(\mu \in I_n) = P\left\{ \frac{|\overline{X}_n - \mu|}{\left[\frac{\sigma^2\{(1-\varrho^*)+n\varrho^*\}}{n}\right]^{\frac{1}{2}}} \leq \frac{d}{\left[\frac{\sigma^2\{(1-\varrho^*)+n\varrho^*\}}{n}\right]^{\frac{1}{2}}} \right\}$$

$$\geq P\left\{ |N(0, 1)| \leq \frac{d}{\left[\frac{\sigma^2(1-\varrho^*)}{n}\right]^{\frac{1}{2}}} \right\} \tag{2.2.7}$$

$$\geq 1 - \alpha,$$

if and only if

$$n \geq a^2\sigma^2(1-\varrho^*)/d^2 = n^*,$$

say, where a = upper $100(\alpha/2)\%$ point of standard normal distribution. Notice that the inequality in (2.2.7) is valid since $-1 < \varrho^* < 0$. Choose $\theta = \mu$, $\xi = \sigma(1-\varrho^*)^{\frac{1}{2}}$, $T_n = \overline{X}_n$, $S_n^2 = (n-1)^{-1}\Sigma_{i=1}^n(X_i - \overline{X}_n)^2$, $U_n = S_n$ for $n \geq 2$. In this case $\omega = \frac{1}{2}$, and $g(x) = x(>0)$. As in (2.2.1), we define

$$N = \max\left\{m, \left[\frac{b_m^2 S_m^2}{d^2}\right]^* + 1\right\} \tag{2.2.8}$$

where $m(\geq 2)$ is the first stage sample size, and b_m = upper $100(\alpha/2)\%$ point of Student's t–distribution with $(m-1)$ degrees of freedom, and we propose the fixed–width confidence interval $I_N = [\overline{X}_N - d, \overline{X}_N + d]$ for μ. One can easily show that the condition (a) holds, and $(n-1)S_n^2 \sim \sigma^2(1-\varrho^*)\chi^2_{n-1}$. See for example p. 197 of Rao (1973).

As in (2.2.3) – (2.2.4), we write [utilizing (2.2.7)]

$$\begin{aligned} P\{\mu \in I_N\} &= \Sigma_{n=m}^{\infty} P\,\{|\overline{X}_n - \mu| \leq d\}P(N = n) \\ &\geq E\,\{2\Phi(d\left[\frac{\sigma^2(1-\varrho^*)}{N}\right]^{-\frac{1}{2}}) - 1\} \\ &\geq E\,[2\Phi(b_m S_m\{\sigma^2(1-\varrho^*)\}^{-\frac{1}{2}}) - 1] \\ &= 1 - \alpha, \end{aligned}$$

which follows from the latter part of the proof of Theorem 2.2.1.

2.2.1 ASYMPTOTIC PROPERTIES OF THE TWO–STAGE PROCEDURE

We will continue using the same notations as introduced for Theorem 2.2.1. From (2.2.3) it follows that $\lim_{d\to 0} N$ is infinite w.p. 1. Again from (2.2.3), it is easy to see that

$$\liminf_{d\to 0}\,(dN^{\omega}) \geq b_m g(U_m) \text{ w.p. } 1 \tag{2.2.9}$$

and

$$b_m g(U_m) \geq \limsup_{d \to 0} d(N-m)^{\omega} = \limsup_{d \to 0} (dN^{\omega}) \text{ w.p. 1.} \quad (2.2.10)$$

Combining (2.2.9) and (2.2.10) we conclude that

$$\lim_{d \to 0} (dN^{\omega}) = b_m g(U_m) \text{ w.p. 1.} \quad (2.2.11)$$

Now,

$$P\{\theta \in I_N\} = E\left\{F\left[\frac{dN^{\omega}}{g(\xi)}\right]\right\}, \quad (2.2.12)$$

which combined with (2.2.11) and the dominated convergence theorem will lead to

$$\lim_{d \to 0} P\{\theta \in I_N\} = E\left\{F\left[\frac{b_m g(U_m)}{g(\xi)}\right]\right\}, \quad (2.2.13)$$

which is equal to $(1-\alpha)$ as we proved in Theorem 2.2.1. So, now we have proved the following theorem.

Theorem 2.2.2. *For the two-stage procedure (2.2.1), under the assumptions (a) and (b), for all* $(\theta, \xi) \in R \times R^{+}$, *we have as* $d \to 0$:

(i) $dN^{\omega} \to b_m g(U_m)$ w.p. 1; (2.2.14)

(ii) $E(N/n^*) \to (b_m/a)^{1/\omega} E[\{g(U_m)/g(\xi)\}^{1/\omega}]$, *provided*

$E[\{g(U_m)\}^{1/\omega}]$ *is finite;* (2.2.15)

(iii) $P\{\theta \in I_N\} \to (1-\alpha)$. (2.2.16)

There are many examples to suit the set-up, where $E[\{g(U_m)/g(\xi)\}^{1/\omega}] = 1$. In these cases, (2.2.15) will become:

$$E(N/n^*) \to (b_m/a)^{1/\omega} \text{ as } d \to 0. \tag{2.2.17}$$

A natural question is to ask whether always $b_m > a$. We strongly feel that this is so, but we cannot justify such a feeling analytically. Studying monotonicity properties of percentiles of various distributions is a difficult problem, and also one needs the form of the distribution to derive any such result. One is referred to Ghosh (1973), DasGupta and Perlman (1974).

However, together with (a) and (b) if some additional conditions are satisfied, then we can conclude that $b_m > a$ in a fairly general set–up. We will elaborate this point as follows. Suppose that conditions (a) and (b) are satisfied and for the moment, pretend that ξ is known. We will utilize the theory of *uniformly most powerful unbiased* (UMPU) tests and *uniformly most accurate unbiased* (UMAU) confidence intervals (or sets) for a parameter. One may refer to Neyman (1937) and Lehmann (1986). One is specially referred to chapter 10 of the book by Zacks (1971). For testing the simple null hypothesis $H_0 : \theta = \theta_0$ against the alternative hypothesis $H_a : \theta \neq \theta_0$, where θ_0 is a specified real number (at the level of significance α), suppose that (c) the acceptance region for H_0, namely $T_n \pm n^{-\omega} a g(\xi)$ is UMPU, and (d) the acceptance region for H_0, namely, $T_n \pm n^{-\omega} b_n g(U_n)$ is unbiased. Thus, under condition (c), $T_n \pm n^{-\omega} a g(\xi)$ is a UMAU confidence interval for θ with confidence coefficient $(1 - \alpha)$. From a result of Pratt (1961), it follows that the confidence interval $T_n \pm n^{-\omega} a g(\xi)$ has the minimum expected length among the two–sided $(1 - \alpha)$ level unbiased confidence intervals for θ.

Hence, under conditions (a) – (d), we obtain

$$2n^{-\omega} a g(\xi) < 2n^{-\omega} b_n E[g(U_m)]. \tag{2.2.18}$$

In case $0 < \omega < 1$, using Jensen's inequality we can write

$$\left[E\{g(U_m)\}\right]^{1/\omega} \le E\left[\{g(U_m)\}^{1/\omega}\right] = \{g(\xi)\}^{1/\omega},$$

the last equality we tacitly assumed right after (2.2.16). Thus, from (2.2.18), one obtains

$$a < b_n, \tag{2.2.19}$$

for all positive integers $n(\geq m)$.

Remark 2.2.1. If conditions (a) – (d) hold and $E[\{g(U_m)/g(\xi)\}^{1/\omega}] = 1$, then for $0 < \omega \leq 1$ we can conclude that $b_m > a$ which implies that the limit in (2.2.17) is bigger than one. That is, the random sample size N associated with the two–stage procedure (2.2.1) overestimates n^* even asymptotically. This property is referred to as *asymptotic inefficiency* or *first–order asymptotic inefficiency* in the sense of Chow and Robbins (1965) and Ghosh and Mukhopadhyay (1981) respectively. The property (2.2.16) is referred to as *asymptotic consistency* in the sense of Chow and Robbins (1965).

Remark 2.2.2. In continuation of example 2.2.2, we take $a = \ln(1/\alpha)$, $n^* = a\sigma/d$ and b_m = upper $100\alpha\%$ point of F–distribution with 2, 2(m–1) degrees of freedom. Since $E(\sigma_m) = \sigma$, we can easily conclude the results of Theorem 2.2.2 (with $\omega = 1$ and $g(x) = x$ for $x \in R^+$). Also, utilizing the corollary 4.2 of Ghosh (1973) we can easily conclude from (2.2.17) that $\lim_{d \to 0} E(Nn^{*-1})$ is strictly bigger than one. There is no doubt that one could also use the types of arguments leading to (2.2.18) to verify that b_m is indeed bigger than a for any fixed $m(\geq 2)$.

Remark 2.2.3. In Birnbaum and Healy (1960), procedures are developed, in a number of problems, for two–stage sampling schemes leading to estimators which are unbiased, having exactly a prescribed variance. But their procedures seem to involve gross waste of information contained in the first stage sample. It may be pointed out that our estimator T_N is also unbiased for θ, if for fixed sample sizes we start with an unbiased estimator T_n for θ.

In Birnbaum and Healy (1960) as well, one has to start with an unbiased fixed–sample estimator for the parameter of interest to arrive at an unbiased estimator at the end. Chatterjee (1991) gives a concise review of the area of multistage estimation procedures.

2.2.2 MODIFIED TWO–STAGE PROCEDURE

Choose and fix a number $\gamma \in R^+$. Let

$$m = m(d) = \max\left\{2, \left[\left[\frac{a}{d}\right]^{\frac{1}{\omega(\gamma+1)}}\right]^* + 1\right\},$$

and we start the experiment with m observations. Then, we define

$$N = N(d) = \max\left\{m, \left[\left\{\frac{b_m g(U_m)}{d}\right\}^{\frac{1}{\omega}}\right]^* + 1\right\}, \tag{2.2.20}$$

and we propose the interval $I_N = [T_N - d, T_N + d]$ for θ. The procedure in (2.2.20) looks exactly like that in (2.2.1), the big difference being the dependence of m, the starting sample size, on d. Also, naturally $m \equiv m(d)\uparrow\infty$ as $d \to 0$, that is, the starting sample size is allowed to grow, but it grows at a slower rate than that of n^*. This type of two–stage procedure was developed in Mukhopadhyay (1980) and further studied in Ghosh and Mukhopadhyay (1981). Cox (1952) is also particularly relevant.

The earlier basic inequality (2.2.3) is changed to

$$\left\{\frac{b_m g(U_m)}{d}\right\}^{\frac{1}{\omega}} \le N \le \left\{\frac{b_m g(U_m)}{d}\right\}^{\frac{1}{\omega}} + \left[\frac{a}{d}\right]^{\frac{1}{\omega(\gamma+1)}} + 2. \tag{2.2.21}$$

The properties of the procedure (2.2.20) can be summarized in the following Theorem.

Theorem 2.2.3. *Suppose the function* $g : R^+ \to R^+$ *is continuous,* U_n is consistent for ξ, $E[\{g(U_m)/g(\xi)\}^{1/\omega}] = 1$. *Then, under conditions (a) and (b), we have:*

(i) $P\{\theta \in I_N\} \geq 1-\alpha$; (2.2.22)

(ii) $\lim (N/n^*) = 1$ in probability, as $d \to 0$; (2.2.23)

(iii) $\lim E(N/n^*) = 1$, as $d \to 0$; (2.2.24)

(iv) $\lim P\{\theta \in I_N\} = 1-\alpha$, as $d \to 0$. (2.2.25)

Proof. Proofs of (2.2.22) and (2.2.25) are exactly the same as those of Theorem 2.2.1 and (2.2.16).

Using condition (b) and consistency of U_n for ξ one can see that the asymptotic (as $n \to \infty$) distribution (which exists) of $n^{\omega}(T_n-\theta)/g(U_n)$ is the same as the asymptotic distribution (which exists) of $n^{\omega}(T_n-\theta)/g(\xi)$ which is independent of n. Hence, $b_n \to a$ as $n \to \infty$. Thus, using (2.2.17) and the fact that $m \to \infty$ as $d \to 0$, we can conclude (2.2.24). (2.2.23) is a direct consequence of (2.2.21), since $\gamma \in R^+$. □

Remark 2.2.4. In further continuation of Example 2.2.2 and Remark 2.2.2, let us now use the modified two–stage procedure (2.2.20) in the negative exponential case. Using an approximation due to Scheffe and Tukey (see p. 84 of Johnson and Kotz, 1970), we obtain

$$b_m = a + a^2(m-1)^{-1} + 0(m^{-2}).$$

From (2.2.24) it follows that this procedure is *asymptotically first–order efficient* in the Ghosh and Mukhopadhyay (1981) sense. However, it follows that $\liminf_{d\to 0} E(N-n^*) \geq \lim_{d\to 0} 0(d^{-\gamma/(\gamma+1)}) = \infty$. It may be noted that Mukhopadhyay's (1974) purely sequential procedure had the property that $E(N) \leq n^* + 0(1)$, which naturally gave the corresponding purely sequential scheme a definite edge over the modified two–stage procedure (2.2.20). For later developments, see Mukhopadhyay (1988, 1993b).

Remark 2.2.5. Under some reasonable conditions, a purely sequential procedure can be proposed along the lines of Chow and Robbins (1965) to achieve (2.2.23) – (2.2.25). We conjecture that a result like that of Simons (1968) can be obtained to achieve exact consistency [(2.2.22)]. One may note that the two–stage procedure (2.2.8) can also be modified accordingly for the problem in Example 2.2.3.

2.3 THREE–STAGE PROCEDURE

One realizes that multistage methods of sampling techniques are used in statistical inference when no suitable fixed sample size procedure is available. Such is the case when the "*optimal*" fixed sample size needed to meet some types of specific goals depends on unknown nuisance parameters. The optimal fixed sample size thus refers to the minimum fixed sample size needed to satisfy certain predetermined criteria, had the nuisance parameters been known. Cox (1952) extended the double sampling techniques of Stein (1945, 1949) to cover a wider class of problems. However, his analysis also revealed that, on the average, the difference between the optimal and second stage sample sizes is asymptotically unbounded, especially when the pilot sample size, that is the starting sample size, happens to be chosen much smaller than the optimal sample size. This remains true even when the starting sample size is allowed to "grow" at a certain rate. Recall the discussions from section 2.2.2. Anscombe (1953), Robbins (1959) and Chow and Robbins (1965) studied the properties of one–by–one sequential sampling plans which are to be discussed in section 2.4. Subsequently many authors have focused on developing further sequential sampling techniques to tackle more difficult problems. Usually, purely sequential procedures enjoy many desirable asymptotic properties. From the practical standpoint, however, sequential procedures suffer, especially when money and particularly time are important design inputs. Triple sampling techniques proposed by Hall (1981) were designed to combine the operational savings made possible by sampling in batches and the efficiency of purely sequential procedures. The basic idea of triple sampling was also put forth in Mukhopadhyay (1976) for constructing fixed–width confidence intervals for the mean of a normal distribution when the variance is unknown, and thereby obtained first–order asymptotic results. For an up to date account, see also Chatterjee (1991).

Hall (1981) mentioned possibilities of applying triple sampling in point estimation problems. But to study the *regret* associated with point estimation in general requires further refinements concerning the asymptotic higher order moments of the stopping variable associated with the three-stage procedure. Since the publication of Hall's paper, the literature on triple sampling for estimation problems has been somewhat limited to the work of Mukhopadhyay (1985a), Mukhopadhyay and Al-Mousawi (1986), Mukhopadhyay and Mauromoustakos (1987), Mukhopadhyay et al. (1987), Woodroofe (1987), Hamdy et al. (1988), Hamdy (1988) and Lohr (1990). These papers pertain to normal or exponential models and rely heavily on the types of results available in Hall (1981). A unified framework is established in Mukhopadhyay (1990) to define a general triple sampling procedure and study the asymptotic second order expansions of both positive and negative moments of the associated stopping variable.

Let $Y_1, Y_2, \ldots$ be a sequence of i.i.d. *positive continuous random variables* having $E(Y_1) = \theta$, say. Write $\overline{Y}_n = n^{-1} \Sigma_{i=1}^n Y_i$, $n \geq 1$, and let m be the starting sample size. We should add that m has to exceed a certain "critical number" in several particular applications. Let $[u]^*$ stand for the largest integer smaller than u as before, and define

$$T = T(\lambda) = \max\{m, [\rho\lambda\overline{Y}_m]^* + 1\} \tag{2.3.1}$$

where $0 < \rho < 1$ and $0 < \lambda < \infty$ are known constants. We sample the difference $(T-m)$ in the second stage if $T > m$. Now, based on $Y_1, \ldots, Y_T$, we define

$$N = N(\lambda) = \max\{T, [\lambda\overline{Y}_T]^* + 1\}. \tag{2.3.2}$$

In the third stage, the sample size T is further augmented, if needed, with $(N-T)$ extra observations. At the stopped stage, θ is estimated by $\overline{Y}_N$.

Let $n^* = \lambda\theta$ and we assume that $\lambda = \lambda(m) \to \infty$, $\limsup (m/n^*) < \rho$, and $n^* = 0(m^r)$ for $r > 1$, as $m \to \infty$. Let $W_i = (Y_i - \theta)/\theta$, $i = 1, 2, \ldots$

Now, we state and prove the following results.

Theorem 2.3.1. *Let* ω *be a positive integer such that* $E(\overline{Y}_T^{\omega}) < \infty$. *Then, as* $m \to \infty$, *we have for the three-stage procedure (2.3.1) – (2.3.2):*

$$E(\overline{Y}_T^{\omega}) = \theta^{\omega} + \tfrac{1}{2}\theta^{\omega}(\rho n^*)^{-1}\,\omega(\omega-3)\mathrm{Var}(W_1) + o(\lambda^{-1}).$$

Proof. $E(\overline{Y}_T^{\omega}) = E\{E(\overline{Y}_T^{\omega} \mid Y_1,\ldots,Y_m)\}$

$$= E[T^{-\omega}\Sigma_{j=0}^{\omega}\tbinom{\omega}{j}(\Sigma_{i=1}^{m}Y_i)^{j}E\{(\Sigma_{i=m+1}^{T}Y_i)^{\omega-j} \mid Y_1,\ldots,Y_m\}]. \qquad (2.3.3)$$

Now, given $Y_1,\ldots,Y_m$, for large m, one gets

$$E\{(\Sigma_{i=m+1}^{T}Y_i)^{\omega-j} \mid Y_1,\ldots,Y_m\} = (T-m)^{\omega-j}\theta^{\omega-j}\{1+0(T^{-1})\}, \qquad (2.3.4)$$

and combining (2.3.3) and (2.3.4), it follows that

$$E(\overline{Y}_T^{\omega}) = \theta^{\omega}\Sigma_{j=0}^{\omega}\tbinom{\omega}{j}\,E\,\{(\Sigma_{i=1}^{m}W_i/T)^{j}\} + o(\lambda^{-1}). \qquad (2.3.5)$$

Now, we expand T^{-j} in a Taylor series around $(\rho n^*)^{-j}$ with a remainder term $R(n^*)$ to write

$$T^{-j} = (\rho n^*)^{-j} - j\rho\lambda(\overline{Y}_m-\lambda)(\rho n^*)^{-j-1} + R(n^*)$$

$$= (\rho n^*)^{-j} - j(m^{-1}\Sigma_{i=1}^{m}W_i)(\rho n^*)^{-j} + R(n^*). \qquad (2.3.6)$$

Combining (2.3.5) and (2.3.6) yields

$$E(\overline{Y}_T^{\omega}) = \theta^{\omega}\Sigma_{j=0}^{\omega}\tbinom{\omega}{j}(\rho n^*)^{-j}\,E\{(\Sigma_{i=1}^{m}W_i)^{j}\}$$

$$-\, m^{-1}\theta^{\omega}\Sigma_{j=0}^{\omega}\binom{\omega}{j}(\rho n^*)^{-j}\, jE\{(\Sigma_{i=1}^{m}W_i)^{j+1}\} + o\,(\lambda^{-1}), \quad (2.3.7)$$

since $E\{(\Sigma_{i=1}^{m}W_i)R(n^*)\} = o(\lambda^{-1})$. Hence, from (2.3.7), we obtain

$$E(\overline{Y}_T^{\omega}) = \theta^{\omega} + \tfrac{1}{2}\theta^{\omega}\omega(\omega-1)m\mathrm{Var}(W_1)(\rho n^*)^{-2} - \theta^{\omega}\omega\mathrm{Var}(W_1)(\rho n^*)^{-1} + o(\lambda^{-1}),$$

which leads to Theorem 2.3.1 since $m/n^* \approx \rho$ for large m. □

Theorem 2.3.2. *Suppose that* ω *is a positive integer such that* $E(\overline{Y}_T^{p}) < \infty$ *where* $p \geq \max(\omega, 2\omega-2)$. *Then, as* $m \to \infty$, *we have for the three-stage procedure (2.3.1) – (2.3.2):*

$$E(N^{\omega}) = n^{*\omega} + \tfrac{1}{2}\omega n^{*\omega-1}\{(\omega-3)\mathrm{Var}(W_1) + \rho\}/\rho + o(\lambda^{\omega-1}).$$

Proof. As in Hall (1981), notice that $N = [\lambda\overline{Y}_T]^* + 1$ w.p. 1 except perhaps on a set $A = \{T \leq m\} \cup \{N \leq [\rho\lambda\overline{Y}_m]^* + 1\}$ of measure zero. Hence, it follows that

$$E(N^{\omega}) = E\{([\lambda\overline{Y}_T]^* + 1)^{\omega}\} + o(\lambda^{\omega-1})$$

$$= E\{(\lambda\overline{Y}_T + Z_T)^{\omega}\} + o(\lambda^{\omega-1}), \quad (2.3.8)$$

where $Z_T = 1 - \{\lambda\overline{Y}_T - [\lambda\overline{Y}_T]^*\} \xrightarrow{\mathscr{L}} \text{Uniform } (0, 1)$ as $m \to \infty$. See, for example, Hall (1981), Nagaev and Mukhin (1966) and Holewijn (1969). Now, (2.3.8) leads to

$$E(N^{\omega}) = \Sigma_{j=0}^{\omega}\binom{\omega}{j}\lambda^{\omega-j}\, E\{\overline{Y}_T^{\omega-j}Z_T^{j}\} + o(\lambda^{\omega-1})$$

$$= \lambda^{\omega}E(\overline{Y}_T^{\omega}) + \omega\lambda^{\omega-1}\, E(\overline{Y}_T^{\omega-1}Z_T^{j}) + o(\lambda^{\omega-1}). \quad (2.3.9)$$

From Theorem 2.3.1, one immediately observes that $\text{Var}(\overline{Y}_T^{\omega-1}) = \theta^{2(\omega-1)}(\rho n^*)^{-1}\text{Var}(W_1)(\omega-1)^2 + o(\lambda^{-1})$ if $E(\overline{Y}_T^{2\omega-2}) < \infty$, and also $\text{Var}(Z_T - \frac{1}{2}) = \frac{1}{12} + o(1)$ as $m \to \infty$. Thus,

$$\lim_{m\to\infty} \text{Cov}^2(\overline{Y}_T^{\omega-1}, Z_T - \tfrac{1}{2}) \leq \lim_{m\to\infty} \{\text{Var}(\overline{Y}_T^{\omega-1})\text{Var}(Z_T - \tfrac{1}{2})\}$$

which equals zero, if $E(\overline{Y}_T^{2\omega-2}) < \infty$. In other words, asymptotically $\overline{Y}_T^{\omega-1}$ and $Z_T - \frac{1}{2}$ are uncorrelated, and hence we have (as $m \to \infty$)

$$E\{\overline{Y}_T^{\omega-1}(Z_T - \tfrac{1}{2})\} = E(\overline{Y}_T^{\omega-1})E(Z_T - \tfrac{1}{2}) = o(1). \tag{2.3.10}$$

Thus,

$$E(\lambda^{\omega-1}\,\overline{Y}_T^{\omega-1}\,Z_T) = E(\tfrac{1}{2}\lambda^{\omega-1}\,\overline{Y}_T^{\omega-1}) + E\{\lambda^{\omega-1}\,\overline{Y}_T^{\omega-1}\,(Z_T - \tfrac{1}{2})\}$$

$$= \tfrac{1}{2}\lambda^{\omega-1}\,E(\overline{Y}_T^{\omega-1}) + o(\lambda^{\omega-1}), \tag{2.3.11}$$

in view of (2.3.10). Now, we combine (2.3.9) and (2.3.11) and use Theorem 2.3.1 (assuming that $E(\overline{Y}_T^{\omega})$ is finite as well) to estimate $E(\overline{Y}_T^{i})$ for $i = \omega-1, \omega$. The proof is complete. □

Theorem 2.3.3. *We have the following for the three-stage procedure (2.3.1) – (2.3.2) as* $m \to \infty$*:*

(i) *Let* $E(Y_1^p) < \infty$ *for some* $p \geq 2$. *Then, as* $m \to \infty$, $P(N \leq \xi n^*) = O(\lambda^{-\frac{1}{2}p/r})$ *for every real number* $\xi \in (0, 1)$, *where* $m = O(n^{*1/r})$ *with* $r > 1$;

(ii) *Let* $p^* = \begin{cases} 2 & \text{if } 1 < r < 2 \\ r+\epsilon & \text{if } r \geq 2 \end{cases}$ *for some* $\epsilon > 0$, *and assume that*

$E(Y_1^p) < \infty$ *for* $p \geq p^*$. *Then,* $(N-n^*)^2/n^* \xrightarrow{\mathscr{L}} \rho^{-1}\mathrm{Var}(W_1)\chi_1^2$ *as* $m \to \infty$, *and* $(N-n^*)^2/n^*$ *is also uniformly integrable;*

(iii) *Let* p^* *be the same as in part (ii), and assume that* $E(Y_1^p) < \infty$ *for* $p \geq \max\{2, 2(\omega+3)r - 2(\omega+2), p^*\}$ *where* ω *is a positive number.*

Then,

$$E(N^{-\omega}) = n^{*-\omega} - \tfrac{1}{2}\omega n^{*-\omega-1}\{\rho-(\omega+3)\mathrm{Var}(W_1)\}/\rho + o(\lambda^{-\omega-1}).$$

Proof. (i) $P(N \leq \xi n^*) \leq P(\lambda\overline{Y}_T \leq \xi n^*) \leq P\{|\overline{Y}_T - \theta| > (1-\xi)\theta\}$

$$\leq P\{\max_{m \leq n \leq [\xi n^*]^*} |\overline{Y}_n - \theta| > (1-\xi)\theta\},$$

which is $0(m^{-p/2})$, by Lemma 2.2 of Sen and Ghosh (1981) and Hajek–Renyi inequality.

(ii) First note that $T(\rho n^*)^{-1} \xrightarrow{P} 1$ as $m \to \infty$, and hence Anscombe's (1952) result leads to

$$(\rho n^*)^{\frac{1}{2}}(\overline{Y}_T - \theta)/\{\theta^2\mathrm{Var}(W_1)\}^{\frac{1}{2}} \xrightarrow{\mathscr{L}} N(0, 1) \tag{2.3.12}$$

as $m \to \infty$. Let x be an arbitrary real number. From (2.3.2) one can immediately see that $N \geq \lambda\overline{Y}_T$ w.p. 1 and thus

$$\limsup_{m \to \infty} P\{N-n^* \leq x[n^*\mathrm{Var}(W_1)/\rho]^{\frac{1}{2}}\}$$

$$\leq \limsup_{m \to \infty} P\{\overline{Y}_T \leq x\theta[\mathrm{Var}(W_1)\{n^*\rho\}^{-1}]^{\frac{1}{2}} + \theta\}$$

$$= \limsup_{m \to \infty} P[(\rho n^*)^{\frac{1}{2}}(\overline{Y}_T - \theta)/\{\theta \mathrm{Var}^{\frac{1}{2}}(W_1)\} \le x] = \Phi(x)$$

in view of (2.3.12). Here and elsewhere one writes $\Phi(x) = \int_{-\infty}^{x} \phi(y)dy$ where $\phi(y) = (2\pi)^{-\frac{1}{2}} \exp(-\frac{1}{2}y^2)$ for $x, y \in R$. Again note that $N \le \lambda \overline{Y}_T + TI(\lambda \overline{Y}_T < T)$ w.p. 1 and thus

$$\liminf_{m \to \infty} P[N - n^* - TI(\lambda \overline{Y}_T < T) \le x\{n^* \mathrm{Var}(W_1)/\rho\}^{\frac{1}{2}}]$$

$$\ge \liminf_{m \to \infty} P[\lambda \overline{Y}_T \le x\{n^* \mathrm{Var}(W_1)/\rho\}^{\frac{1}{2}} + n^*] = \Phi(x), \qquad (2.3.13)$$

in view of (2.3.12). Now, it is easy to see that the first line in (2.3.13) will reduce to $\liminf_{m \to \infty} P[N - n^* \le x\{n^* \mathrm{Var}(W_1)/\rho\}^{\frac{1}{2}}]$ if we verify that as $m \to \infty$,

$$TI(T > \lambda \overline{Y}_T) n^{*-\frac{1}{2}} \xrightarrow{P} 0,$$

and so it suffices to show that

$$I(T > \lambda \overline{Y}_T) n^{*\frac{1}{2}} \xrightarrow{P} 0. \qquad (2.3.14)$$

Let ϵ be an arbitrary positive number. Then,

$$P\{n^{*\frac{1}{2}} I(T > \lambda \overline{Y}_T) > \epsilon\} \le n^{*\frac{1}{2}} \epsilon^{-1} P(T > \lambda \overline{Y}_T)$$

$$\le n^{*\frac{1}{2}} \epsilon^{-1} P\{\lambda \overline{Y}_T \le m + \rho \lambda \overline{Y}_m\}$$

$$\le n^{*\frac{1}{2}} \epsilon^{-1} [P\{|\overline{Y}_m - \theta| > \tfrac{1}{2} \frac{(1-\rho)\theta - m\lambda^{-1}}{\rho}\}$$

$$+ P\{|\overline{Y}_T - \theta| > \tfrac{1}{2}\frac{(1-\rho)\theta - m\lambda^{-1}}{\rho}\}]$$

$$= 0(\lambda^{\frac{1}{2}})0(\lambda^{-\frac{1}{2}p/r}) = o(1),$$

if $p \geq 2$ and $p > r$. This verifies (2.3.14) and hence

$$(N - n^*)\{n^*\mathrm{Var}(W_1)/\rho\}^{-\frac{1}{2}} \xrightarrow{\mathscr{L}} N(0, 1)$$

as $m \to \infty$, if $p \geq 2$ and $p > r$, which is the first part of (ii). From Theorem 2.3.2, we get

$$E\{(N - n^*)^2/n^*\} = n^{*-1}E(N^2) - 2E(N) + n^*$$

$$= \mathrm{Var}(W_1)\rho^{-1} + o(1), \tag{2.3.15}$$

since $E(\overline{Y}_T^2) < \infty$ under the given assumptions. Now, the first part of (ii) together with (2.3.15) show uniform integrability of $(N - n^*)^2/n^*$ under the stated moment conditions.

(iii) Here, we expand $E(N^{-\omega})$ up to $o(\lambda^{-\omega-1})$ where ω is a positive number. Use Taylor expansion to write

$$E(N^{-\omega}) = n^{*-\omega} - \omega n^{*-\omega-1}\{E(N) - n^*\}$$

$$+ \tfrac{1}{2}\omega(\omega + 1)n^{*-\omega-1}E\{W^*(n^*/M)^{\omega+2}\}, \tag{2.3.16}$$

where M is a suitable random variable between n^* and N, and we define $W^* = (N - n^*)^2/n^*$. Since $E(\overline{Y}_T) < \infty$, Theorem 2.3.2 leads to

$$E(N - n^*) = \tfrac{1}{2}\{\rho - 2\mathrm{Var}(W_1)\}/\rho + o(1). \tag{2.3.17}$$

Now, $W^*(n^*/M)^{\omega+2}I(N > \tfrac{1}{2}n^*) \leq 2^{\omega+2}W^*$, since $M > \tfrac{1}{2}n^*$ on the set $N >$

$\frac{1}{2}n^*$. Hence, from part (ii), we can immediately claim uniform integrability of $W^*(n^*/M)^{\omega+2}I(N > \frac{1}{2}n^*)$, under the assumption that $E(Y_1^p)$ is finite for $p \geq p^*$, and thus

$$E\{W^*(n^*/M)^{\omega+2}I(N > \tfrac{1}{2}n^*)\} = \mathrm{Var}(W_1)\rho^{-1} + o(1). \qquad (2.3.18)$$

Again, since $M > m$ on the set $N \leq \frac{1}{2}n^*$, we get

$$\begin{aligned} E\{W^*(n^*/M)^{\omega+2}I(N \leq \tfrac{1}{2}n^*)\} &\leq (n^*/m)^{\omega+2}E\{(N^2n^{*-1} + n^*)I(N \leq \tfrac{1}{2}n^*)\} \\ &\leq \tfrac{5}{4}(n^{*\omega+3}/m^{\omega+2})P(N \leq \tfrac{1}{2}n^*) \\ &= O(\lambda^{\omega+3})O(\lambda^{-(\omega+2)/r})O(\lambda^{-\frac{1}{2}p/r}), \qquad (2.3.19) \end{aligned}$$

which follows from part (i). Now, the order given in (2.3.19) will be $o(1)$ if $p \geq \max\{2, 2r(\omega+3) - 2(\omega+2)\}$. Thus,

$$E\{W^*(n^*/M)^{\omega+2}\} = \mathrm{Var}(W_1)\rho^{-1} + o(1), \qquad (2.3.20)$$

by combining (2.3.18) and (2.3.19). Now, we put (2.3.17), (2.3.20) and (2.3.16) together to complete the proof. □

2.3.1 APPLICATIONS OF THE THREE–STAGE PROCEDURE

Let $\underset{\sim}{X}_1\ \underset{\sim}{X}_2, \ldots$ be a sequence of $q(\geq 1)$ dimensional i.i.d. random variables having continuous distributions with unknown parameters $(\underset{\sim}{\nu}^*, \theta^*) \in \mathscr{A} \subseteq R^q \times R^+$. The domain space $\mathscr{A}$ is permitted to depend upon $\underset{\sim}{\nu}^*$, but not on θ^*. The primary focus here is on estimating $\underset{\sim}{\nu}^*$ by means of a fixed–size confidence region, and this development parallels that of Mukhopadhyay (1982), summarized in section 2.2.

Having recorded $\underset{\sim}{X}_1, \ldots, \underset{\sim}{X}_n$, let $\underset{\sim}{U}_n = \underset{\sim}{U}_n(\underset{\sim}{X}_1, \ldots, \underset{\sim}{X}_n)$ and $V_n =$

$V_n(\underset{\sim}{X}_1,\ldots,\underset{\sim}{X}_n)$ be estimators of $\underset{\sim}{\nu}^*$ and θ^* respectively, and suppose that they satisfy the following conditions:

(i) For any fixed $n(\geq 2)$, $\underset{\sim}{U}_n$ is independent of $(V_2, V_3,\ldots,V_n)$ and $E(V_n) = \theta^*$;

(ii) For some $\zeta(>0)$ and positive definite $q \times q$ matrix H, the distribution function of $n^\zeta(\underset{\sim}{U}_n - \underset{\sim}{\nu}^*)' H^{-1}(\underset{\sim}{U}_n - \underset{\sim}{\nu}^*)/\theta^{*\zeta}$ does not depend upon n, $\underset{\sim}{\nu}^*$ and θ^*, and we write

$$G^*(x) = P\{n^\zeta(\underset{\sim}{U}_n - \underset{\sim}{\nu}^*)' H^{-1}(\underset{\sim}{U}_n - \underset{\sim}{\nu}^*) \leq \theta^{*\zeta} x\},\ x > 0;$$

(iii) The two infinite dimensional sequences $\{V_n : n \geq 2\}$ and $\{\overline{Y}_n : n \geq 1\}$ are identically distributed for suitably defined i.i.d. positive continuous random variables $Y_1, Y_2, \ldots$.

Given $L(>0)$ and $0 < \alpha < 1$, we propose the following fixed-size confidence region for $\underset{\sim}{\nu}^*$. Let

$$\mathscr{R}_n = \{\underset{\sim}{x} \in R^q : (\underset{\sim}{U}_n - \underset{\sim}{x})' H^{-1}(\underset{\sim}{U}_n - \underset{\sim}{x}) \leq L\}; \qquad (2.3.21)$$

however, we also require that the associated confidence coefficient is approximately $(1-\alpha)$. Let "a" be the upper $100\alpha\%$ point of G^*, and now $P\{\underset{\sim}{\nu}^* \in \mathscr{R}_n\} = G^*(n^\zeta L/\theta^{*\zeta}) \geq 1 - \alpha$ if n is the smallest integer $\geq (a/L)^{1/\zeta}\theta^*$ $= n^*$, say. We write $\lambda = (a/L)^{1/\zeta}$ and carry out the asymptotics as $L \to 0$. The three-stage procedure (2.3.1) – (2.3.2) transforms to

$$T = T(\lambda) = \max\{m, [\rho\lambda V_m]^* + 1\} \qquad (2.3.22)$$

and

$$N = N(\lambda) = \max\{T, [\lambda V_T]^* + 1\}, \qquad (2.3.23)$$

thereby proposing the region $\mathcal{R}_N$ for $\underset{\sim}{\nu}^*$ defined in (2.3.21).

Let $W_1 = (Y_1 - \theta^*)/\theta^*$, and then under appropriate moment conditions on Y_1, Theorems 2.3.1 and 2.3.2 lead to

$$E(V_T) = \theta^* - \theta^* \mathrm{Var}(W_1)(\rho n^*)^{-1} + o(L^{1/\zeta}), \tag{2.3.24}$$

$$E(N) = n^* + \tfrac{1}{2} - \rho^{-1}\mathrm{Var}(W_1) + o(1), \tag{2.3.25}$$

$$\mathrm{Var}(N) = \rho^{-1} n^* \mathrm{Var}(W_1) + o(L^{-1/\zeta}). \tag{2.3.26}$$

Example 2.2.1 (Continued). In the formulation of Hall (1981), we have $q = 1$, $\zeta = 1$, $L = d^2$, $m \geq 2$, $\underset{\sim}{\nu}^* = \mu$, $\theta^* = \sigma^2$, $U_n = \overline{X}_n (= n^{-1}\Sigma_{i=1}^n X_i)$, $V_n = (n-1)^{-1}\Sigma_{i=1}^n (X_i - \overline{X}_n)^2$, $Y_1 \sim \sigma^2\chi_1^2$, and thus $\mathrm{Var}(W_1) = 2$. Hence, (2.3.24), (2.3.25) and (2.3.26) agree with Hall's (1981) equations (4.1), (2.2) and (2.3), respectively, in this special case.

Example 2.3.1. Suppose that $\underset{\sim}{X}$'s are i.i.d. $N_p(\underset{\sim}{\mu}, \sigma^2 H)$, as in the formulation of Mukhopadhyay and Al–Mousawi (1986), and we have $q = p$, $\zeta = 1$, $L = d^2$, $m \geq 2$, $\underset{\sim}{\nu}^* = \underset{\sim}{\mu}$, $\theta^* = \sigma^2$, $\underset{\sim}{U}_n = \overline{\underset{\sim}{X}}_n (= n^{-1}\Sigma_{i=1}^n \underset{\sim}{X}_i)$, $V_n = (pn-p)^{-1}\Sigma_{j=1}^n (\underset{\sim}{X}_j - \overline{\underset{\sim}{X}}_n)' H^{-1} (\underset{\sim}{X}_j - \overline{\underset{\sim}{X}}_n)$, $Y_1 \sim \sigma^2 p^{-1}\chi_p^2$, and thus $\mathrm{Var}(W_1) = 2/p$. Hence, (2.3.24), (2.3.25) and (2.3.26) agree with Mukhopadhyay and Al–Mousawi's Theorem 2 (parts (a) and (b)) and (5.5), respectively, in this special case as well.

In general, the confidence coefficient associated with the region $\mathcal{R}_N$ is given by

$$P\{\underset{\sim}{\nu}^* \in \mathcal{R}_N\} = E\{G^*(N^\zeta L/\theta^{*\zeta})\} \tag{2.3.27}$$

under the assumptions (i) – (iii). It is quite easy to see that $\lim_{L\to 0} P\{\underset{\sim}{\nu}^* \in \mathcal{R}_N\} = 1 - \alpha$; however, to guarantee that the coverage probability is $(1 - \alpha) + o(L^{1/\zeta})$, the sample size needed in the third stage (see (2.3.23)) will require some "fine–tuning." The following Theorem gives a precise statement regarding the magnitude of such "fine–tuning." *Assume*, for simplicity, that *all moments of* Y's *are finite.*

Theorem 2.3.4. *Let* $G^{*'''}(x)$ *be finite for all* $x > 0$ *and suppose that* $|G^{*'''}(x)|$ *is bounded for* $x > x_0$ *for any arbitrary* x_0 (>0). *Assume that* $\mathrm{Var}(W_1)$ *is known and let*

$$\epsilon^* = \tfrac{1}{2}\rho^{-1}\mathrm{Var}(W_1)\{3 - \zeta - \frac{\zeta a g^{*'}(a)}{g^*(a)}\} - \tfrac{1}{2} \tag{2.3.28}$$

where $g^*(x) = G^{*'}(x)$ *and* $g^{*'}(x) = G^{*''}(x)$. *Let* T *be the same as in (2.3.22), but define*

$$N^* = N^*(\lambda) = \max\{T, [\lambda V_T + \epsilon^*]^* + 1\}, \tag{2.3.29}$$

and propose the confidence region $\mathcal{R}_{N^*}$ *for* $\underset{\sim}{\nu}^*$. *For this modified three–stage procedure, we have:*

$$P\{\underset{\sim}{\nu}^* \in \mathcal{R}_{N^*}\} = (1 - \alpha) + o(L^{1/\zeta})$$

as $m \to \infty$.

Proof. We continue working with N given in (2.3.23) and show that the "fine–tuning" needed is indeed ϵ^* given by (2.3.28). Let $\ell = L/\theta^{*\zeta}$ and $\psi^*(x) = G^*(x^\zeta)$ for $x > 0$, so that

$$P\{\underset{\sim}{\nu}^* \in \mathcal{R}_N\} = E\{\psi^*(N\ell^{1/\zeta})\}. \tag{2.3.30}$$

Note that $\psi^*(a^{1/\zeta}) = 1 - \alpha$, $\psi^{*\prime}(a^{1/\zeta}) = \zeta a^{1-\frac{1}{\zeta}} g^*(a)$, and $\psi^{*\prime\prime}(a^{1/\zeta}) = \zeta\{\zeta a^{2-\frac{2}{\zeta}} g^{*\prime}(a) + (\zeta-1)a^{1-\frac{2}{\zeta}} g^*(a)\}$. Now we obtain

$$P\{\underset{\sim}{\nu}^* \in \mathscr{R}_N\} = E\{\psi^*(E(N)\ell^{1/\zeta})\}$$

$$+\tfrac{1}{2}\ell^{2/\zeta}\mathrm{Var}(N)\psi^{*\prime\prime}(E(N)\ell^{1/\zeta})+r_1(L), \tag{2.3.31}$$

where $|r_1(L)| \le K^*\ell^{3/\zeta}E|N - E(N)|^3 = o(L^{1/\zeta})$ by Theorem 2.3.2. Here, K^* stands for a positive generic constant independent of L. Now, let $\epsilon = \epsilon(L)$ be a sequence of constants converging to zero, to be determined explicitly very shortly. We have

$$\psi^*(E(N)\ell^{1/\zeta}) = \psi^*(a^{1/\zeta}) + \{\ell^{1/\zeta}E(N) - a^{1/\zeta}\}\psi^{*\prime}(a^{1/\zeta})$$

$$+ \tfrac{1}{2}\{\ell^{1/\zeta}E(N) - a^{1/\zeta}\}^2\psi^{*\prime\prime}(z), \tag{2.3.32}$$

for some suitable z (>0). Thus, we can write

$$\psi^*(E(N)\ell^{1/\zeta}) = (1-\alpha) + \psi^{*\prime}(a^{1/\zeta})\{\ell^{1/\zeta}E(N) - a^{1/\zeta}\} + r_2(L)$$

where $r_2(L)$ is defined to be the remainder term from (2.3.32). Let us now suppose that $\lambda = \lambda(L) = (a/L)^{1/\zeta}(1+\epsilon)$. Then, $r_2(L) = o(L^{1/\zeta} + |\epsilon|)$ as we know that $\{\ell^{1/\zeta}E(N) - a^{1/\zeta}\}^2 = o(L^{1/\zeta} + |\epsilon|)$. Also,

$$\ell^{1/\zeta}E(N)-a^{1/\zeta} = a^{1/\zeta}\epsilon+[\tfrac{1}{2}-\rho^{-1}\mathrm{Var}(W_1)]L^{1/\zeta}/\theta^*+o(L^{1/\zeta}) \tag{2.3.33}$$

and from (2.3.26), we get

$$\tfrac{1}{2}\ell^{2/\zeta}\mathrm{Var}(N) = \tfrac{1}{2}(aL)^{1/\zeta}(\rho\theta^*)^{-1}\mathrm{Var}(W_1) + o(L^{1/\zeta}). \tag{2.3.34}$$

Now, combining (2.3.31) – (2.3.32), we obtain

$$P\{\underset{\sim}{\nu}^* \in \mathscr{R}_N\} = (1-\alpha) + \psi^{*\prime}(a^{1/\zeta})\{a^{1/\zeta}\epsilon$$

$$+ [\tfrac{1}{2} - \rho^{-1}\mathrm{Var}(W_1)]L^{1/\zeta}\theta^{*-1} + o(L^{1/\zeta})\}$$

$$+o(L^{1/\zeta}+|\epsilon|)+\tfrac{1}{2}\psi^{*\prime\prime}(a^{1/\zeta})(aL)^{1/\zeta}(\rho\theta^*)^{-1}\mathrm{Var}(W_1) + o(L^{1/\zeta})$$

$$= (1-\alpha) + o(L^{1/\zeta}),$$

provided that we define $\epsilon^* = \epsilon n^*$ where ϵ satisfies the equation

$$\psi^{*\prime}(a^{1/\zeta})a^{1/\zeta}\epsilon + \psi^{*\prime}(a^{1/\zeta})\{\tfrac{1}{2} - \mathrm{Var}(W_1)\rho^{-1}\}L^{1/\zeta}\theta^{*-1}$$

$$+ \tfrac{1}{2}\psi^{*\prime\prime}(a^{1/\zeta})(aL)^{1/\zeta}\mathrm{Var}(W_1)(\rho\theta^*)^{-1} = 0,$$

which implies that ϵ^* must be the same as in (2.3.28). □

Examples 2.2.1 and 2.3.1 (Continued).

We remark that in the special setups of Hall (1981) and Mukhopadhyay and Al–Mousawi (1986), the "fine–tuning" ϵ^* given here agrees completely with theirs.

Example 2.2.2 (Continued).

Let $X_1, X_2, \ldots$ be i.i.d. with the probability density function $f(x; \nu^*, \theta^*)$ given by (2.2.5). Let $\mathscr{R}_n = \{x \in R : U_n - L < \nu^* < U_n\}$ where $U_n = \min\{X_1, \ldots, X_n\}$ and $V_n = (n-1)^{-1}\Sigma_{i=1}^n (X_i - U_n)$. The notation used here is slightly different from what one had in example 2.2.2. Here the support of the distribution depends only on ν^* and the corresponding fixed–width confidence interval problem for ν^* has the same basic setup as in (i) – (iii) of section 2.3.1 where $m \geq 2$, $\zeta = 1$, $G^*(x) = 1 - e^{-x}$, Y's are i.i.d. having the p.d.f. $f(y; 0, \theta^*)$, and $\mathrm{Var}(W_1) = 1$. Theorem 2.3.4 now implies Theorem 2 of Mukhopadhyay and Mauromoustakos (1987).

2.4 PURELY SEQUENTIAL PROCEDURE

Consider a sequence $\{N_\nu : \nu \geq 1\}$ of positive integer valued random variables defined as follows:

$$N = N_\nu = \inf\{n \geq m \; : \; n \geq \psi_\nu T_n\} \tag{2.4.1}$$

where m is the starting sample size, ψ_ν is a sequence of positive constants, $\to \infty$ as $\nu \to \infty$, and $\{T_n \; : \; n \geq m\}$ are statistics such that $P(T_n \leq 0) = 0$ for all $n \geq m$. The purely sequential procedures of Robbins (1959), Chow and Robbins (1965), Starr and Woodroofe (1972), Mukhopadhyay (1974) and many others have the same form given in (2.4.1). Under certain special circumstances, Siegmund (1968) and Bhattacharyya and Mallik (1973) studied asymptotic distributions of properly standardized stopping variables arising from specific sequential processes. We now mention a few analogous results under the general framework (2.4.1) along the lines of Ghosh and Mukhopadhyay's (1975) developments.

Theorem 2.4.1. *For the purely sequential procedure (2.4.1), if both*

$$N_\nu^{\frac{1}{2}}(T_{N_\nu} - a)/b \text{ and } N_\nu^{\frac{1}{2}}(T_{N_\nu - 1} - a)/b \tag{2.4.2}$$

converge to N(0, 1) *in distribution as* $\nu \to \infty$, *where* a(>0) *and* b(>0) *are constants, then we have:*

$$a^{\frac{1}{2}}(N_\nu - a\psi_\nu)/(b\psi_\nu^{\frac{1}{2}}) \xrightarrow{\mathscr{L}} N(0, 1) \text{ as } \nu \to \infty.$$

Proof. Consider the basic inequality

$$\psi_\nu T_{N_\nu} \leq N_\nu < m + \psi_\nu T_{N_\nu - 1}. \tag{2.4.3}$$

From (2.4.1), $N_\nu \xrightarrow{P} \infty$ as $\nu \to \infty$, hence from (2.4.2) we have $T_{N_\nu} \xrightarrow{P} a$

and $T_{N_\nu - 1} \xrightarrow{P} a$ as $\nu \to \infty$. Dividing now all sides of (2.4.3) by $a\psi_\nu$ and using the fact that $\psi_\nu \to \infty$ as $\nu \to \infty$, we get $N_\nu/(a\psi_\nu) \xrightarrow{P} 1$ as $\nu \to \infty$. Thus, (2.4.2) can alternatively be expressed as

$$(a\psi_\nu)^{\frac{1}{2}}(T_{N_\nu} - a)/b \xrightarrow{\mathscr{L}} N(0,1) \text{ and } (a\psi_\nu)^{\frac{1}{2}}(T_{N_\nu - 1} - a)/b \xrightarrow{\mathscr{L}} N(0,1) \qquad (2.4.4)$$

when $\nu \to \infty$. Hence, for $x \in R$, in view of (2.4.3),

$$\Phi(x) = \lim_{\nu \to \infty} P[(a\psi_\nu)^{\frac{1}{2}}(T_{N_\nu - 1} - a)/b \le x]$$

$$\le \liminf_{\nu \to \infty} P[(a\psi_\nu)^{\frac{1}{2}}\{(N_\nu - m)\psi_\nu^{-1} - a\}/b \le x]$$

$$= \liminf_{\nu \to \infty} P[a^{\frac{1}{2}}(N_\nu - a\psi_\nu)/(b\psi_\nu^{\frac{1}{2}}) \le x]$$

$$\le \limsup_{\nu \to \infty} P[a^{\frac{1}{2}}(N_\nu - a\psi_\nu)/(b\psi_\nu^{\frac{1}{2}}) \le x]$$

$$\le \limsup_{\nu \to \infty} P[a^{\frac{1}{2}}(\psi_\nu T_{N_\nu} - a\psi_\nu)/(b\psi_\nu^{\frac{1}{2}}) \le x]$$

$$= \limsup_{\nu \to \infty} P[(a\psi_\nu)^{\frac{1}{2}}(T_{N_\nu} - a)/b \le x] = \Phi(x).$$

Now, the proof is complete. □

Remark 2.4.1 It is clear from the proof that Φ can be replaced by any non–degenerate d.f. $F(\cdot)$. Also, one should note that the usual sufficient conditions to ensure (2.4.4) are the ones due to Anscombe (1952). These are:

(i) for a sequence n_ν of positive integers converging to infinity as $\nu \to \infty$, one has $N_\nu/n_\nu \xrightarrow{P} 1$ as $\nu \to \infty$,

(ii) $n^{\frac{1}{2}}(T_n - a)/b \xrightarrow{\mathscr{L}} N(0, 1)$ as $n \to \infty$, and

(iii) given $\epsilon > 0$, $\eta^* > 0$, there exist $\delta(>0)$ and a positive integer n_0 such that

$$P\Big[\sup_{|n'_\nu - n_\nu| \le \delta n_\nu} \Big| T_{n'_\nu} - T_{n_\nu} \Big| > \epsilon n_\nu^{-\frac{1}{2}}\Big] < \eta^* \tag{2.4.5}$$

for $n_\nu \geq n_0$. The condition (2.4.5) is known in this literature as Anscombe's "Uniform Continuity in Probability" condition. It is also often referred as the *tightness condition.*

Remark 2.4.2 Anscombe (1952) had shown that (2.4.5) is satisfied when T_n is replaced by the largest (or smallest) order statistics or the maximum likelihood estimators. Kolmogorov's inequality immediately shows that (2.4.5) holds when T_n is replaced by the sample mean of i.i.d. random variables.

One will find several applications of Theorem 2.4.1 in Mukhopadhyay (1988) and Mukhopadhyay and Al–Mousawi (1986), among several others. Now, let us discuss a few "rate of convergence" problems associated with the results of Theorem 2.4.1 in a special case, having however a very broad range of applications.

2.4.1 RATES OF CONVERGENCES FOR STOPPING TIMES BASED ON U–STATISTICS

Problems related to rates of convergences for randomly stopped U–statistics have received much attention lately. Some basic results were due to Landers and Rogge (1976), Ghosh and DasGupta (1980) and Ghosh (1980). Csenki (1980) used the results of Landers and Rogge (1976) to derive the rate of convergence of the coverage probability for Chow and Robbins' (1965) fixed–width sequential confidence interval procedure. Mukhopadhyay (1981) generalized Csenki's results by applying the tools of Ghosh and DasGupta (1980) in the framework of Sproule's (1969, 1974) sequential fixed–width confidence intervals for the means of U–statistics. Much of these findings had been unified later in Mukhopadhyay and Vik (1985), and other related citations can also be found in the references provided. See Wijsman (1991) for an up to date account. In the following, we summarize some of the results of Mukhopadhyay and Vik (1985) for completeness.

Let $X_1, X_2, \ldots$ be a sequence of i.i.d. random variables having d.f. $F(\cdot)$, where $F(\cdot)$ belongs to a family $\mathcal{F}$ of d.f.'s. Let $\phi^*(X_1,\ldots,X_r)$ be a symmetric kernel of degree r. Now for $n \geq r$, U–statistics are defined as follows:

$$U_n = \binom{n}{r}^{-1} \sum_{n,r} \phi^*(X_{\alpha_1},\ldots,X_{\alpha_r})$$

where $\sum_{n,r}$ is the summation over all combinations $\{\alpha_1 < \cdots < \alpha_r\}$ formed from the integers $\{1,2,\ldots,n\}$. Furthermore, we define

$$\theta = E\{\phi^*(X_1,\ldots,X_r)\},$$

$$\phi^*_c(x_1,\ldots,x_c) = E\{\phi^*(X_1,\ldots,X_r) \,\big|\, X_1 = x_1,\ldots,X_c = x_c\},$$

$$\xi_c = \mathrm{Var}\{\phi^*_c(X_1,\ldots,X_c)\}, \quad c = 1,2,\ldots,r.$$

As illustrations, we give two examples on U–statistics. First let $\mathcal{F}$ be a class of distribution functions F for which the variance exists. Let $\theta = \theta(F) = E(X_1)$ and $\sigma^2 = \sigma^2(F) = \mathrm{Var}(X_1)$ for all $F \in \mathcal{F}$. Then

$$U_n = \binom{n}{1}^{-1} \sum_{n,1} X_{\alpha_i} = n^{-1} \sum_{i=1}^{n} X_i = \overline{X}_n$$

is a U–statistic, where $\phi^*(X_i) = X_i$, $\phi_1^*(X_1) = X_1$, and $\xi_c = \text{Var}[\phi_1^*(X_1)] = \sigma^2$.

Next, let $\mathscr{F}$ be a class of distributions for which $E|X_1|^4$ exists. Consider $\phi^*(X_i, X_j) = \frac{1}{2}(X_i - X_j)^2$. This is a symmetric kernel of degree 2, and $E\{\phi^*(X_i, X_j)\} = \sigma^2$ for all $F \in \mathscr{F}$. Now,

$$U_n = \binom{n}{2}^{-1} \sum_{n,2} \tfrac{1}{2}(X_{\alpha_1} - X_{\alpha_2})^2$$

$$= \tfrac{1}{2}\binom{n}{2}^{-1} \sum_{1 \le i < j \le n} (X_i - X_j)^2$$

$$= \frac{1}{(n-1)n}\{n \Sigma_{i=1}^n X_i^2 - (\Sigma_{i=1}^n X_i^2 + 2\sum_{1 \le i < j \le n} X_i X_j)\}$$

$$= (n-1)^{-1} \Sigma_{i=1}^n (X_i - \overline{X}_n)^2,$$

which shows that the familiar sample variance is also a U–statistic. Here $\theta = \sigma^2$, with

$$\phi_1^*(x_1) = E\{\tfrac{1}{2}(x_1 - X_2)^2\} = \tfrac{1}{2}\{x_1^2 - 2x_1 E(X_2) + E(X_2^2)\}, \text{ and}$$

$$\xi_1 = \text{Var}\{\phi_1^*(X_1)\} = \text{Var}\{\tfrac{1}{2}(X_1^2 - 2\mu X_1)\}, \text{ where } \mu = E(X_2).$$

Now, let us return to the general theory. Hoeffding (1948) proved in his classic paper that $n^{\frac{1}{2}}(U_n - \theta) \xrightarrow{\mathscr{L}} N(0, r^2\xi_1)$ as $n \to \infty$, if $\xi_1 > 0$ and $E(\phi^{*2}) < \infty$. For a similar result when the sample size is a random variable, we state the following theorem which is due to Sproule (1969, 1974). Throughout, we write $\sup_x$ or Sup_x instead of $\sup_{x \in R}$ or $\text{Sup}_{x \in R}$ respectively.

Theorem 2.4.2 *Let $\{n_\nu\}$ be a nondecreasing sequence of positive integers tending to ∞ as $\nu \to \infty$, and let $\{N_\nu\}$ be a sequence of proper positive integer valued random variables. Assume that $\xi_1 > 0$ and $E\{\phi^{*2}\} < \infty$. Also, if*

$n_\nu^{-1} N_\nu \xrightarrow{P} 1$ *as* $\nu \to \infty$, *then we have as* $\nu \to \infty$:

$$N_\nu^{\frac{1}{2}}(U_{N_\nu} - \theta) \xrightarrow{\mathscr{L}} N(0, r^2\xi_1) \text{ and } N_\nu^{\frac{1}{2}}(U_{N_\nu - 1} - \theta) \xrightarrow{\mathscr{L}} N(0, r^2\xi_1).$$

Assume the following:

$\{U_n: n \geq r\}$ is a sequence of U–statistics and $U_n \geq 0$ w.p. 1;
$E\{U_n\} = \theta \in R^+$ and $E\{\phi^{*2}\} < \infty$;
$\alpha(>0)$, $\gamma(>0)$, $\delta(>0)$ and $t(\geq 0)$ are all constants;
$\{\psi_\nu: \nu = 1,2,...\}$ is a sequence of nondecreasing real numbers (>0) where $\psi_\nu \to \infty$ as $\nu \to \infty$.

Now, define a stopping time as

$$N_\nu = \inf\{n \geq m: (n + t)^\delta \geq \psi_\nu(U_n^\alpha + n^{-\gamma})\}, \tag{2.4.6}$$

$\nu = 1,2,...$; $m(\geq r)$ is an integer and can be thought as the starting sample size. One can prove that

$P\{N_\nu < \infty\} = 1$ *for* $\nu = 1,2,...$;
$N_{\nu_1} \leq N_{\nu_2}$ *w.p. 1 if* $\nu_1 \leq \nu_2$;
$N_\nu \to \infty$ *w.p. 1 as* $\nu \to \infty$; and
$U_{N_\nu} \to \theta$ *w.p. 1*, $U_{N_\nu - 1} \to \theta$ *w.p. 1*, and $N_\nu/(\psi_\nu^{1/\delta}\theta^{\alpha/\delta}) \to 1$ *w.p. 1 as* $\nu \to \infty$.

Also, if $E\{|\phi^*|^{2\xi}\} < \infty$ for some $\xi > n_0(\geq 1)$, then

$$\lim_{\nu \to \infty} E\{[N_\nu/(\psi_\nu^{1/\delta}\,\theta^{\alpha/\delta})]^{n_0}\} = 1.$$

Now, we state a result that is a direct generalization of Theorem 2.4.1:

Theorem 2.4.3. *Consider a sequence* $\{N^*_\nu\}$ *of positive integer valued random variables defined by* $N^*_\nu = \inf\{n \geq m: n \geq b_\nu T_n - t\}$ *where* $m \geq 1$, $\{b_\nu: \nu = 1,2,...\}$ *is a sequence of positive real numbers tending to infinity as* $\nu \to \infty$, $\{T_n: n \geq m\}$ *is a sequence of statistics such that* $P\{T_n > 0\} = 1$ *for all* $n \geq m$, *and* $t \in R$ *is a constant. Now, if for all* $x \in R$,

$$\lim_{\nu\to\infty} P\{N^{*\frac{1}{2}}_\nu (T_{N^*_\nu} - c) \leq dx\} = \Phi(x), \quad \textit{and}$$

$$\lim_{\nu\to\infty} P\{N^{*\frac{1}{2}}_\nu (T_{N^*_\nu - 1} - c) \leq dx\} = \Phi(x)$$

for some constants $c(>0)$ *and* $d(>0)$, *then we have:*

$$\lim_{\nu\to\infty} P\{c^{\frac{1}{2}}(N^*_\nu - cb_\nu) \leq xdb^{\frac{1}{2}}_\nu\} = \Phi(x) \quad \textit{for all } x \in R.$$

Theorem 2.4.4. *Assume* $\gamma > \frac{1}{2}$ *in the definition of* N_ν *given in (2.4.6). Then we have:*

$$(c/b_\nu)^{\frac{1}{2}} (N_\nu - cb_\nu) \xrightarrow{\mathscr{L}} N(0, d^2),$$

as $\nu \to \infty$, *where* $c = \theta^{\alpha/\delta}$, $b_\nu = \psi_\nu^{1/\delta}$ *and* $d = r\xi_1^{\frac{1}{2}}\alpha\delta^{-1}\theta^{(\alpha-\delta)/\delta}$.

Theorem 2.4.5. *Let* N^*_ν, b_ν *and* T_n *be defined as in Theorem 2.4.3 and let* $\{\epsilon_\nu\}$ *be a sequence of positive real numbers such that* $\epsilon_\nu \geq b_\nu^{-1}$. *Now, if*

$$P\{|N^*_\nu (cb_\nu)^{-1} - 1| > \epsilon_\nu\} = 0(\epsilon_\nu^{\frac{1}{2}}),$$

$$\sup_x |P\{N^{*\frac{1}{2}}_\nu (T_{N^*_\nu} - c) \leq dx\} - \Phi(x)| = 0(\epsilon_\nu^{\frac{1}{2}}), \quad \textit{and}$$

$$\sup_x |P\{N_\nu^{*\frac{1}{2}}(T_{N_\nu^*-1} - c) \le dx\} - \Phi(x)| = 0(\epsilon_\nu^{\frac{1}{2}}),$$

for some constants c(>0) *and* d(>0), *then we have:*

$$\sup_x |P\{c^{\frac{1}{2}}(N_\nu^* - cb_\nu) \le xdb_\nu^{\frac{1}{2}}\} - \Phi(x)| = 0(\epsilon_\nu^{\frac{1}{2}}).$$

Theorem 2.4.6. *Consider the stopping time* N_ν *defined in (2.4.6) and let* $\gamma > \frac{1}{2}$. *Also let* $\lambda \in (0, \frac{1}{2})$. *Now if* $E\{|\phi^*|^{2\xi}\} < \infty$ *for* $\xi \ge \max\{(5-2\lambda)/8\lambda, 1 + (\delta+\gamma)(1-2\lambda)/4\delta\}$, *then we have:*

$$P\{|N_\nu n_\nu^{-1} - 1| > c\psi_\nu^{-(\frac{1}{2}-\lambda)/\delta}\} = 0(\psi_\nu^{-(\frac{1}{2}-\lambda)/2\delta}),$$

for arbitrary c(>0) *and* $n_\nu = \psi_\nu^{1/\delta}\theta^{\alpha/\delta}$.

Theorem 2.4.7. *Consider* N_ν *as defined in (2.4.6), and assume that* $\gamma > \frac{3}{4}$. *Let* $\lambda \in (0, \frac{1}{2})$. *If* $E\{|\phi^*|^{2\xi}\} < \infty$ *for* $\xi \ge \max\{(5-2\lambda)/(1+2\lambda), (5-2\lambda)/8\lambda, 1 + (\delta+\gamma)(1-2\lambda)/4\delta\}$, *then we have:*

$$\sup_x |P\{c^{\frac{1}{2}}(N_\nu - cb_\nu) \le xdb_\nu^{\frac{1}{2}}\} - \Phi(x)| = 0(\psi_\nu^{-(\frac{1}{2}-\lambda)/2\delta}),$$

where $c = \theta^{\alpha/\delta}$, $b_\nu = \psi_\nu^{1/\delta}$ *and* $d = r\xi_1^{\frac{1}{2}}\alpha\delta^{-1}\theta^{(\alpha-\delta)/\delta}$.

2.4.2 NONLINEAR RENEWAL THEORY

The *nonlinear renewal theory* was developed and fully exploited in the area of sequential analysis in order to derive asymptotic second–order expansions of the risk functions associated with certain specific problems in the area of estimation and testing of hypotheses. Major breakthroughs came from Lai and Siegmund (1977, 1979) and Woodroofe (1977). See also the comprehensive accounts in Woodroofe (1982, 1991) and Siegmund (1985). In what follows, we consider the setup discussed in Woodroofe (1977); however, we also present some of the materials along the lines of Mukhopadhyay (1988). Suppose that $W_1, W_2, \ldots$ are i.i.d. *positive* and *continuous* random variables having *all positive moments finite*. Define

$$N = N(h^*) = \inf\{n \geq m : \Sigma_{i=1}^{n} W_i < h^* n^{\delta} L(n)\}, \tag{2.4.7}$$

where $\delta > 1$, $h^* > 0$, $m \geq 1$ and $L(u) = 1 + L_0 u^{-1} + o(u^{-1})$ as $u \to \infty$ with $-\infty < L_0 < \infty$. Write $\theta = E(W_1)$, $\tau^2 = E(W_1^2) - \theta^2$, $\beta^* = 1/(\delta-1)$, $n_0^* = (\theta/h^*)^{\beta^*}$ and $p = \beta^{*2}\tau^2/\theta^2$. Assume that

$$P(W_1 \leq u) \leq B^* u^b \text{ for all } u > 0 \tag{2.4.8}$$

for some $B^* > 0$ and $b > 0$. Let

$$\nu = \frac{\beta^*}{2\theta}\{(\delta-1)^2\theta^2 + \tau^2\} - \Sigma_{n=1}^{\infty} n^{-1} E\{\max(0, \Sigma_{i=1}^{n} W_i - n\delta\theta)\}, \tag{2.4.9}$$

$$\eta = \beta^*\theta^{-1}\nu - \beta^* L_0 - \tfrac{1}{2}\delta\beta^{*2}\tau^2\theta^{-2}. \tag{2.4.10}$$

Let $U^* = \{N(h^*) - n_0^*\}/n_0^{*\frac{1}{2}}$.

Theorem 2.4.8. *For the stopping variable* $N(h^*)$ *defined in (2.4.7), we have as* $h^* \to 0$:

(i) $P\{N(h^*) \leq \epsilon n_0^*\} = O(h^{*mb})$ *if* $0 < \epsilon < 1$ *and* $m \geq 1$;

(ii) $U^* \xrightarrow{\mathscr{L}} N(0, p)$ *if* $m \geq 1$;

(iii) $\sup_x |P(U \leq p^{\frac{1}{2}} x) - \Phi(x)| = O(h^{*(\frac{1}{4}-\frac{1}{2}\lambda)\beta^*})$ *if* $0 < \lambda < \frac{1}{2}$ *and* $m \geq 1$;

(iv) $|U^*|^{\omega}$ *is uniformly integrable if* $mb > \frac{1}{2}\beta^*\omega$ *where* ω *is an arbitrary, but fixed positive number;*

(v) $E\{[N(h^*)/n_0^*]^{\omega}\} = 1 + \{\omega\eta + \frac{1}{2}\omega(\omega-1)p\}n_0^{*-1} + o(n_0^{*-1})$ *if* (a) $m > (3-\omega)\beta^*/b$ *for* $\omega \in (-\infty, 2) - \{-1, 1\}$, (b) $m > \beta^*/b$ *for* $\omega = 1$ *and* $\omega \geq 2$, *and* (c) $m > 2\beta^*/b$ *for* $\omega = -1$. *Here* b *and* η *come from (2.4.8) and (2.4.10) respectively, while* $p = \beta^{*2}\tau^2/\theta^2$ *and* ω *is an arbitrary, but fixed real number.*

Proof. Parts (i) and (iv) were proved in Woodroofe's (1977) Lemma 2.3 and Theorem 2.3 respectively. Part (v) with $\omega = 1$ was proved in Theorem 2.4 in Woodroofe (1977). Part (v) in its fullest generality was proved in Theorem 3 of Mukhopadhyay (1988). We will now utilize the tools summarized in section 2.4.1 to prove parts (ii) – (iii). The stopping variable $N(h^*)$ can be rewritten as

$$N(h^*) = \inf\{n \geq m \ : \ n \geq (h^*)^{-\beta^*}[\overline{W}_n/L(n)]^{\beta^*}\} \tag{2.4.11}$$

where $\overline{W}_n = n^{-1}\Sigma_{i=1}^{n}W_i$. Now, (2.4.11) agrees with (2.4.1) where one writes $\nu = [1/n^*]^* + 1$, $\psi_\nu = (h^*)^{-\beta^*}$ and $T_n = \{\overline{W}_n/L(n)\}^{\beta^*}$. It is easy to see that T_{N_ν} and $T_{N_\nu - 1}$ satisfy (2.4.2) with "a" $= \theta^{\beta^*}$ and "b" $= \tau\beta^*\theta^{\beta^*-1}$. Hence, $a^{\frac{1}{2}}(N_\nu - a\psi_\nu)/(b\psi_\nu^{\frac{1}{2}}) = [N(h^*) - n_0^*]/[n_0^{*\frac{1}{2}}\tau\beta^*\theta^{-1}]$ and now part (ii) follows from Theorem 2.4.1

Also (2.4.11) "agrees" with (2.4.6) where $t = 0$, $\delta = 1$, $\alpha = \beta^*$, $\psi_\nu = h^{*-\beta^*}$. Note that $U_n = \overline{W}_n/L(n)$, but $L(n) \to 1$ as $n \to \infty$ and also that one does not quite need the "fudge" factor $n^{-\gamma}$ in (2.4.6) in our special case since W's are i.i.d. positive and continuous random variables. Also, it is assumed that W's have all moments finite. Hence, part (iii) follows immediately from Theorem 2.4.7. □

2.4.2.1 Application of Nonlinear Renewal Theory

Suppose that $X_1, Y_1, X_2, Y_2, \ldots$ are independent random variables where X's are i.i.d. $N(\mu_1, \sigma^2)$, Y's are i.i.d. $N(\mu_2, \sigma^2)$ with $(\mu_1, \mu_2, \sigma) \in R \times R \times R^+$. Given $d(>0)$ and $\alpha \in (0, 1)$, we wish to construct a confidence interval for $\Delta = \mu_1 - \mu_2$, having length 2d and the confidence coefficient *approximately* $(1 - \alpha)$. Having recorded $X_1, Y_1, \ldots, X_n, Y_n$, let us write $\overline{X}_n = n^{-1}\Sigma_{i=1}^n X_i$, $\overline{Y}_n = n^{-1}\Sigma_{i=1}^n Y_i$, $S_{1n}^2 = (n-1)^{-1}\Sigma_{i=1}^n (X_i - \overline{X}_n)^2$, $S_{2n}^2 = (n-1)^{-1}\Sigma_{i=1}^n (Y_i - \overline{Y}_n)^2$ and $S^2(n) = \frac{1}{2}(S_{1n}^2 + S_{2n}^2)$, for $n \geq 2$. Consider the confidence interval

$$I_n = [\overline{X}_n - \overline{Y}_n \pm d]$$

for Δ. Now, $P\{\Delta \in I_n\} = 2\Phi(d[n/2]^{\frac{1}{2}}\sigma^{-1}) - 1 \geq 1 - \alpha$, provided that $d(n/2)^{\frac{1}{2}}\sigma^{-1} \geq a_0$, that is $n \geq 2a_0^2\sigma^2/d^2 = n_1^*$, say, where $\Phi(a_0) = 1 - \frac{1}{2}\alpha$. In reality, all the parameters will be unknown and hence n_1^* is indeed unknown. So, one starts with $m(\geq 2)$ observations on both X's and Y's and then takes one pair of (X, Y) at a time according to the following purely sequential stopping rule. Let

$$N = N(d) = \inf\{n \geq m : n \geq 2a_0^2S^2(n)/d^2\}.$$

$P(N < \infty)$ is, of course, unity and hence one proposes the fixed–width confidence interval I_N for Δ. It is easy to see that $P\{\Delta \in I_N\} = 2E[\Phi(d[N/2]^{\frac{1}{2}}\sigma^{-1})] - 1$. Now, N can be equivalently written as $Q + 1$ where

$$Q = \inf\{n \geq m-1 \ : \ 2n(n+1)n_1^{*-1} \geq \Sigma_{i=1}^n W_i\},$$

the W's being i.i.d. χ_2^2. Now, Q agrees with (2.4.7) where $\delta = 2$, $L_0 = 1$, $h^* = 2/n_1^*$, $\theta = 2$, $\tau^2 = 4$, $\beta^* = 1$, $n_0^* = n_1^*$, $b = 1$ and $p = 1$. Now, Theorem 2.4.8 certainly becomes very relevant for finding asymptotic (as $d \to 0$) second–order expansions of E(N) and $P\{\Delta \in I_N\}$. See problems 2.2 and 2.17 in this context.

2.5 Accelerated Sequential Procedure

In principle, we now try to accelerate the purely sequential procedure (2.4.7). Suppose then we have W's which are i.i.d. *positive* and *continuous* random variables having *all positive moments finite.* The basic idea is to go forward sequentially to a particular step along the lines of (2.4.7) and then augment that by observing a few more W's in one single batch. To be specific, we first choose and fix $0 < \rho < 1$, $q \geq 0$ and let

$$t(h^*) = \inf\{n \geq m \ : \ h^* n^\delta \geq \rho \overline{W}_n\} \tag{2.5.1}$$

with $h^* > 0$, $m \geq 1$, $\delta > 0$ and $\overline{W}_n = n^{-1}\Sigma_{i=1}^n W_i$. Now, based on $W_1, \ldots, W_m, \ldots, W_{t(h^*)}$, we define

$$N_1(h^*) = [\{h^{*-1}\overline{W}_{t(h^*)}\}^{1/\delta} + q]^* + 1, \tag{2.5.2}$$

$$N(h^*) = \max\{t(h^*), N_1(h^*)\}. \tag{2.5.3}$$

If $t(h^*) \geq N_1(h^*)$, then sampling terminates at stage $t(h^*)$, that is, with $W_1, \ldots, W_{t(h^*)}$. On the other hand, we sample the difference $N_1(h^*) - t(h^*)$ in *one single batch* if $N_1(h^*) > t(h^*)$. We refer to $N(h^*)$ given by (2.5.3) as the *accelerated stopping time.* Hall (1983) has studied second–order asymptotic properties of $N(h^*)$ in a particular situation involving the construction of fixed–width confidence intervals for the mean of a normal population that also had an unknown variance. The present structure is very general and it covers, as we shall see in Chapters 3 – 7, other examples of interest. This unified theory is summarized from Mukhopadhyay and Solanky (1991).

2.5.1 SECOND–ORDER EXPANSIONS

Let $E(W_1) = \theta$, $Var(W_1) = \tau^2$ and $n_0 = (\rho\theta/h^*)^{1/\delta}$. It is easy to see that $P\{t(h^*) < \infty\} = 1$, $E\{t(h^*)\} < \infty$ for all fixed $h^* > 0$. As $h^* \to 0$, $t(h^*) \to \infty$, $t(h^*)/n_0 \to 1$ w.p. 1, and $E\{t(h^*)/n_0\} \to 1$, so that $h^{*1/\delta}t(h^*)$ is uniformly integrable in $h^* < h_0$, for some $h_0 > 0$. It is known that $\overline{W}_{t(h^*)} \to \theta$ w.p. 1 as $h^* \to 0$. As before, $I(\cdot)$ stands for the indicator function of $(\cdot)$. We assume the smoothness condition (2.4.8) and note that "ν" given by (2.4.9) reduces to

$$\nu=\tfrac{1}{2}[\delta^2\theta^2+\tau^2](\delta\theta)^{-1}-\Sigma_{n=1}^{\infty}n^{-1}E[\max\{0,\Sigma_{i=1}^{n}W_i-n\theta(1+\delta)\}]. \quad (2.5.4)$$

Define $Q^*(h^*) = (n_0)^{-\frac{1}{2}}\{t(h^*)-n_0\}$ and $p = \tau^2(\delta^2\theta^2)^{-1}$. From Theorem 2.4.1 and along the lines of Theorem 2.4.8 (ii), it follows that

$$Q^*(h^*) \xrightarrow{\mathscr{L}} N(0, p) \text{ as } h^* \to 0,$$

and Theorem 2.4.8 (iv) – (v) respectively lead to:

$$|Q^*(h^*)|^{\omega} \text{ is uniformly integrable if } m>\tfrac{1}{2}\omega(b\delta)^{-1}, \quad (2.5.5)$$

$$E\{t(h^*)\}=n_0+\nu(\delta\theta)^{-1}-\tfrac{1}{2}p(1+\delta)+o(1) \text{ if } m>(b\delta)^{-1}, \quad (2.5.6)$$

for $\omega > 0$.

Let r, s, r' and s' be all positive numbers such that $r^{-1} + s^{-1} = r'^{-1} + s'^{-1} = 1$. We define

$$m_0(\delta)=\begin{cases}\max\{\tfrac{1}{2}sr'(b\delta)^{-1},\ ss'(b\delta)^{-1},\ 2/(\delta^3 b),\ \delta+1\} \\ \qquad \text{if } \delta = 1 \text{ or } \delta \geq 2; \\ \max\{\tfrac{1}{2}sr'(b\delta)^{-1},\ ss'(b\delta)^{-1},\ (3-\delta)(b\delta)^{-1}, \\ \qquad 2/(\delta^3 b),\ \delta+1\} \\ \qquad \text{if } 0 < \delta < 1 \text{ or } 1 < \delta < 2; \end{cases} \tag{2.5.7}$$

where "b" comes from (2.4.8).

Theorem 2.5.1. *For the accelerated stopping time* $N(h^*)$ *defined in (2.5.3), we have as* $h^* \to 0$:

(i) $N(h^*)/n_0^* \to 1$ w.p. 1;

(ii) $E\{N(h^*)/n_0^*\} \to 1$;

(iii) $n_0^{*-\frac{1}{2}}\{N(h^*) - n_0^*\} \xrightarrow{\mathscr{L}} N(0, p\rho^{-1/\delta})$;

where $n_0^* = (\theta/h^*)^{1/\delta}$ *and* $p = \tau^2(\delta^2\theta^2)^{-1}$.

Proof. Since $\overline{W}_{t(h^*)} \to \theta$ w.p. 1, $t(h^*)/(\rho^{1/\delta}n_0^*) \to 1$ w.p. 1 and $N_1(h^*)/n_0^* \to 1$ w.p. 1 as $h^* \to 0$, part (i) immediately follows from (2.5.2) – (2.5.3). Note therefore that $P\{N(h^*) = N_1(h^*)\ \forall\ h^* < h_0\} = P\{N_1(h^*) \geq t(h^*)$ * for all $h^* < h_0\} \to 1$ as $h_0 \to 0$. Now, $t(h^*)I(t(h^*) > N_1(h^*))/n_0^* \leq t(h^*)/n_0^*$, and hence the left hand side is uniformly integrable since $h^{*1/\delta}t(h^*)$ is uniformly integrable. Thus,

$$E\{t(h^*)I(t(h^*) > N_1(h^*))/n_0^*\} = o(1), \tag{2.5.8}$$

and since

$$N(h^*) = t(h^*)I(t(h^*) > N_1(h^*)) + N_1(h^*)I(N_1(h^*) \geq t(h^*)),$$

to prove part (ii), it suffices to show that

$$\limsup_{h^*\to 0} E\{N_1(h^*)I(N_1(h^*) \geq t(h^*))/n_0^*\} \leq 1. \tag{2.5.9}$$

Now, $N_1(h^*) \leq \{h^{*-1}\overline{W}_{t(h^*)}\}^{1/\delta} + q + 1$, and hence

$$E\{N_1(h^*)I(N_1(h^*) \geq t(h^*))/n_0^*\} \leq E\{(\overline{W}_{t(h^*)}\theta^{-1})^{1/\delta}\} + h^{*1/\delta}o(1).$$

From (2.5.1), we have $(\theta^{-1}\overline{W}_{t(h^*)})^{1/\delta} \leq t(h^*)/n_0$ which immediately implies $\limsup_{h^*\to 0} E\{(\theta^{-1}\overline{W}_{t(h^*)})^{1/\delta}\} \leq 1$. Hence, (2.5.9) is proved.

For part (iii), note that the asymptotic distributions of $h^{*1/\delta}(\rho^{1/\delta}n_0^*)^{\frac{1}{2}}\{N_1(h^*) - n_0^*\}$ and $(\rho^{1/\delta}n_0^*)^{\frac{1}{2}}(\overline{W}_{t(h^*)}^{1/\delta} - \theta^{1/\delta})$ are identical and then appeal to Anscombe's (1952) result. □

For the next Theorem, we define

$$M^*(h^*) = \max\{t(h^*), (h^{*-1}\overline{W}_{t(h^*)})^{1/\delta} + q\}. \tag{2.5.10}$$

Theorem 2.5.2. *For* $N(h^*)$, $M^*(h^*)$ *and* $m_0(\delta)$ *defined respectively in (2.5.3), (2.5.10) and (2.5.7), we have as* $h^* \to 0$:

(i) $E\{M^*(h^*)\} = n_0^* - p\rho^{-1/\delta} + \frac{1}{2}(1-\delta)\tau^2\delta^{-2} + q + o(1)$ *if* $m > m_0(\delta)$;

(ii) $\mathrm{Var}\{N(h^*)\} = \rho^{-1/\delta}pn_0^* + o(h^{*-1/\delta})$ *if* $m > (b\delta)^{-1}$;

(iii) $E\{|N(h^*) - E(N(h^*))|^r\} = 0(h^{*-\frac{1}{2}r/\delta})$ *for* $r > 0$ *if* $m > \frac{1}{2}r(b\delta)^{-1}$.

Proof. In order to prove part (i), we first note that we may indeed do so with

$\{h^{*-1}\overline{W}_{t(h^*)}\}^{1/\delta} + q = Q'(h^*)$, say, instead of $M^*(h^*)$, that is, when $m > \max\{2/(\delta^3 b), (\delta+1)\}$,

$$E\{M^*(h^*)\} = E\{Q'(h^*)I(M^*(h^*) = Q'(h^*))\} + o(1). \qquad (2.5.11)$$

Let $Y(h^*) = h^* t^{\delta+1}(h^*)\rho^{-1} - \Sigma_{i=1}^{t(h^*)} W_i$, the excess over the boundary for the stopping time in (2.5.1) in the Woodroofe (1977) sense, and thus $\overline{W}_{t(h^*)} = \rho^{-1}h^* t^{\delta}(h^*) - Y(h^*)t^{-1}(h^*)$. From Theorem 2.4.8 (v), with $\omega = 1$, we obtain

$$E\{h^* t^{\delta}(h^*)/\rho\} = \theta + \theta\{\delta\eta^* + \tfrac{1}{2}\delta(\delta-1)p\}n_0^{-1} + o(n_0^{-1}) \qquad (2.5.12)$$

if (i) $m > (b\delta)^{-1}$ when $\delta = 1$ or $\delta \geq 2$, and (ii) $m > (3-\delta)(b\delta)^{-1}$ when $0 < \delta < 1$ or $1 < \delta < 2$, where $\eta^* = \nu(\delta\theta)^{-1} - \tfrac{1}{2}p(1+\delta)$. Recall that ν was defined in (2.5.4). Now, choose r, r', s, s' such that they are all positive and $r^{-1} + s^{-1} = r'^{-1} + s'^{-1} = 1$. From Woodroofe's (1977) Theorem 2.2, we obtain $E\{Y(h^*)\} = \nu + o(1)$, and hence

$$E\{Y(h^*)/t(h^*)\} = \nu n_0^{-1} + o(n_0^{-1}) - E\{Y(h^*)[\frac{t(h^*)-n_0}{n_0 t(h^*)}]\}. \qquad (2.5.13)$$

Now,

$E\{|Y(h^*)(\frac{t(h^*)-n_0}{t(h^*)})|\} \leq n_0^{-\frac{1}{2}s}O(1)E^{1/sr'}\{|Q'(h^*)|^{sr'}\}E^{1/ss'}\{|\frac{n_0}{t(h^*)}|^{ss'}\}$ $= o(1)$, if $m > \max\{\tfrac{1}{2}sr'(b\delta)^{-1}, ss'(b\delta)^{-1}\}$. Here, we appealed to Theorem 2.4.8 (iv) and used the fact that $\{n_0/t(h^*)\}^{ss'}$ is uniformly integrable if $m > ss'(b\delta)^{-1}$. Thus, from (2.5.13), we can write

$$E\{Y(h^*)/t(h^*)\} = \nu n_0^{-1} + o(n_0^{-1}) \qquad (2.5.14)$$

if $m > \max\{\tfrac{1}{2}sr'(b\delta)^{-1}, ss'(b\delta)^{-1}\}$, and hence we immediately obtain from

(2.5.12) and (2.5.14)

$$E\{(\overline{W}_{t(h^*)}-\theta)h^{*-1/\delta}\} = -\ \delta p\theta(\rho\theta)^{-1/\delta} + o(1) \qquad (2.5.15)$$

if (i) $m > \max\{(b\delta)^{-1}, \frac{1}{2}sr'(b\delta)^{-1}, ss'(b\delta)^{-1}\}$ when $\delta = 1$ or $\delta \geq 2$, and (ii) $m > \max\{(3-\delta)(b\delta)^{-1}, \frac{1}{2}sr'(b\delta)^{-1}, ss'(b\delta)^{-1}\}$ when $\delta \in (0, 2) - \{1\}$.

From Theorem 2 of Chow et al. (1979), it is immediate to observe that $n_0(\overline{W}_{t(h^*)}-\theta)^2/\tau^2$ is uniformly integrable if $m > (b\delta)^{-1}$. Hence, for $m > (b\delta)^{-1}$, we have

$$E[n_0(\overline{W}_{t(h^*)}-\theta)^2] = \tau^2 + o(1). \qquad (2.5.16)$$

Now, with appropriate conditions on m, we can write by combining (2.5.15) – (2.5.16),

$$E\{h^{*-1/\delta}\overline{W}^{1/\delta}_{t(h^*)}\} = (h^{*-1}\theta)^{1/\delta} - p\rho^{-1/\delta} + \tfrac{1}{2}(1-\delta)\tau^2\delta^{-2} + o(1) + r(h^*) \quad (2.5.17)$$

where $|r(h^*)| \leq K^* n_0^{-\frac{1}{2}} E\{|\frac{n_0^{\frac{1}{2}}(\overline{W}_{t(h^*)}-\theta)}{\theta}|^3\} = o(1)$ if $m > (b\delta)^{-1}$. Here K^* is a generic positive constant independent of h^*. Since $ss'(b\delta)^{-1} > (b\delta)^{-1}$, we obtain from (2.5.17)

$$E\{h^{*-1/\delta}\overline{W}^{1/\delta}_{t(h^*)}\} = (h^{*-1}\theta)^{1/\delta} - p\rho^{-1/\delta} + \tfrac{1}{2}(1-\delta)\tau^2\delta^{-2} + o(1) \quad (2.5.18)$$

if (i) $m > \max\{\frac{1}{2}sr'(b\delta)^{-1}, ss'(b\delta)^{-1}\}$ when $\delta = 1$ or $\delta \geq 2$, and also (ii) $m > \max\{(3-\delta)(b\delta)^{-1}, \frac{1}{2}sr'(b\delta)^{-1}, ss'(b\delta)^{-1}\}$ when $0 < \delta < 1$ or $1 < \delta < 2$. Hence part (i) follows in view of (2.5.11).

Now, we turn to part (ii). Note that $\rho^{-1/\delta}t(h^*) - \{h^{*-1}\overline{W}_{t(h^*)}\}^{1/\delta} \leq$

$(\delta+1)/\delta$. Hence, $\mathrm{Var}\{(h^{*-1}\overline{W}_{t(h^*)})^{1/\delta}\} \approx \rho^{-2/\delta}\mathrm{Var}[t(h^*)]$, and now apply (2.5.5) with $\omega = 2$. Thus, part (ii) follows, and part (iii) is very similar. □

Corollary 2.5.1. *For* $N(h^*)$ *defined in (2.5.3), we have as* $h^* \to 0$:

$$\eta'' \le \liminf \{E(N(h^*)) - n_0^*\} \le \limsup \{E(N(h^*)) - n_0^*\} \le \eta'' + 1$$

if $m > m_0(\delta)$, *where* $m_0(\delta)$ *comes from (2.5.7) and* $\eta'' = \frac{1}{2}(1-\delta)\tau^2\delta^{-2} + q - p\rho^{-1/\delta}$.

Proof. The result is obvious from of Theorem 2.5.2 (i) since $M^*(h^*) \le N(h^*) \le M^*(h^*) + 1$. □

Remark 2.5.1: Suppose that $\rho^{-1/\delta}$ is an integer. Then, one can show that

$$E\{N(h^*)\} = n_0^* + \eta' + o(1) \tag{2.5.19}$$

if $m > m_0(\delta)$ where $\eta' = -p\rho^{-1/\delta} + \frac{1}{2}(1-\delta)\tau^2\delta^{-2} + q + q^*$, and $q^* = E\{[q + \rho^{-1/\delta}(1-Z)]^* - 1 + [q + \rho^{-1/\delta}(1-Z)]\}$. Here, Z is a suitable random variable having a continuous distribution on $(0, 1)$. The proof is similar to what is given in section 2 (page 221) of Hall (1983).

Theorem 2.5.3. *Assume that* $\rho^{-1/\delta}$ *is an integer. Then, for* $N(h^*)$ *defined in (2.5.3) and* $0 < \epsilon < \rho^{1/\delta}$, *we have as* $h^* \to 0$:

(i) $P(N(h^*) \le \epsilon n_0^*) = O(h^{*mb})$;

(ii) $E\{[N(h^*)/n_0^*]^l\} = 1 + \{l\eta' + \frac{1}{2}l(l-1)p\rho^{-1/\delta}\}n_0^{*-1} + o(n_0^{*-1})$

for fixed $l < 0$ *if* (i) $m > \max\{m_0(\delta), 2(b\delta)^{-1}\}$ *when* $l = -1$, and also (ii) $m > \max\{m_0(\delta), (3-l)(b\delta)^{-1}\}$ *when* $l \in (-\infty, 0) - \{-1\}$, *where* η' *and* $m_0(\delta)$ *respectively come from (2.5.19) and (2.5.7).*

Proof. $P(N(h^*) \le \epsilon n_0^*) \le P\{t(h^*) \le \epsilon n_0^*\} = P\{t(h^*) \le \epsilon' n_0\}$ with ϵ' defined as $\epsilon\rho^{-1/\delta}$. Now, part (i) follows from Theorem 2.4.8 (i). Part (ii) follows from Theorem 2.4.8 (v). □

Remark 2.5.2. One will note that $E[\{N(h^*) - n_0^*\}^2/N(h^*)] = p\rho^{-1/\delta} + o(1)$ if $m > \max\{m_0(\delta), 2(b\delta)^{-1}\}$ *for all values of* ρ *in* $(0, 1)$. Also, Theorem 2.5.3 (i) holds *for all values of* ρ *in* $(0, 1)$.

Corollary 2.5.2. *Assume that* $0 < \rho < 1$ *is arbitrary. Then, for* $N(h^*)$ *defined in (2.5.3), we have as* $h^* \to 0$:

$$1 + \{l(\eta''+1) + \tfrac{1}{2}l(l-1)p\rho^{-1/\delta}\}n_0^{*-1} + o(n_0^{*-1}) \le E\{[N(h^*)/n_0^*]^l\}$$

$$\le 1 + \{l\eta'' + \tfrac{1}{2}l(l-1)p\rho^{-1/\delta}\}n_0^{*-1} + o(n_0^{*-1})$$

for fixed $l < 0$ *if* (i) $m > \max\{m_0(\delta), 2(b\delta)^{-1}\}$ *when* $l = -1$, and also (ii) $m > \max\{m_0(\delta), (3-l)(b\delta)^{-1}\}$ *when* $l \in (-\infty, 0) - \{-1\}$, *where* η'' is defined as $\frac{1}{2}(1-\delta)\tau^2\delta^{-2} + q - p\rho^{-1/\delta}$ *and* $m_0(\delta)$ *comes from (2.5.7).*

2.6 NOTES AND OVERVIEW

In section 2.4.2, most of the proofs have been left out for brevity since the nonlinear renewal theory of purely sequential procedures has already been well documented in Lai and Siegmund (1977, 1979), Woodroofe (1977, 1982, 1991), Siegmund (1985) and Mukhopadhyay (1988), among others.

There is much need to consider the three–stage analogue of (2.3.1) – (2.3.2) where n^* is of the form $\lambda\theta^\delta$ for fixed but arbitrary $\delta > 0$ and θ is estimated by a sample mean of i.i.d. positive and continuous random variables. In section 2.3, we of course had a special case, namely $\delta = 1$. In the general case, we conjecture that for $k \in \{1,2,3,\ldots\} \cup (-\infty, 0)$,

$$E(N^k) = n^{*k} + \tfrac{1}{2}kn^{*k-1}[\delta(\delta k - 2\delta - 1)\mathrm{Var}(W_1) + \rho]\rho^{-1} + o(n^{*k-1}), \quad (2.6.1)$$

under appropriate moment conditions on Y's. Such generalizations will be of considerable practical interest in solving some otherwise very difficult estimation and related problems.

One should also investigate second–order properties of a three–stage procedure defined along the lines of (2.3.1) – (2.3.2) where θ is estimated instead by something like $\overline{Y}_n + Y_n^*$ where Y_n^* is an appropriate sequence of slowly changing nonnegative random variables.

Second–order properties should be investigated for the accelerated sequential procedure defined along the lines of (2.5.1) – (2.5.3) when $\overline{W}_n$ is replaced by $\overline{W}_n + Y_n^*$, Y_n^* being an appropriate sequence of slowly changing nonnegative random variables.

Complements and Problems

2.1 Suppose that $X_1,\dots,X_n$ are i.i.d. $N(\mu, \sigma^2)$, $(\mu, \sigma^2) \in R \times R^+$. Let $\overline{X}_m = m^{-1}\Sigma_{i=1}^m X_i$, $S_m^2 = (m-1)^{-1}\Sigma_{i=1}^m (X_i - \overline{X}_m)^2$, $n \geq m \geq 2$. Show that $\overline{X}_n$ and $(S_2^2,\dots,S_n^2)$ are independent.

[*Hints*: For arbitrary but fixed $\sigma \in R^+$, note that $\overline{X}_n$ is both complete and sufficient for μ and $(S_2^2,\dots,S_n^2)$ is ancillary for μ. Hence, Basu's (1955) Theorem gives the result.]

2.2 In Problem 2.1, let us write $W_1 = n^{-\frac{1}{2}}\Sigma_{i=1}^n X_i$ as well as $W_j = \{(j-1)j\}^{-\frac{1}{2}} \{\Sigma_{i=1}^{j-1} X_i - (j-1)X_j\}$, $j = 2,\dots,n$. This orthogonal transformation from $(X_1,\dots,X_n)$ to $(W_1,\dots,W_n)$ is called *Helmert's transformation.* Obviously $\overline{X}_n = n^{-\frac{1}{2}}W_1$. Show that $\Sigma_{i=1}^m (X_i - \overline{X}_m)^2 = \Sigma_{j=2}^m W_j^2$ and hence prove that $\overline{X}_n$ and $(S_2^2,\dots,S_n^2)$ are independent. Note that one can now write $S_n^2 = (n-1)^{-1}\Sigma_{i=2}^n W_i^2$ for all $n \geq 2$.

2.3 Suppose that $X_1,...,X_n$ are i.i.d. with the p.d.f. $f(x; \mu, \sigma)$ given in (2.2.5). Let $X_{n:1} \leq X_{n:2} \leq \cdots \leq X_{n:n}$ be the corresponding order statistics. Let $T_n = X_{n:1}$ and $U_m = (m-1)^{-1}\Sigma_{i=1}^m(X_i - X_{m:1})$, $n \geq m \geq 2$. Show that T_n is independent of $(U_2,...,U_n)$. Here, $(\mu, \sigma) \in \mathbb{R} \times \mathbb{R}^+$.

[*Hints*: For an arbitrary but fixed $\sigma \in \mathbb{R}^+$, note that T_n is both complete and sufficient for μ and $(U_2,...,U_n)$ is ancillary for μ. Then, use Basu's (1955) Theorem. See also Ghurye (1958).]

2.4 In Problem 2.3, let $W_{in} = (n-i+1)\{X_{n:i} - X_{n:i-1}\}$ $i = 2,...,n$. Show that $U_n = (n-1)^{-1}\Sigma_{i=2}^n W_{in}$ where the W's are i.i.d. having the exponential p.d.f. $f(x; 0, \sigma)$ given in (2.2.5). Compare the representation of U_n with that of S_n^2 derived in Problem 2.2. Any comments?

2.5 Let $Z_1, Z_2,...$ be i.i.d. and N be a positive integer valued random variable such that $I(N > n)$ depends only on $(Z_1,...,Z_n)$ for every $n = 1, 2,...$ and assume that $E(N) < \infty$.

(a) Assuming that $E|Z_1| < \infty$, show that $E\{\Sigma_{i=1}^N Z_i\} = E(Z_1)E(N)$. This result is known as Wald's first equation.

(b) Assuming that $E(Z_1^2) < \infty$, show that $E\{[\Sigma_{i=1}^N Z_i - N\mu]^2\} = Var(Z_1)E(N)$ where $\mu = E(Z_1)$. This result is known as Wald's second equation.

[*Hints*: See Chow, Robbins and Teicher (1965).]

2.6 Let $X_1, X_2, \ldots$ be i.i.d. and $N = N(\lambda) = \inf\{n \geq m: n \geq \lambda S_n^2\}$ where we consider $\lambda > 0$, $S_n^2 = (n-1)^{-1}\Sigma_{i=1}^{n}(X_i - \overline{X}_n)^2$ and $m(\geq 2)$ is the starting sample size. Assume that $0 < \sigma^2 = \mathrm{Var}(X_1) < \infty$ and $n^* = \lambda\sigma^2$.

(a) Show that $P(N < \infty) = 1$.

[*Hints*: $P(N = \infty) = 1 - \lim_{n\to\infty} P(N > n) \leq 1 - \lim_{n\to\infty} P\{n < \lambda S_n^2\} = 1$.]

(b) Show that $N/n^* \to 1$ w.p. 1 as $\lambda \to \infty$.

[*Hints*: Note that $N \uparrow \infty$ w.p. 1 as $\lambda \to \infty$. Hence, S_N^2 and S_{N-1}^2 both tend to σ^2 w.p. 1 as $\lambda \to \infty$. Now, use the basic inequality $\lambda S_N^2 \leq N \leq m + \lambda S_{N-1}^2$, then divide throughout by n^* and take limits as $\lambda \to \infty$.]

(c) Show that $E(N/n^*) \to 1$ as $\lambda \to \infty$.

[*Hints*: Let $\mu = E(X_1)$ and first assume that $E(N) < \infty$. Then, from the basic inequality's upper bound, one obtains $(N - m)(N - 2) \leq \lambda \Sigma_{i=1}^{N-1} (X_i - \overline{X}_{N-1})^2 \leq \lambda \Sigma_{i=1}^{N-1} (X_i - \mu)^2 \leq \lambda \Sigma_{i=1}^{N} (X_i - \mu)^2$. Hence, $E^2(N) - (m+2)E(N) \leq E\{(N - m)(N - 2)\} \leq \lambda E(N)\sigma^2$, in view of Problem 2.5 (a), which implies $E(N) \leq n^* + (m+2)$. In general, consider $N_\tau = \min\{N, \tau\}$, $\tau = 1,2,\ldots$. $N_\tau \uparrow N$ w.p. 1 as $\tau \to \infty$ and $E(N_\tau) \leq n^* + (m+2)$. The Monotone Convergence Theorem implies $E(N) \leq n^* + (m+2)$ and hence, note that $E(N)$ is indeed finite. Now, $\limsup_{\lambda\to\infty} E(N/n^*) \leq 1$. Next, Fatou's Lemma and (a) will complete the proof.]

(d) Show that $E(N^2/n^{*2}) \to 1$ as $\lambda \to \infty$.

[*Hints*: From (c), note that $E(N^2) - (m+2)E(N) \leq \lambda E(N)\sigma^2$. Hence

$E(N^2/n^{*2}) \le (m+2)E(N)/n^{*2} + E(N/n^*)$, that is $\limsup_{\lambda\to\infty} E(N^2/n^{*2}) \le 1$. Again, Fatou's Lemma and (a) will complete the proof.]

2.7 Consider the problem discussed in Example 2.2.2 and suppose that one proposes the following *three-stage procedure*. Start with $m(\ge 2)$ samples and let $T = \max\{m, [\frac{1}{2}\rho a U_m d^{-1}]^* + 1\}$, $N = \max\{T, [\frac{1}{2}a U_T d^{-1}]^* + 1\}$ where $0 < \rho < 1$ is kept fixed and $a = \ln(1/\alpha)$. Propose the fixed-width confidence interval $[\tilde{X}_N - 2d, \tilde{X}_N]$ for μ. Give particular attention to sufficient conditions involving m.

(a) Derive the asymptotic second-order expansions of $E(N)$ and $Var(N)$, as $d \to 0$.

[*Hints*: Use Theorem 2.3.2 and Theorem 2.3.3.]

(b) Show that $P\{\tilde{X}_N - 2d < \mu < \tilde{X}_N\} = E\{1 - \exp(-2Nd/\sigma)\}$.

(c) Use Theorem 2.3.4 to determine the "fine-tuning" ϵ^*, needed to modify N to N^*, so that one can claim $P\{\tilde{X}_{N^*} - 2d < \mu < \tilde{X}_{N^*}\} = (1-\alpha) + o(d)$.

[*Hints*: See Mukhopadhyay and Maruomoustakos (1987).]

2.8 Consider the set up of Problem 2.7 and suppose that one proposes the following *accelerated sequential procedure*. Start with $m(\ge 2)$ samples and let $t = \inf\{n \ge m : n \ge \frac{1}{2}a\rho U_n^* d^{-1}\}$, $N_1 = [\frac{1}{2}a U_t^* d^{-1} + q]^* + 1$, $N = \max\{N_1, t\}$ and propose the fixed-width confidence interval $[\tilde{X}_N - 2d, \tilde{X}_N]$ for μ, where $0 < \rho < 1$ and $q \ge 0$ are chosen and fixed in advance. Give particular attention to sufficient conditions involving m. Here, $U_n^* = n(n-1)^{-1}U_n$ and $C = \frac{1}{2}a\sigma/d$.

(a) Show that $N/C \to 1$ w.p. 1 as $d \to 0$.

(b) Show that $E(N/C) \to 1$ as $d \to 0$.

(c) Show that $(C/N)^{\omega}$ is uniformly integrable for all $\omega > 0$.

(d) Show that $E(N - C)$ is bounded as $d \to 0$.

(e) Show that $\lim_{d \to 0} P\{\tilde{X}_N - 2d < \mu < \tilde{X}_N\} = 1 - \alpha$.

(f) Determine q as a function α and ρ such that one can claim $P\{\tilde{X}_N - 2d < \mu < \tilde{X}_N\} \geq (1 - \alpha) + o(d)$.

[*Hints*: In (c), use Theorem 2.5.3 (i). Use Corollary 2.5.1 in (d). See section 3.2 in Mukhopadhyay and Solanky (1991) for (f).]

2.9 Let $X_1, X_2, \ldots$ be i.i.d. $N(\mu, \sigma^2)$, $\mu \in R$, $\sigma \in R^+$. The loss in estimating μ by $\overline{X}_n$ is given by $L_n = A(\overline{X}_n - \mu)^2$ where $A(>0)$ is a known constant. In order to achieve the associated risk that is bounded from above by W^*, a preassigned positive number, one considers the following purely sequential procedure. Let $N = \inf\{n \geq m : n \geq AS_n^2/W^*\}$ where $m(\geq 2)$ is the starting sample size and $S_n^2 = (n-1)^{-1}\Sigma_{i=1}^n (X_i - \overline{X}_n)^2$. Let $n^* = A\sigma^2/W^*$. Give particular attention to sufficient conditions on m.

(a) Show that $N/n^* \to 1$ w.p. 1 as $A \to \infty$.

(b) Show that $E(N/n^*) \to 1$ as $A \to \infty$.

(c) Derive the asymptotic second-order expansions of $E(N)$, $Var(N)$ and $E(L_N)$.

[*Hints*: Note that $E(L_N) = W^*E(n^*/N)$ and use Theorem 2.4.8. See also Mukhopadhyay (1985a)].

2.10 In (2.4.6), assume that $\delta = 1$ and $t = 0$. Then, show that $N_\nu \geq \psi_\nu^{1/(\gamma+1)}$ w.p. 1.

[*Hints*: Observe that $N_\nu \geq \psi_\nu(U_N^\alpha + N^{-\gamma}) \geq \psi_\nu N^{-\gamma}$ w.p. 1. In other words, $N^{\gamma+1} \geq \psi_\nu$ w.p. 1. This result motivates the expression for the starting sample size m in section 2.2.2.]

2.11 Let $N = \inf\{n \geq m(>a) : n > \lambda Y_n\}$ where m, a, b are positive integers, $\lambda > 0$ and $A_n = (n-a)bY_n/\theta - \chi^2_{(n-a)b}$ for some $\theta > 0$. Define $n^* = \lambda\theta$, $p(n, n^*) = n(n-a)bn^{*-1}$.

(a) Show that $N/n^* \to 1$ w.p. 1 as $\lambda \to \infty$.

(b) Show that $P(N = m) = 0_e(n^{*\,-\frac{1}{2}b(m-a)})$, where 0_e stands for exact order.

[*Hints*: $P(N = m) = P\{A_m \leq p(m, n^*)\} = \{2^{\frac{1}{2}q}\,\Gamma(\frac{1}{2}q)\}^{-1} \int_0^{p(m,n^*)} \exp(-\frac{1}{2}x) \times x^{\frac{1}{2}q-1} \leq 0(1)[x^{\frac{1}{2}q}]_0^{p(m,n^*)} = 0(n^{*\,-\frac{1}{2}b(m-a)})$, where $q = b(m-a)$. Also, note that $P(N = m) \geq 0(1)e^{-\frac{1}{2}p(m,n^*)}[x^{\frac{1}{2}q}]_0^{p(m,n^*)} = 0(n^{*\,-\frac{1}{2}b(m-a)})$. The orders in the lower and upper bounds being same, one obtains the exact order of $P(N = m)$.]

(c) Show that $P(N = n) \leq P\{A_n \leq n(n-a)bn^{*-1}\}$ for $n > m$.

[*Hints*: $P(N = n) = P(m < \lambda Y_m \cap \cdots \cap n-1 < \lambda Y_{n-1} \cap n > \lambda Y_n) \le P(n > \lambda Y_n)$.]

(d) Show that $P(N = n) \le \{\frac{n}{n^*} e^{1-\frac{n}{n^*}}\}^{\frac{b(n-a)}{2}}$ for $m < n \le \epsilon n^*$ with $0 < \epsilon < 1$.

[*Hints*: Use (c) to write $P(N = n) \le \inf_{h>0} P[e^{-hA_n} \ge e^{-hp(n,n^*)}] = \inf_{h>0}\{(1 + 2h)^{-\frac{1}{2}b(n-a)} e^{hp(n,n^*)}\} = (1 + 2h_0)^{-\frac{1}{2}b(n-a)} e^{h_0 p(n,n^*)}$ where one lets $h_0 = \frac{1}{2}[\frac{b(n-a)}{p(n,n^*)} - 1] = \frac{1}{2}\{n^* n^{-1} - 1\}$, which is positive in the range of n.]

(e) Let $\epsilon \in (0, 1)$. Show that $P(N \le \epsilon n^*) = 0(n^{*-\frac{1}{2}b(m-a)})$.

[*Hints*: $P(m+1 \le N \le \epsilon n^*) = \Sigma_{n=m+1}^{d} P(N = n)$ where d = largest integer $\le \epsilon n^*$. Use (d) to write $P(m+1 \le N \le \epsilon n^*) \le \Sigma_{n=m+1}^{d}\{\frac{n}{n^*} e^{1-\frac{n}{n^*}}\}^{\frac{1}{2}b(n-a)} = \Sigma_{n=m+1}^{d} \{\frac{n}{n^*} e^{1-\frac{n}{n^*}}\}^{\frac{1}{2}b(n-m-a)} \{\frac{n}{n^*} e^{1-\frac{n}{n^*}}\}^{\frac{1}{2}bm} \le n^{*-\frac{1}{2}bm} e^{\frac{1}{2}bm} \Sigma_{n=m+1}^{d} \epsilon^{\frac{1}{2}b(n-m-a)} n^{\frac{1}{2}bm} = 0(n^{*-\frac{1}{2}bm})$. In the last step, we used the facts that $xe^{1-x} \uparrow$ in $x \in (0, 1)$ and that the infinite series is convergent (via root test). Now, use (b) to complete the proof.]

2.12 Consider the example and the purely sequential procedure given in section 2.4.2.1. Give particular attention to sufficient conditions involving m.

(a) Show that $P\{\Delta \in I_N\} = 2E[\Phi(d\{N/2\}^{\frac{1}{2}}\sigma^{-1})] - 1$.

[*Hints*: Use Problems 2.2 – 2.3.]

(b) Show that $N/n_1^* \to 1$ w.p. 1 and $E(N/n_1^*) \to 1$ as $d \to 0$.

(c) Show that $(N - n_1^*)^2/n_1^* \xrightarrow{\mathscr{L}} \chi_1^2$ as $d \to 0$.

(d) Derive the asymptotic second–order expansions of $E(N)$ and the confidence coefficient, $P\{\Delta \in I_N\}$.

[*Hints*: See Mukhopadhyay and Al–Mousawi's (1986) Theorem 4.]

2.13 Consider the fixed–width confidence interval problem discussed in section 2.4.2.1.

(a) Propose a suitable two–stage procedure so that one can claim $P\{\Delta \in I_N\} \geq 1 - \alpha$. Is this procedure asymptotically first–order efficient?

(b) Propose the corresponding modified two–stage procedure for the same problem. Is this procedure asymptotically first– or second–order efficient?

(c) Propose the corresponding three–stage procedure. Is this procedure asymptotically second–order efficient? Also derive the second–order expansion of $P\{\Delta \in I_N\}$.

[*Hints*: See Mukhopadhyay and Al–Mousawi (1986) and improvise.]

2.14 In section 2.2, suppose that $E(T_n) = \theta$ for all $n \geq m$. Then, show that $E(T_N) = \theta$ for the two–stage procedure (2.2.1).

2.15 Refer to (2.3.27) and show directly that $P\{\underset{\sim}{\nu}^* \in \mathscr{R}_N\} \to 1 - \alpha$ as $L \to 0$.

2.16 Suppose $X_1, X_2, \ldots$ are i.i.d. random variables. Let $\theta = E(X_1)$, $\sigma^2 = E\{(X_1-\theta)^2\}$, $\mu_4 = E\{(X_1-\theta)^4\}$, and assume that $\mu_4 > \sigma^4$ and $\mu_4 \in R^+$. Define $N = N(\lambda) = \inf\{n \geq 2 : n \geq \lambda T_n\}$ where $T_n = S_n^2 + n^{-\gamma}$ with $S_n^2 = (n-1)^{-1}\Sigma_{i=1}^n (X_i - \overline{X}_n)^2$, $0 < \gamma < \frac{1}{2}$. Let $n^* = \lambda\sigma^2$. Derive the asymptotic (as $\lambda \to \infty$) distribution of $N^{-\frac{1}{2}}(N - n^*)$.

[*Hints*: Observe that S_n^2 is a U–statistic and use Theorem 2.4.2 with $\theta = \sigma^2$ and $\xi_1 = \frac{1}{4}(\mu_4 - \sigma^2)$. Then, apply Theorem 2.4.1.]

2.17 In (2.4.7) suppose that W's are i.i.d. χ_p^2 random variables. Then, equation (2.4.8) holds with $b = \frac{1}{2}p$.

[*Hints*: $P(W_1 \leq u) = k_o \int_0^u e^{-x/2}\, x^{\frac{1}{2}p-1}\, dx \leq k_o \int_0^u x^{\frac{1}{2}p-1}\, dx$, $u > 0$ where $k_o = \{2^{p/2}\Gamma(p/2)\}^{-1}$.]

REFERENCES

Anscombe (1952, 1953), Basu (1971, 1991), Basu (1955), Bhattacharyya and Mallik (1973), Birnbaum and Healy (1960), Chatterjee (1991), Chow, Hsiung and Lai (1979), Chow and Robbins (1965), Chow, Robbins and Teicher (1965), Cox (1952), Csenki (1980), DasGupta and Perlman (1974), Ghosh (1973), Ghosh (1980), Ghosh and DasGupta (1980), Ghosh and Mukhopadhyay (1975, 1981), Ghurye (1958), Hall (1981, 1983), Hamdy (1988), Hamdy, Mukhopadhyay, Costanza and Son (1988), Hoeffding (1948), Holewijn (1969), Johnson and Kotz (1970), Lai and Siegmund (1977, 1979), Landers and Rogge (1976), Lehmann (1986), Lohr (1990), Mahalanobis (1940), Mukhopadhyay (1974, 1976, 1980, 1981, 1982, 1985a, 1988, 1990, 1993b), Mukhopadhyay and Al–Mousawi (1986), Mukhopadhyay, Hamdy, Al–Mahmeed and Costanza

(1987), Mukhopadhyay and Mauromoustakos (1987), Mukhopadhyay and Solanky (1991), Mukhopadhyay and Vik (1985), Nagaev and Mukhin (1966), Neyman (1937), Pratt (1961), Rao (1973), Robbins (1959), Sen and Ghosh (1981), Siegmund (1968, 1985), Simons (1968), Sproule (1969, 1974), Starr and Woodroofe (1972), Stein (1945, 1949), Wald (1947), Wijsman (1991), Woodroofe (1977, 1982, 1987, 1991), Zacks (1971).

3

SELECTING THE BEST NORMAL POPULATION

3.1 INTRODUCTION

Suppose we have π_i, $i = 1,\ldots,k$ (≥ 2), *independent*, normally distributed populations, having *unknown mean* μ_i and *common unknown variance* σ^2. We assume that $\mu_i \in R$ and $\sigma \in R^+$, $i = 1,\ldots,k$. Let $\mu_{[1]} \leq \cdots \leq \mu_{[k]}$ be the ordered μ–values. Since the variance is same in all populations, we sample equally from each population and at any point of time, whenever we need "new" samples from π's, we take a certain equal number, to be made specific later, of samples from each π at that step. Having recorded i.i.d. random variables $X_{i1},\ldots,X_{in}$ from π_i, we write

$$\overline{X}_{in} = n^{-1}\Sigma_{j=1}^{n}X_{ij},\ S_{in}^2 = (n-1)^{-1}\Sigma_{j=1}^{n}(X_{ij} - \overline{X}_{in})^2,$$

$$U(n) = k^{-1}\Sigma_{i=1}^{k}S_{in}^2,\ U^*(n) = n(n-1)^{-1}U(n) \tag{3.1.1}$$

for $i = 1,\ldots,k$ and $n \geq 2$. One will note that μ_i is to be estimated by $\overline{X}_{in}$. The common unknown variance σ^2 is often estimated by the pooled unbiased estimator $U(n)$. For technical reasons, in section 3.6, σ^2 will be estimated by $U^*(n)$. It should be added that both $U(n)$ and $U^*(n)$ are consistent estimators of σ^2, that is, $U(n) \xrightarrow{P} \sigma^2$ as well as $U^*(n) \xrightarrow{P} \sigma^2$ as $n \to \infty$.

If the ith population π_i is characterized by X_i which is, for example, perhaps the yield of a particular *variety* of some crop or the "strength" of certain material etc., then from a practical point of view, it may be worthwhile to select the variety "j" or the brand "j" of the material, in such a way that

$$P(X_j \geq a^*) = \max_{1 \leq i \leq k} P(X_i \geq a^*), \tag{3.1.2}$$

for all $a^* \in R^+$. In fact, it is very easy to check that (3.1.2) holds for all $a^* \in R$ when X_j corresponds to the population having its mean equal to $\mu_{[k]}$. Hence, our aim is to *select the population associated with the largest mean*, that is $\mu_{[k]}$, and such a population is called the *best population* in this context. We assume that there is just one population associated with $\mu_{[k]}$. We do not, however assume any knowledge about the association of the μ_i's with $\mu_{[i]}$'s. The dual problem of selecting the population associated with $\mu_{[1]}$ is not given any attention here.

In what follows, we adopt Bechhofer's (1954) *indifference zone* approach, and given $\delta^* > 0$, we let

$$\Omega(\delta^*) = \{\underline{\mu} = (\mu_1, \ldots, \mu_k) : \mu_{[k]} - \mu_{[k-1]} \geq \delta^*\}. \tag{3.1.3}$$

See section 1.2.1.1. For a given $P^* \in (k^{-1}, 1)$, we are interested in selection procedures such that the probability of *correct selection* (CS) of the population associated with $\mu_{[k]}$ is at least P^* or *asymptotically* (as $\delta^* \to 0$) at least P^* whenever the true parameter $\underline{\mu} \in \Omega(\delta^*)$. In other words, we are interested to identify the best population having certain minimum or approximately minimum probability P^* that our final decision is the right one when the best population's mean is at least δ^*–unit ahead of the mean of the second best. The parameter space $\Omega(\delta^*)$ is called, as one will recall from section 1.2.1.1, the *preference zone*, while $\Omega^c(\delta^*)$ is called the *indifference zone* in the sense that the experimenter is *not* willing to pick the best population when $\mu_{[k]} - \mu_{[k-1]} < \delta^*$, that is the experimenter expresses indifference in the parameter space $\Omega^c(\delta^*)$, because then the best population is apparently not that much better than the second best.

3.2 PRELIMINARIES

For a moment let us pretend that σ is known and we have recorded $X_{i1},\ldots,X_{in}$ from π_i, $i = 1,\ldots,k$. The natural selection rule is simply to pick the population associated with the largest sample mean, that is, one implements the following selection procedure.

$$\mathcal{P}_n : \text{Select } \pi_j \text{ as the best population if } \overline{X}_{jn} = \underset{1\leq i\leq k}{\text{Max}} \overline{X}_{in}. \tag{3.2.1}$$

We should add that the selection rule (3.2.1) indeed turns out to be *natural*, having justified through likelihood comparisons along the lines of Bechhofer et al. (1968). See also Gupta and Panchapakesan (1979) as well as Mukhopadhyay (1983a) in this context. Such a selection procedure also turns out to coincide with the Bayes rule under an appropriate setup. See Gupta and Miescke (1984). Throughout this chapter, we work only with the selection procedure $\mathcal{P}_n$ given by (3.2.1). Recall that $\phi(y) = (2\pi)^{-\frac{1}{2}}\exp(-y^2/2)$ and $\Phi(y) = \int_{-\infty}^{y} \phi(x)dx$, for all $y\epsilon R$.

Let $\pi_{(i)}$ stand for the population associated with the mean $\mu_{[i]}$ and let us write $\overline{X}_{(in)}$ for the sample mean arising from the population $\pi_{(i)}$, $i = 1,\ldots,k$. Also, write $\delta_i = \mu_{[k]} - \mu_{[i]}$ and note that

$$\delta_i \geq 0 \text{ for arbitrary } \underset{\sim}{\mu}\epsilon\Omega(\delta^*)\cup\Omega^c(\delta^*)$$

and

$$\delta_i \geq \delta^* \text{ for arbitrary } \underset{\sim}{\mu}\epsilon\Omega(\delta^*).$$

Then, along the lines of (1.2.2) one notes that the probability of correct selection is given by

$$P_n(CS) = P\{\overline{X}_{(kn)} > \overline{X}_{(in)},\ i = 1,\ldots,k-1\}$$

$$= P\Big[n^{\frac{1}{2}}\{\overline{X}_{(kn)} - \mu_{[k]}\}\sigma^{-1} + n^{\frac{1}{2}}\sigma^{-1}\{\mu_{[k]} - \mu_{[i]}\}$$

$$> n^{\frac{1}{2}}\{\overline{X}_{(in)} - \mu_{[i]}\}\sigma^{-1},\ i = 1,\ldots,k-1\Big]$$

$$= \int_{-\infty}^{\infty} \prod_{i=1}^{k-1} \Phi\,[y + n^{\frac{1}{2}}\delta_i\sigma^{-1}]\ \phi(y)\,dy. \tag{3.2.2}$$

This integral is really $f^*(n, \delta_1,\ldots,\delta_{k-1}, \sigma)$ as introduced before equation (1.2.3). From (3.2.2), it is easy to see that

$$\operatorname*{Inf}_{\underset{\sim}{\mu}\in\Omega(\delta^*)\cup\Omega^c(\delta^*)} P_n(CS) = \int_{-\infty}^{\infty} \Phi^{k-1}(y)\phi(y)\,dy = 1/k,$$

which compares with (1.2.4). Hence, one works in the light of (1.2.6) and notes that for all $\underset{\sim}{\mu}\in\Omega(\delta^*)$, that is when $\delta_i \geq \delta^*$,

$$P_n(CS) \geq \int_{-\infty}^{\infty} \prod_{i=1}^{k-1} \Phi\,[y + n^{\frac{1}{2}}\delta^*\sigma^{-1}]\ \phi(y)dy,$$

and the equality holds if and only if $\delta_i = \delta^*$, for all $i = 1,\ldots,k-1$. Hence,

$$\operatorname*{Inf}_{\underset{\sim}{\mu}\in\Omega(\delta^*)} P_n(CS) = \int_{-\infty}^{\infty} \Phi^{k-1}(y + n^{\frac{1}{2}}\delta^*\sigma^{-1})\ \phi(y)\,dy, \tag{3.2.3}$$

the infimum being attained when $\mu_{[1]} = \cdots = \mu_{[k-1]} = \mu_{[k]} - \delta^*$, a parameter configuration that is known as the *least favorable configuration* (LFC). Let $h = h(k, P^*)$ be such that

$$\int_{-\infty}^{\infty} \{\Phi(y+h)\}^{k-1}\ \phi(y)\,dy = P^*, \tag{3.2.4}$$

and then it follows from (3.2.3) that

$$P_n(CS) \geq P^* \text{ for all } \underset{\sim}{\mu} \in \Omega(\delta^*)$$

when $n \geq h^2\sigma^2/\delta^{*2} = C$, say, and note that this C corresponds to $\tilde{n}^*$ defined in section 1.2.1.1. In other words, had σ been known, one would take $[C]^* + 1$ samples from each π and implement the corresponding selection rule (3.2.1). Thus, C is known to be the *optimal fixed sample size* needed from each π, in conjunction with (3.2.1), in order to meet the P^*–requirement for $P(CS)$ in $\Omega(\delta^*)$, had σ been known. In fact, however, C is unknown. Moreover, one can show that there exists no fixed–sample size procedure that will meet the P^*–requirement for $P(CS)$ in $\Omega(\delta^*)$, uniformly in $\sigma \in R^+$. See Dudewicz (1971) in this context. The values of h had been tabulated in Bechhofer (1954), Gupta (1963), Gupta et al. (1973) and these are also available in Gibbons et al.'s (1977) Table A.1. In what follows, we develop two–stage, purely sequential, accelerated sequential and three–stage selection methodologies for this problem.

Suppose, in general, one devises a statistical methodology giving rise to a positive integer valued random variable N estimating C and selects the population associated with the largest sample mean, that is to implement the selection procedure $\mathscr{P}_N$ given by (3.2.1).

Theorem 3.2.1. *Suppose that* N *satisfies the following conditions. The distribution of* N does not depend on $\underset{\sim}{\mu}$; $I(N = n)$ *is independent of* $(\overline{X}_{1n}, \ldots, \overline{X}_{kn})$ *for every fixed* $n \geq 2$; *and* $P(N < \infty) = 1$. *Then,*

$$\underset{\underset{\sim}{\mu} \in \Omega(\delta^*)}{\operatorname{Inf}} P(CS) = E\left[\int_{-\infty}^{\infty} \{\Phi(y + N^{\frac{1}{2}}\delta^*\sigma^{-1})\}^{k-1} \phi(y)dy\right]$$

and this infimum is attained when $\mu_{[1]} = \cdots = \mu_{[k-1]} = \mu_{[k]} - \delta^*$.

Proof. As before, write $\delta_i = \mu_{[k]} - \mu_{[i]}$ and $\delta_i \geq \delta^*$ whenever $\underset{\sim}{\mu} \in \Omega(\delta^*)$, defined in (3.1.3). An experiment will start with a few pilot samples from each π. Suppose that $m_1(\geq 2)$ is the starting sample size from each π that ultimately gives rise to N. The condition $P(N < \infty)$ guarantees that the selection procedure (3.2.1) remains well–defined when "n" is replaced by N which is a random variable. Now,

$$P(CS) = \Sigma_{n=m_1}^{\infty} P(CS \mid N = n)P(N = n)$$

$$= \Sigma_{n=m_1}^{\infty} P\{\overline{X}_{(kn)} > \overline{X}_{(in)},\ i = 1,\dots,k-1\}P(N = n),$$

since $I(N = n)$ and $(\overline{X}_{1n},\dots,\overline{X}_{kn})$ are independent for all $n \geq m_1$. That is,

$$P(CS) = \Sigma_{n=m_1}^{\infty} P_n(CS)P(N = n) \qquad (3.2.5)$$

where $P_n(CS)$ was defined in (3.2.2). Since the distribution of N does not involve $\underset{\sim}{\mu}$ and $P_n(CS)$ attains the minimum value in $\Omega(\delta^*)$ when $\mu_{[1]} = \cdots = \mu_{[k-1]} = \mu_{[k]} - \delta^*$, whatever n, it follows from (3.2.5) that

$$\underset{\underset{\sim}{\mu}\in\Omega(\delta^*)}{\mathrm{Inf}} P(CS) = E[\int_{-\infty}^{\infty} \{\Phi(y + N^{\frac{1}{2}}\delta^*\sigma^{-1})\}^{k-1}\phi(y)dy] \qquad (3.2.6)$$

and this infimum is attained when $\mu_{[1]} = \cdots = \mu_{[k-1]} = \mu_{[k]} - \delta^*$. □

3.3 Two–stage procedure

Bechhofer et al. (1954) had proposed a two–stage procedure for this problem along the lines of Stein (1945). Let $m_1(\geq 2)$ be the starting sample size from each π and $\tau = \tau(m_1, k, P^*)$ be a positive real number, to be determined specifically a little later. Define

$$N = N(\delta^*) = \max\{m_1, [\tau^2 \mathsf{U}(m_1)\delta^{*-2}]^* + 1\}. \tag{3.3.1}$$

This procedure looks like (2.2.1) where $m = m_1$, $\omega = \frac{1}{2}$, $d = \delta^*$, $U_m = \mathsf{U}(m_1)$, $g(x) = x$ and $b_m = \tau$. If $N = m_1$, we do not take any more samples from any population. However, if $N > m_1$ then we sample the difference in the second stage, that is, we take $(N - m_1)$ more samples from each π in the second stage. Based on the totality of all samples, that is having $X_{i1},\ldots,X_{iN}$ from π_i, $i = 1,\ldots,k$, one now implements the usual selection procedure $\mathscr{P}_N$ given by (3.2.1). Throughout, $\underset{\sim}{\mu}$ and σ are kept fixed.

Theorem 3.3.1. *For the two-stage procedure (3.3.1), with* τ *defined in (3.3.4), we have:*

(i) $P(CS) \geq P^*$ *for all* $\underset{\sim}{\mu} \in \Omega(\delta^*)$;

(ii) $\tau^2\sigma^2\delta^{*-2} \leq E(N) \leq m_1 + \tau^2\sigma^2\delta^{*-2}$;

(iii) $E(N/C) \to \tau^2/h^2 (>1)$ *as* $\delta^* \to 0$;

(iv) $\liminf P(CS) \geq P^*$, *for all* $\underset{\sim}{\mu} \in \Omega(\delta^*)$, *as* $\delta^* \to 0$;

where $C = h^2\sigma^2/\delta^{*2}$ *and* h *comes from (3.2.4).*

Proof. The basic inequality (2.2.3) turns out to be

$$\tau^2 \mathsf{U}(m_1)\delta^{*-2} \leq N \leq m_1 + \tau^2 \mathsf{U}(m_1)\delta^{*-2}, \tag{3.3.2}$$

and now taking expectations throughout leads to part (ii), since $E(\mathsf{U}(m_1)) = \sigma^2$. Next, dividing throughout the expressions in part (ii) by C and then taking limits as $\delta^* \to 0$ leads to part (iii). The fact that τ^2 is larger than h^2 follows from appropriate tables. See Dunnett and Sobel (1955), Gupta and Sobel (1957), and Krishnaiah and Armitage (1965).

In order to prove part (i), first observe that Theorem 3.2.1 holds for the two–stage selection procedure (3.3.1) and thus $\underset{\underset{\sim}{\mu}\in\Omega(\delta^*)}{\mathrm{Inf}}\ P(CS)$ is given by the expression in (3.2.6) and the LFC turns out to be $\mu_{[1]} = \cdots = \mu_{[k-1]} = \mu_{[k]} - \delta^*$. Thus,

$$\underset{\underset{\sim}{\mu}\in\Omega(\delta^*)}{\mathrm{Inf}}\ P(CS) \geq E[\int_{-\infty}^{\infty} \{\Phi(y + \tau\, U^{\frac{1}{2}}(m_1)\sigma^{-1})\}^{k-1} \phi(y)dy]$$

$$= E\{P[Y_i < Y_k + \tau\, U^{\frac{1}{2}}(m_1)\sigma^{-1},\ i = 1,\ldots,k-1 \,\big|\, U(m_1)]\} \qquad (3.3.3)$$

where $Y_1,\ldots,Y_k$ are i.i.d. $N(0, 1)$ and they are chosen independently of $U(m_1)$, since $N^{\frac{1}{2}}\delta^* \geq \tau\, U^{\frac{1}{2}}(m_1)$ w.p. 1, which follows from the left hand side of (3.3.2). Now,

$$(Y_1 - Y_k,\ldots,Y_{k-1} - Y_k) \sim N_{k-1}(\underset{\sim}{0}, \Sigma)$$

where $\Sigma = (\sigma_{ij})$ with $\sigma_{ii} = 2$, $\sigma_{ij} = 1$, $1 \leq i \neq j \leq k-1$. Let $T_i = (Y_i - Y_k)/\{2^{\frac{1}{2}}U^{\frac{1}{2}}(m_1)\sigma^{-1}\}$, $i = 1,\ldots,k-1$. Then, $(T_1,\ldots,T_{k-1})$ is distributed as $(k-1)$ dimensional multivariate t with equicorrelation $= \frac{1}{2}$ and the degrees of freedom $= k(m_1 - 1)$. Now, (3.3.3) can be rewritten as

$$\underset{\underset{\sim}{\mu}\in\Omega(\delta^*)}{\mathrm{Inf}}\ P(CS) \geq P\{T_i \leq \tau/2^{\frac{1}{2}},\ i = 1,\ldots,k-1\},$$

in other words, we determine $\tau = \tau(m_1, k, P^*)$ in such a way that

$$P[T_i \leq \tau/2^{\frac{1}{2}},\ i = 1,\ldots,k-1] = P^*, \qquad (3.3.4)$$

so that part (i) will then follow.

In order to prove part (iv), first notice from (3.3.2) that obviously $N^{\frac{1}{2}}\delta^* \to \tau\, U^{\frac{1}{2}}(m_1)$ w.p. 1 as $\delta^* \to 0$. From (3.2.6) and the dominated convergence theorem one obtains

$$\liminf_{\delta^*\to 0} P(CS) \geq E[\int_{-\infty}^{\infty} \{\Phi(y + \tau U^{\frac{1}{2}}(m_1)\sigma^{-1})\}^{k-1} \phi(y)dy], \qquad (3.3.5)$$

for all $\underset{\sim}{\mu}\in\Omega(\delta^*)$. Hence, part (iv) is obvious from (3.3.3) – (3.3.4). □

Remark 3.3.1. Table A.4 in Gibbons et al. (1977) provides the values of τ for $k = 2(1)10$, $P^* = .95, .99$ and $\nu = k(m_1 - 1)$. See also Chapter 37 in Johnson and Kotz (1972) for further citations.

In the two–stage procedure (3.3.1), the choice of m_1 is very crucial. From a practical point of view, both "very large" as well as "very small", but fixed, m_1 values will, more than likely, "hurt" the experimenter, because then E(N) will tend to overestimate C by wide margin under both scenarios. In fact, from Theorem 3.3.1 (iii), it is obvious that

$$\liminf_{\delta^*\to 0} E(N - C) = \infty, \qquad (3.3.6)$$

whatever fixed k, P^* and m_1. That is, as long as m_1 is held fixed, N will *overestimate* C even *asymptotically*. For moderate values of δ^* as well, the magnitude of overestimation of C by means of N can be quite severe. For further discussions, see Mukhopadhyay and Judge (1991) and particularly refer to the simulation results reported in Table 3 there. Also, see section 3.8 for such details.

3.4 Modified two–stage procedure

In this procedure, proposed along the lines of section 2.2.2, we determine m_1 in such a way that m_1 grows as $\delta^* \to 0$, but m_1 still stays fairly small compared with C. In other words, the idea is to choose m_1 large when C is perhaps large, but one is free to pick a smaller m_1 when one suspects that C is perhaps small. But, then the problem is that C is unknown. We pick $m_0^*(\geq 2)$, $\gamma \in R^+$ and define

$$m_1 = m_1(\delta^*) = \max\{m_0^*, [(h^2\delta^{*-2})^{1/(\gamma+1)}]^* + 1\}, \qquad (3.4.1)$$

where recall that h came from (3.2.4). This m_1 is then the experimenter's starting sample size from each population. Now, let

$$N = N(\delta^*) = \max\{m_1, [\tau^2 U(m_1)\delta^{*-2}]^* + 1\}, \qquad (3.4.2)$$

and from a practical point of view, we implement the modified two–stage procedure (3.4.1) – (3.4.2) along the lines of the original two–stage procedure (3.3.1) in conjunction with the customary selection procedure $\mathscr{P}_N$ given by (3.2.1). The procedure (3.4.2) looks like (2.2.20) where $m = m_1$, $\omega = \frac{1}{2}$, $d = \delta^*$, $U_m = U(m_1)$ and $g(x) = x$. Again, we hold $\underset{\sim}{\mu}$ and σ fixed.

Theorem 3.4.1. *For the modified two–stage procedure (3.4.1) – (3.4.2), with* $0 < \gamma < \infty$ *arbitrary, but fixed* and τ *defined in (3.3.4), we have:*

(i) $P(CS) \geq P^*$ *for all* $\underset{\sim}{\mu} \in \Omega(\delta^*)$;

(ii) $N/C \to 1$ *w.p.* 1 *and* $E(N/C) \to 1$ *as* $\delta^* \to 0$;

(iii) $\liminf P(CS) \geq P^*$, *for all* $\underset{\sim}{\mu} \in \Omega(\delta^*)$, *as* $\delta^* \to 0$;

where $C = h^2\sigma^2/\delta^{*2}$ *and* h *comes from (3.2.4).*

Proof. Part (i) is nonasymptotic in nature and it follows immediately along the lines of Theorem 3.3.1 (i). Now, we turn to part (ii). Notice that the starting sample size $m_1(\delta^*) \to \infty$ and $m_1(\delta^*)/C \to 0$ as $\delta^* \to 0$, whatever $\gamma \in \mathbb{R}^+$. The basic inequality (2.2.21) is now given by

$$\tau^2 U(m_1)\delta^{*-2} \le N \le m_0^* + \{h^2\delta^{*-2}\}^{1/(\gamma+1)} + \tau^2 U(m_1)\delta^{*-2}, \quad (3.4.3)$$

where note that $\tau = \tau(m_1, k, P^*)$ is actually a function of δ^* since k and P^* are held fixed. Recall the T_i's and Y_i's defined in the proof of Theorem 3.3.1. It is obvious that as $\delta^* \to 0$,

$$(T_1,\ldots,T_{k-1}) \xrightarrow{\mathscr{L}} 2^{-\frac{1}{2}}(Y_1 - Y_k,\ldots,Y_{k-1} - Y_k)$$

and hence $\tau^2 = \tau^2(\delta^*) \to h^2$. Now, from (3.4.3) we get

$$(\tau^2/h^2)(U(m_1)/\sigma^2) \le N/C \le m_0^* C^{-1} + o(\delta^{*2-2(\gamma+1)^{-1}})$$

$$+ (\tau^2/h^2)(U(m_1)/\sigma^2) \quad (3.4.4)$$

which implies $N/C \to 1$ w.p. 1 as $\delta^* \to 0$ since $U(n) \to \sigma^2$ w.p. 1 as $n \to \infty$. Taking expectation throughout (3.4.4), one gets

$$\tau^2 h^{-2} \le E(N/C) \le m_0^* C^{-1} + o(\delta^{*2\gamma/(\gamma+1)}) + \tau^2 h^{-2}$$

and hence $\lim_{\delta^* \to 0} E(N/C) = 1$.

Now, in view of part (ii) and the dominated convergence theorem, one immediately obtains along the lines of (3.2.6)

$$\liminf_{\delta^* \to 0} P(CS) \ge E\left[\int_{-\infty}^{\infty} \{\Phi(y+h)\}^{k-1}\phi(y)dy\right] = P^*,$$

for all $\underset{\sim}{\mu}\in\Omega(\delta^*)$. □

One can see that the modified two–stage procedure (3.4.1) – (3.4.2) has the same properties given in Theorem 3.3.1 (i) and (iv) which are enjoyed by the two–stage procedure (3.3.1). Also, the sample size N required by the modified methodology from each population additionally gets to be in the close proximity of C in the following senses: As $\delta^* \to 0$, $N/C \to 1$ w.p. 1 and $E(N/C) \to 1$. Now, contrast these with Theorem 3.3.1 (iii). At this point, we introduce certain definitions of phrases which are used hence forth repeatedly.

Consider an arbitrary multistage or sequential selection methodology for the problem on hand that gives rise to a random number of random samples $N = N(\delta^*)$ from each population π. Such a procedure is called

(i) *consistent* or *exactly consistent* in the Stein (1945) or Mukhopadhyay (1980) sense respectively if

$$P(CS) \geq P^* \text{ for all } \underset{\sim}{\mu}\in\Omega(\delta^*); \tag{3.4.5}$$

(ii) *asymptotically consistent* or *asymptotically first–order consistent* in the Chow and Robbins (1965) or Mukhopadhyay (1981) sense respectively if

$$\liminf_{\delta^*\to 0} P(CS) \geq P^* \text{ for all } \underset{\sim}{\mu}\in\Omega(\delta^*); \tag{3.4.6}$$

(iii) *asymptotically efficient* or *asymptotically first–order efficient* in the Chow and Robbins (1965) or Ghosh and Mukhopadhyay (1981) sense respectively if

$$\lim_{\delta^*\to 0} E(N/C) = 1 \text{ for all } \underset{\sim}{\mu}\in\Omega(\delta^*); \tag{3.4.7}$$

(iv) *asymptotically second–order efficient* in the Ghosh and Mukhopadhyay (1981) sense if

$$\lim_{\delta^* \to 0} E(N - C) \text{ is bounded.} \tag{3.4.8}$$

In our terminologies then, both the two–stage procedure (3.3.1) and modified two–stage procedure (3.4.1) – (3.4.2) are *consistent* and *asymptotically consistent*. On the other hand, the two–stage procedure is not even asymptotically first–order efficient while the modified version is indeed *asymptotically first–order efficient*. Let us now address the question whether the modified two–stage procedure (3.4.1) – (3.4.2) is asymptotically second–order efficient. Follow (3.4.3) and combine with (3.4.4) to write

$$E(N - C) \geq \sigma^2 \delta^{*-2}(\tau^2 - h^2). \tag{3.4.9}$$

Now, in order to take the limit on the right hand side of (3.4.9) as $\delta^* \to 0$, we first need the expansion of $\tau = \tau(m_1(\delta^*))$ in terms of h. The Cornish–Fisher expansion of $\tau(m_1)$ in terms of h for large m_1 can be obtained from equation (5.1) of Gupta and Sobel (1957). One may also refer to section 26.2.49 of Abramowitz and Stegun (1965). It is known that for a fixed positive real number q^*, we have

$$\tau(m_1) = h + q^*(m_1 - 1)^{-1} + 0(m_1^{-2}).$$

Here, q^* does not depend on m_1. Thus, from (3.4.9), one gets

$$\liminf_{\delta^* \to 0} E(N - C) \geq \lim_{\delta^* \to 0} 0(\delta^{*-2} m_1^{-1})$$

$$= \lim_{\delta^* \to 0} 0(\delta^{*-2\gamma/(\gamma+1)}) = \infty, \tag{3.4.10}$$

since $\gamma \epsilon R^+$. Thus it follows that the modified two–stage procedure (3.4.1) – (3.4.2) is *not asymptotically second–order efficient* in the sense of (3.4.8).

One may like to speculate the primary reasons behind the asymptotic second–order inefficiency property of the modified two–stage methodology. It is clear that σ^2 had been estimated only once in the determination of N, the final sample size. On the other hand, this estimator can be quite unstable and this instability may lead to the eventual blowup of $E(N - C)$. Intuitively, it seems reasonable to guess that $E(N - C)$ will perhaps remain bounded (as $\delta^* \to 0$) if the selection methodology has the built–in provision of suitably updating the estimators of σ^2 in the course of collecting samples from each population, while estimating C. The next three selection methodologies are of this nature.

3.5 Purely Sequential Procedure

The purely sequential procedure discussed here was originally proposed by Robbins et al. (1968). Recall that $C = h^2\sigma^2/\delta^{*2}$, which is the optimal fixed sample size required from each population, had σ been known. Since σ is unknown, the basic idea is to keep updating estimators of σ^2 until we arrive at some kind of "stability", that is the sample size n and the corresponding estimator $U(n)$ of σ^2 satisfies certain relationship which is to be made precise very shortly.

We start with $m_1(\geq 2)$ samples from each π. Then, we keep taking one more sample at a time from each population according to the following stopping rule:

$$N = N(\delta^*) = \inf\{n \geq m_1 \ : \ n \geq h^2 U(n)\delta^{*-2}\}. \tag{3.5.1}$$

For fixed $\underset{\sim}{\mu}$, σ^2, m_1 and δ^*, we have $P(N < \infty) = 1 - \lim_{n\to\infty} P(N > n) \geq 1 - \lim_{n\to\infty} P\{n < h^2U(n)\delta^{*-2}\} = 1$, since $U(n) \to \sigma^2$ w.p. 1 as $n \to \infty$. That is, one has $P(N < \infty) = 1$, in other words, the purely sequential procedure (3.5.1) terminates with probability one. Based on the totality of all samples, that is having $X_{i1},\ldots,X_{iN}$ from π_i, $i = 1,\ldots,k$, one now implements the customary selection procedure $\mathscr{P}_N$ given by (3.2.1). Observe that the procedure (3.5.1) looks like (2.4.1) where $m = m_1$, $T_n = U(n)$ and $\psi_\nu = h^2/\delta^{*2}$. Throughout,

$\underset{\sim}{\mu}$ and σ are kept fixed.

Theorem 3.5.1. *For the purely sequential procedure (3.5.1), we have as* $\delta^* \to 0$:

(i) $N/C \to 1$ *w.p.* 1

(ii) $E(N/C) \to 1$;

(iii) $C^{-\frac{1}{2}}(N - C) \xrightarrow{\mathscr{L}} N(0, 2/k)$;

(iv) $\liminf P(CS) \geq P^*$ *for all* $\underset{\sim}{\mu} \in \Omega(\delta^*)$;

(v) $\lim P(CS) = P^*$ *under the LFC, namely* $\mu_{[1]} = \cdots = \mu_{[k-1]} = \mu_{[k]} - \delta^*$;

where $C = h^2\sigma^2/\delta^{*2}$ *and* h *comes from (3.2.4).*

Proof. Utilizing Lemma 1 of Chow and Robbins (1965), it follows that as $\delta^* \to 0$, we have $N \to \infty$ w.p. 1, $U(N) \to \sigma^2$ w.p. 1 and $U(N-1) \to \sigma^2$ w.p. 1. Recall that the sequential procedure (3.5.1) agrees with (2.4.1) where $\nu = [\delta^{*-1}]^* + 1$, $\psi_\nu = h^2/\delta^{*2}$, $T_n = U(n)$ and $m = m_1$. The basic inequality (2.4.3) simplifies to

$$h^2 U(N)\delta^{*-2} \leq N \leq m_1 + h^2 U(N-1)\delta^{*-2}. \tag{3.5.2}$$

Now divide throughout (3.5.2) by C and take limits as $\delta^* \to 0$. This leads to part (i).

Invoke Helmert's orthogonal transformation (Problem 2.2) to construct $W_1', W_2', \ldots$ which are i.i.d. $k^{-1}\sigma^2\chi_k^2$ such that

$$U(n) = (n-1)^{-1} \Sigma_{i=1}^{n-1} W_i'. \tag{3.5.3}$$

Use Anscombe's (1952) results to claim that the sufficient conditions given in (2.4.2) hold with $a = \sigma^2$ and $b = (2/k)^{\frac{1}{2}}\sigma^2$. Now, part (iii) follows from Theorem 2.4.1.

In order to prove parts (iv) – (v), first observe that Theorem 3.2.1 holds for the purely sequential selection procedure (3.5.1). Hence, for all $\underset{\sim}{\mu}\in\Omega(\delta^*)$,

$$P(CS) \geq E[\int_{-\infty}^{\infty} \{\Phi\,(y + N^{\frac{1}{2}}\delta^*\sigma^{-1})\}^{k-1}\phi(y)dy]. \tag{3.5.4}$$

Also, from part (i), one gets $N^{\frac{1}{2}}\delta^*\sigma^{-1} \to h$ w.p. 1 as $\delta^* \to 0$, and hence (3.5.4) together with the dominated convergence theorem will lead to part (iv). On the other hand, note that the LFC is given by $\mu_{[1]} = \cdots = \mu_{[k-1]} = \mu_{[k]} - \delta^*$ and under the LFC, equality holds in (3.5.4). Thus, similar arguments then lead to part (v).

Now, we turn to prove part (ii). From the right hand side of the basic inequality (3.5.2) it follows that

$$N \leq m_1 + h^2\delta^{*-2}W^*,$$

that is $NC^{-1} \leq m_1 + \sigma^{-2}W^*$ for sufficiently small δ^* such that C^{-1} becomes smaller than unity, where $W^* = \sup_{n\geq 2}\{(n-1)^{-1}\Sigma_{i=1}^{n-1}W_i'\}$. By Wiener's (1939) dominated ergodic theorem one concludes that $E(W^*) < \infty$. Now, the dominated convergence theorem and part (i) together imply part (ii). □

Remark 3.5.1. We mention that the purely sequential selection procedure (3.5.1) is *asymptotically* both *first–order efficient* and *first–order consistent.* However, one should note that the exact consistency property (3.4.5), which holds for the two–stage as well as the modified two–stage versions, can no longer be claimed for the purely sequential methodology. In the proof of Theorem 3.5.1 (ii), one can avoid using Wiener's (1939) result. Along the lines of Simons (1968) and Ghosh and Mukhopadhyay (1976), one can first prove instead that $E(N) \le C + 0(1)$. Hence, Theorem 3.5.1 (i) and Fatou's Lemma together will complete the proof. See Problem 3.2.

Remark 3.5.2. Suppose that the stopping rule (3.5.1) is modified slightly along the following lines:

$$N = N(\delta^*) = \inf\{n \ge m_1 : n \ge h^2\delta^{*-2}[U(n)+n^{-\gamma}]\} \tag{3.5.5}$$

for $0 < \gamma < \infty$. By looking at (3.5.3), one can easily claim that the stopping rule (3.5.5) is indeed of the form given in (2.4.6) with $\delta = 1$, $t = 0$, $\alpha = 1$, $m = m_1$ and $\psi_\nu = h^2/\delta^{*2}$ with $\nu = [\delta^{*-1}]^* + 1$. Hence, parts (i), (ii), (iv) and (v) of Theorem 3.5.1 holds for (3.5.5) for all $\gamma \in R^+$. Theorem 2.4.4 on the other hand implies that Theorem 3.5.1 (iii) holds when $\gamma > \frac{1}{2}$. Additionally, one can now obtain the rate of convergence also in Theorem 3.5.1 (iii) when the stopping variable N is taken from (3.5.5). For $\gamma > \frac{3}{4}$, Theorem 2.4.7 gives the following:

$$\sup_x \mid P[(2C/k)^{-\frac{1}{2}}(N - C) \le x] - \Phi(x) \mid = 0(\delta^{*\frac{1}{2}-\lambda})$$

for $0 < \lambda < \frac{1}{2}$.

3.5.1 SECOND–ORDER ASYMPTOTICS

In this section, we strengthen the results given in Theorem 3.5.1 (ii) and (v) by providing expansions of (a) $E(N)$ for all $\underset{\sim}{\mu} \in \Omega(\delta^*)$ and (b) $P(CS)$ under the LFC, for the purely sequential selection methodology (3.5.1). Use the representation given in (3.5.3) to rewrite (with $W_i = k\sigma^{-2}W_i'$)

$$N = \inf\{n \ge m_1 \ : \ kn(n-1)C^{-1} \ge \Sigma_{i=1}^{n-1} W_i\}$$

and hence $N = Q + 1$ where

$$Q = \inf\{n \geq m_1 - 1 : kn^2(1 + n^{-1})C^{-1} \geq \Sigma_{i=1}^{n} W_i\}, \tag{3.5.6}$$

W's being i.i.d. χ_k^2. The stopping variable Q is of the form given in (2.4.7) with $\delta = 2$, $L_0 = 1$, $m = m_1 - 1$, $h^* = k/C$, $\theta = k$, $\tau^2 = 2k$, $\beta^* = 1$, $n_0^* = C$, $p = 2/k$, $b = \frac{1}{2}k$ and

$$\nu = \nu(k) = \tfrac{1}{2}(k+2) - \Sigma_{n=1}^{\infty} n^{-1} E[\max\{0, \chi_{nk}^2 - 2nk\}]. \tag{3.5.7}$$

See (2.4.7) – (2.4.9) in this context. The second–order theory summarized here was originally developed in Mukhopadhyay and Judge (1989) for general $k \geq 2$. The case when $k = 2$ was considered in Mukhopadhyay (1983b). Let $g_0(y) = \int_{-\infty}^{\infty} \{\Phi(x+y)\}^{k-1} \phi(x)dx$ and $f(y) = g_0(y^{\frac{1}{2}})$ for $y > 0$. Now, we have the following theorem.

Theorem 3.5.2. *For the purely sequential procedure (3.5.1), we have as* $\delta^* \to 0$:

(i) $E(N) = C + \beta + o(1)$ *for all* $\underset{\sim}{\mu} \in \Omega(\delta^*)$ *if* (a) $m_1 \geq 3$ *when* $k = 2$, (b) $m_1 \geq 2$ *when* $k \geq 3$;

(ii) $P(CS) = P^* + h^2C^{-1}\{\beta f'(h^2) + h^2k^{-1}f''(h^2)\} + o(\delta^{*2})$ *under the LFC if* (a) $m_1 \geq 4$ *when* $k = 2$, (b) $m_1 \geq 3$ *when* $k = 3, 4, 5$, (c) $m_1 \geq 2$ *when* $k \geq 6$;

where $\beta = \beta(k) = k^{-1}\{\nu(k) - 2\}$ *with* $\nu(k)$ *defined in(3.5.7).*

Before we prove this theorem, we first prove a lemma. Let us define $U^* = (N - C)^2/C$. Then, Theorem 2.4.8 (iv) implies that

$$U^* \text{ is uniformly integrable when } m_1 > 1+2k^{-1}. \tag{3.5.8}$$

Lemma 3.5.1. *Suppose that (i)* $M(x)$ *is a real valued function of* $x(>0)$ *such that there exists* $a_i > 0$, $\alpha_i > 0$, $i = 1,\ldots,r$, *for which* $|M(x)| \le \Sigma_{i=1}^r a_i/x^{\alpha_i}$; *(ii)* Z *is a positive random variable such that* $\min\{h^2, Nh^2/C\} \le Z \le \max\{h^2, Nh^2/C\}$ w.p. 1 *for* N *defined in (3.5.1). Then, we have*

$$E\{M(Z)(N-C)^2/C\} = (2/k)M(h^2) + o(1) \quad \textit{as} \quad \delta^* \to 0,$$

if $m_1 > (2/k)(\max_{1\le i\le r} \alpha_i + 1) + 1$.

Proof. Write

$$M(Z)(N-C)^2/C = Q^* + R^*,$$

where $Q^* = M(Z)U^*I(N>C/2)$ and $R^* = M(Z)U^*I(N\le C/2)$. From assumption (i), we have

$$|Q^*| \le (\Sigma_{i=1}^r a_i/Z^{\alpha_i})U^*I(N>C/2).$$

Now assumption (ii) gives us $Z \ge \min\{h^2, Nh^2/C\}$ and thus note that on the set $\{N > C/2\}$, we have $N/C > \frac{1}{2}$ and $Z > h^2/2$. Therefore,

$$a_i/Z^{\alpha_i} < 2^{\alpha_i}a_i/h^{2\alpha_i} \quad \text{and} \quad |Q^*| < (\Sigma_{i=1}^r 2^{\alpha_i}a_i/h^{2\alpha_i})U^*.$$

By appealing to (3.5.8) under the condition $m > 1 + 2/k$, $|Q^*|$ is uniformly integrable, and since $U^* \xrightarrow{\mathscr{L}} (2/k)\chi_1^2$ as $\delta^* \to 0$ and $Z \xrightarrow{P} h^2$ as $\delta^* \to 0$, we have

$$E(Q^*) = (2/k)M(h^2) + o(1) \quad \text{as } \delta^* \to 0. \tag{3.5.9}$$

Similarly,

$$|R^*| \le (\Sigma_{i=1}^r a_i/Z^{\alpha_i})U^*I(N\le C/2)$$

where $Z \geq \min\{h^2, Nh^2/C\}$. On the set $\{N \leq C/2\}$, we have $N/C \leq \frac{1}{2}$ and thus $Z > Nh^2/C$. From this we obtain the inequality $1/Z \leq C/(h^2N) \leq C/(h^2m_1)$, since $N \geq m_1$, and therefore

$$a_i/Z^{\alpha_i} < a_i\{C/(h^2m_1)\}^{\alpha_i} \text{ and } |R^*| < [\Sigma_{i=1}^r a_i\{C/(h^2m_1)\}^{a_i}]U^*I(N\leq C/2).$$

It follows that $E|R^*| \leq [\Sigma_{i=1}^r a_i\{C/h^2m_1\}^{\alpha_i}]E\{U^*I(N\leq C/2)\}$. Note that on the set $\{N \leq C/2\}$ we also have

$$U^* = (N - C)^2/C \leq (N^2+C^2)/C \leq \{(C^2/4)+C^2\}/C = (5/4)C.$$

This then leads us to conclude

$$E|R^*| \leq [\Sigma_{i=1}^r a_i\{C/(h^2m_1)\}^{\alpha_i}]\,(5C/4)P(N \leq C/2).$$

From Theorem 2.4.8 (i), we have $P(N \leq C/2) = 0(C^{k(1-m_1)/2})$, and thus, as $\delta^*\to 0$, one has

$$[\Sigma_{i=1}^r a_i(C/h^2m_1)^{\alpha_i}](5C/4)P(N\leq C/2) = 0(C^{\{\max \alpha_i+1+k(1-m_1)/2\}}).$$

For this to be $o(1)$ as $\delta^* \to 0$, we need $\max \alpha_i + 1 + k(1-m_1)/2 < 0$, and hence under the condition $m_1 > (2/k)(\max \alpha_i + 1) + 1$, we have $E(R^*) = o(1)$ as $\delta^* \to 0$. Therefore, combining this with (3.5.9), we conclude

$$E\{M(Z)(N-C)^2/C\} = E(Q^*) + E(R^*) = (2/k)M(h^2)+o(1) \text{ as } \delta^* \to 0. \qquad \square$$

Proof of Theorem 3.5.2. Recall that, for $y > 0$, we write $g_0(y) = \int_{-\infty}^{\infty} \{\Phi(x+y)\}^{k-1}\phi(x)dx$ and $f(y) = g_0(y^{\frac{1}{2}})$, and we obtain $f'(y) = \frac{1}{2}y^{-\frac{1}{2}}g_0'(y^{\frac{1}{2}})$,

$$f''(y) = \tfrac{1}{4}\{y^{-1}g_0''(y^{\frac{1}{2}}) - y^{-3/2}g_0'(y^{\frac{1}{2}})\}, \tag{3.5.10}$$

$$|g_0'(y)| \le (k-1)(2\pi)^{-\frac{1}{2}} \tag{3.5.11}$$

$$|g_0''(y)| \le (k-1)(k-2)(2\pi)^{-1} + (k-1)y(2\pi)^{-1}. \tag{3.5.12}$$

First, we prove part (i). Note that η in (2.4.10) corresponding to Q given in (3.5.6) simplifies to $(\nu - 2)k^{-1} - 1$ and hence Theorem 2.4.8 (v) with $\omega = 1$ leads to

$$E(N) = 1 + E(Q) = 1 + C + \eta + o(1)$$

$$= C + (\nu - 2)k^{-1} + o(1),$$

if $m_1 - 1 > 2k^{-1}$, that is, if $m_1 > 1 + 2k^{-1}$. This is part (i).

For part (ii), recall from(3.2.6) and (3.5.4) that under the LFC, we have

$$P(CS) = E\{\int_{-\infty}^{\infty}\{\Phi[z + (N^{\frac{1}{2}}\delta^*/\sigma)]\}^{k-1}\phi(z)dz\} = E\{f(Nh^2/C)\}.$$

Expanding $f(x)$ in a Taylor series around $x = h^2$ gives us

$$f(x) = f(h^2) + f'(h^2)(x-h^2) + f''(Z(x))(x-h^2)^2/2,$$

where $\min(x, h^2) \le Z(x) \le \max(x, h^2)$. Therefore,

$$E\{f(Nh^2/C)\} = f(h^2) + (h^2/C)f'(h^2)E\{N-C\}$$

$$+(h^4/2C)E\{f''(Z(Nh^2/C))(N-C)^2/C\}.$$

Note that $f(h^2) = P^*$ and $E\{N - C\} = \beta + o(1)$ as $\delta^* \to 0$. Now, utilizing (3.5.10) – (3.5.12), we have

$$|f''(x)| \le \tfrac{1}{4}\{|x^{-1}g_0''(x^{\frac{1}{2}})| + |x^{-3/2}g_0'(x^{\frac{1}{2}})|\},$$

and hence, we can immediately claim $|f''(x)| \le \Sigma_{i=1}^{r} a_i/x^{\alpha_i}$ where $r = 3$, $a_1 = \frac{1}{4}(2\pi)^{-\frac{1}{2}}\{(k-1)(k-2)(2\pi)^{-\frac{1}{2}} + (k-1)a'\}$, $a_2 = \frac{1}{4}(2\pi)^{-\frac{1}{2}}(k-1)$, $a_3 = \frac{1}{4}(k-1)(2\pi)^{-\frac{1}{2}}$ and $\alpha_1 = 1$, $\alpha_2 = \frac{1}{2}$, $\alpha_3 = 3/2$. Thus, by the Lemma 3.5.1, for $m_1 > 1 + 5/k$, we have

$$E\{f''(Z(Nh^2/C))(N-C)^2/C\} = (2/k)f''(h^2) + o(1)$$

as $\delta^* \to 0$. Therefore, under the LFC, for $m_1 > 1 + 5/k$, we obtain

$$\begin{aligned} P(CS) &= P^* + (h^2/C)f'(h^2)\{\beta + o(1)\} + (h^4/2C)\{(2/k)f''(h^2) + o(1)\} \\ &= P^* + (h^2/C)\{\beta f'(h^2) + h^2 f''(h^2)/k\} + o(\delta^{*2}) \text{ as } \delta^* \to 0, \end{aligned}$$

which is part (ii). □

One should note that Theorem 3.5.1 (ii) and (iv) show that the purely sequential selection procedure (3.5.1) is *asymptotically* both *first–order efficient* and *first–order consistent.* Theorem 3.5.2 (i), however, shows that the same methodology is also *asymptotically second–order efficient.* Recall (3.4.5) – (3.4.8) in this context. Even so, one can easily feel that the purely sequential rule can be quite cumbersome for implementation, because here the experimenter takes one sample at a time from each population and then this process may take a while to stop. It will be desirable to reduce such sampling operations to a certain extent and yet maintain the asymptotic second–order efficiency property. In the next two sections, the selection methodologies discussed aim at achieving this goal, that is to have more operational convenience in conjunction with asymptotic second–order properties.

3.6 Accelerated sequential procedure

Recall that $C = h^2\sigma^2/\delta^{*2}$ and thus far σ^2 was being estimated by $U(n)$. For technical reasons, however, we will now estimate σ^2 by $U^*(n)$ given in (3.1.1). Asymptotically, that is as $n \to \infty$, there is of course not much difference between $U(n)$ and $U^*(n)$. In fact, the sequential procedure (3.5.1) can *essentially* be replaced by

$$L = L(\delta^*) = \inf\{n \geq m_1 : n \geq h^2 U^*(n)\delta^{*-2}\} \tag{3.6.1}$$

where $m_1(\geq 2)$ is the starting sample size from each π. In the end one would use selection rule $\mathscr{R}_L$ given by (3.2.1). Note that $L = Q + 1$ where

$$Q = \inf\{n \geq m_1 - 1 : kn^2C^{-1} \geq \Sigma_{i=1}^n W_i\}, \tag{3.6.2}$$

W's being i.i.d. χ_k^2. Thus, one can prove along the lines of Theorem 3.5.2 (i) that

$$E(L) = C + (\nu-2)k^{-1} + 1 + o(1) \tag{3.6.3}$$

if $m_1 > 1 + 2k^{-1}$, for all $\underset{\sim}{\mu}\in\Omega(\delta^*)$. Since $U^*(n)$ overestimates σ^2 slightly for each $n \geq 2$, the fact that the purely sequential procedure (3.6.1) takes *on the average* one extra sample compared with the original purely sequential rule (3.5.1) is quite intuitive indeed. The second–order expansion of $P(CS)$ given in Theorem 3.5.2 (ii) also holds for the purely sequential procedure (3.6.1), having replaced β by the expression $(\nu-2)k^{-1} + 1$. In these senses, the two purely sequential selection rules (3.5.1) and (3.6.1) can be considered *essentially same*, and we motivate our accelerated version of the methodology via (3.6.1) along the lines of section 2.5. This procedure was developed in Mukhopadhyay and Solanky (1992a).

We first choose and fix $0 < \rho < 1$ and $q \geq 0$. Along the lines of (2.5.1) – (2.5.3), let us define

$$t = t(\delta^*) = \inf\{n \geq m_1 : n \geq \rho h^2 U^*(n)/\delta^{*2}\}, \tag{3.6.4}$$

$$N_1 = N_1(\delta^*) = [h^2 U^*(t)\delta^{*-2} + q]^* + 1, \tag{3.6.5}$$

$$N = N(\delta^*) = \max\{t(\delta^*), N_1(\delta^*)\}. \tag{3.6.6}$$

We start the experiment with $m_1(\geq 2)$ samples from each π. Then, we take one sample at a time from each π sequentially and end up with $X_{i1},\ldots,X_{it}$ from π_i, $i = 1,\ldots,k$. Note that $t(\delta^*)$ estimates ρC, a fraction of the optimal fixed sample size. Now, $N_1(\delta^*)$ estimates C based on t samples from each π. If $N = t$, we do not take any more samples. If $N > t$, we sample the difference $(N - t)$ from each π in *one single batch* at this stage. Now, based on the totality of all samples, that is having $X_{i1},\ldots,X_{iN}$ from π_i, $i = 1,\ldots,k$, one now implements the customary selection procedure $\mathscr{P}_N$ given by (3.2.1). Throughout, $\underset{\sim}{\mu}$ and σ are kept fixed.

Observe that Theorem 3.2.1 holds for the accelerated sequential selection procedure (3.6.4) – (3.6.6). Hence, for all $\underset{\sim}{\mu} \in \Omega(\delta^*)$,

$$P(CS) \geq E[\int_{-\infty}^{\infty} \{\Phi(y + N^{\frac{1}{2}}\delta^*\sigma^{-1})\}^{k-1}\phi(y)dy] \tag{3.6.7}$$

as in (3.5.4), while the equality in (3.6.7) holds under LFC which is given by $\mu_{[1]} = \cdots = \mu_{[k-1]} = \mu_{[k]} - \delta^*$, as before. The constant $\nu = \nu(k)$ is the same as in (3.5.7) and $f(y)$ as well as $g_0(y)$ are exactly same as these were in the previous section. First, we discuss certain first-order asymptotic results parallel to those in Theorem 3.5.1.

Theorem 3.6.1. *For the accelerated sequential procedure (3.6.4) – (3.6.6), we have as* $\delta^* \to 0$:

(i) $N/C \to 1$ *w.p. 1;*

(ii) $E(N/C) \to 1$;

(iii) $C^{-\frac{1}{2}}(N - C) \xrightarrow{\mathscr{L}} N(0, 2(k\rho)^{-1})$;

(iv) $\liminf P(CS) \geq P^*$ *for all* $\underset{\sim}{\mu} \in \Omega(\delta^*)$;

(v) $\lim P(CS) = P^*$ *under the LFC, namely* $\mu_{[1]} = \cdots = \mu_{[k-1]} = \mu_{[k]} - \delta^*$;

where $C = h^2\sigma^2/\delta^{*2}$ *and* h *comes from (3.2.4).*

Proof. Note that along the lines of (3.6.2), one can write

$$t = t(\delta^*) = 1 + \inf\{n \geq m_1 - 1 : kC^{-1}n \geq \rho\overline{W}_n\}$$

where $\overline{W}_n = n^{-1}\Sigma_{i=1}^n W_i$, the W's being i.i.d χ_k^2, and it is immediate to observe that the accelerated sequential procedure (3.6.4) – (3.6.6) agrees with (2.5.1) – (2.5.3) with $\delta = 1$, $h^* = k/C$, $m = m_1 - 1$, $\theta = k$, $\tau^2 = 2k$, $n_0 = \rho C$, $n_0^* = C$ and $p = 2/k$. Now, parts (i) – (iii) follow from Theorem 2.5.1. Parts (iv) – (v) follow along the lines of proof of Theorem 3.5.1 (iv) – (v). □

Remark 3.6.1. Comparing part (iii) in Theorems 3.5.1 and 3.6.1, one immediately notices that the sample size N obtained from the accelerated sequential procedure is somewhat more variable than that for the purely sequential one. This is clear since one finds that Var(N | Acc.Seq.) "appears" to be $2(k\rho)^{-1}$ instead of $2k^{-1}$ obtained in the purely sequential case and $\rho \in (0, 1)$. A more precise statement is given in Theorem 3.6.2 (ii). In practice, the increase in the asymptotic variance of N defined in (3.6.6) will have to be eventually balanced with the possible significant reduction ($\approx 100(1-\rho)\%$) in sampling operations compared with the purely sequential rule, and hence comments on appropriate choices of ρ would be crucial and these are briefly discussed later in section 3.8.

3.6.1 SECOND–ORDER ASYMPTOTICS

In this section, we strengthen the results given in Theorem 3.6.1 (ii) and (v) by providing expansions of (a) E(N) for all $\underset{\sim}{\mu} \in \Omega(\delta^*)$, and (b) P(CS) under the LFC, for the accelerated sequential methodology (3.6.4) – (3.6.6). Let $g_0(y)$ and $f(y)$ be as before, for $y > 0$. Now, we have the following theorem.

Theorem 3.6.2. *For the accelerated sequential procedure (3.6.4) – (3.6.6), we have as* $\delta^* \to 0$:

(i) $q - 2(k\rho)^{-1} \le \liminf E(N-C) \le \limsup E(N-C) \le q + 1 - 2(k\rho)^{-1}$ *for all* $\underset{\sim}{\mu} \in \Omega(\delta^*)$ *if* $m_1 \ge 4$;

(ii) $\text{Var}(N) = 2(k\rho)^{-1}C + o(\delta^{*-2})$ *for all* $\underset{\sim}{\mu} \in \Omega(\delta^*)$ *if* (a) $m_1 \ge 3$ *when* $k = 2$, (b) $m_1 \ge 2$ *when* $k \ge 3$;

(iii) $P(CS) \ge P^* + o(\delta^{*2})$ *under the LFC if* (a) $m_1 \ge 5$ *when* $k = 2$, (b) $m_1 \ge 4$ *when* $k \ge 3$, *with* $q = \max\{0, 2(k\rho)^{-1} - h^2 f''(h^2)[k\rho f'(h^2)]^{-1}\}$;

where $C = h^2\sigma^2/\delta^{*2}$ *and* h *comes from (3.2.4).*

Proof. Here, we fully exploit the tools developed in section 2.5.1. With $\delta = 1$, $m_0(\delta)$ given in (2.5.7) indeed depends upon k, and thus we may write $m_0(k)$ instead and it is easy to check that (having $r' = 2$, $s = 1.01$)

$$m_0(k) = \begin{cases} 2.02 & \text{if } k = 2 \\ 2 & \text{if } k \ge 3 \end{cases} \tag{3.6.8}$$

Part (i) follows immediately from Corollary 2.5.1 for $m = m_1 - 1 > m_0(k)$

where $m_0(k)$ is defined in (3.6.8). Part (ii) follows from Theorem 2.5.2 (ii) for $m = m_1 - 1 > 2/k$. In part (iii), we restrict attention only to LFC. Now, as before,

$$P(CS) = E\{\int_{-\infty}^{\infty} [\Phi(z + N^{\frac{1}{2}}\delta^* \sigma^{-1})]^{k-1} \phi(z)dz\} = E\{f(Nh^2/C)\}. \quad (3.6.9)$$

Expand $f(x)$ in a Taylor series around $x = h^2$, and for a suitable function $W^*(x)$ such that $\min(x, h^2) \le W^*(x) \le \max(x, h^2)$, we can write from (3.6.9):

$$P(CS) = f(h^2) + (h^2/C)f'(h^2)E\{N-C\}$$
$$+ \tfrac{1}{2}(h^4/C)E[f''(W^*(Nh^2/C))(N-C)^2/C] \quad (3.6.10)$$

One can see that $|f''(x)| \le \Sigma_{i=1}^{3} a_i/x^{\alpha_i}$ where $\alpha_1 = 1$, $\alpha_2 = \frac{1}{2}$, $\alpha_3 = 3/2$ and a_1, a_2, a_3 are positive constants. Let $U^* = (N-C)^2/C$. Combining Remark 2.5.2 and Theorem 3.6.1 (iii), we observe that

$$U^* \text{ is uniformly integrable and } E(U^*) = 2(k\rho)^{-1} + o(1)$$
$$\text{if } m = m_1 - 1 > \max\{m_0(k), 4k^{-1}\}. \quad (3.6.11)$$

Also, Theorem 2.5.3 (i) leads to

$$P\{N \le \epsilon C\} = 0(C^{-\frac{1}{2}(m_1-1)k}) \quad (3.6.12)$$

for all $\epsilon \in (0, \rho)$. In view of our (3.6.11) – (3.6.12), it is not hard to realize that Lemma 3.5.1 holds for N given by (3.6.6), and hence

$$E[f''(W^*(Nh^2/C))U^*] = 2(k\rho)^{-1}f''(h^2) + o(1) \quad (3.6.13)$$

if $m_1 > 1 + \max\{m_0(k), 1 + 5k^{-1}\}$. Now, we combine (3.6.10), (3.6.13) and part (i). Thus, for $m_1 > 1 + \max\{m_0(k), 1 + 5k^{-1}\}$, we can write

$$P(CS) \geq P^* + h^2C^{-1}f'(h^2)\{q - 2(k\rho)^{-1}\} + h^4C^{-1}f''(h^2)(k\rho)^{-1} + o(\delta^{*2})$$

$$= P^* + o(\delta^{*2})$$

if $q = \max\{0, (k\rho)^{-1}\{2 - h^2f''(h^2)\{f'(h^2)\}^{-1}\}$. □

3.7 THREE–STAGE PROCEDURE

The three–stage selection methodology developed here reduces the sampling operations practically to the bare minimum and yet one maintains second–order asymptotic characteristics. This procedure was originally discussed in Mukhopadhyay and Judge (1991).

Again recall that $C = h^2\sigma^2/\delta^{*2}$ and we go back to estimating σ^2 by means of $U(n)$. In this three–stage selection procedure, we start with an initial sample of size $m_1 = m_1(\delta^*)(\geq 2)$ from each of the k populations, where

$$m_1 = m_1(\delta^*) = o(C^{1/v}) \tag{3.7.1}$$

for some $v > 1$. This expression is indeed very much in line with the choice of m_1 given in (3.4.1) while defining the modified two–stage procedure (3.4.1) – (3.4.2). As such, the number "v" in (3.7.1) can be thought of as the number "$\gamma + 1$" used in (3.4.1). We choose and fix $0 < \rho < 1$. The idea is to first estimate ρC, a fraction C, by means of two–stage sampling, and then estimate all of C in the third stage of sampling. In other words, this methodology consists of batch sampling all through. Along the lines of (2.3.1) – (2.3.2), we now write

$$T = T(\delta^*) = \max\{m_1, [\rho h^2U(m_1)\delta^{*-2}]^* + 1\}, \tag{3.7.2}$$

$$N = N(\delta^*) = \max\{T, [h^2U(T)\delta^{*-2}]^* + 1\}. \tag{3.7.3}$$

In terms of implementation, we proceed as follows. We start with X_{i1},

$X_{i2},\ldots,X_{im_1}$ from π_i, $i = 1,\ldots,k$. If $T = m_1$, then we do not sample any more from any population at the second stage. However, if $T > m_1$, then we sample the difference at the second stage to extend $X_{i1}, X_{i2},\ldots,X_{im_1}$ to X_{i1}, $X_{i2},\ldots,X_{iT}$ from π_i, $i = 1,\ldots,k$. Based on T samples from each population, we determine $U(T)$ and thus we find N. If $N = T$, we do not sample any more from any population at the third stage. However, if $N > T$, then we sample the difference at the third stage to extend $X_{i1}, X_{i2},\ldots,X_{iT}$ to X_{i1}, $X_{i2},\ldots,X_{iN}$ from π_i, $i = 1,\ldots,k$. Once we have N samples from each of the k populations, one now implements the customary selection procedure $\mathcal{P}_N$ given by (3.2.1). Throughout, $\underset{\sim}{\mu}$ and σ are kept fixed.

Observe that Theorem 3.2.1 holds for the three–stage selection procedure (3.7.2) – (3.7.3). Hence, for all $\underset{\sim}{\mu}\in\Omega(\delta^*)$,

$$P(CS) \geq E[\int_{-\infty}^{\infty} \{\Phi(y+N^{\frac{1}{2}}\delta^*\sigma^{-1})\}^{k-1}\phi(y)dy] \qquad (3.7.4)$$

as in (3.6.7), while the equality in (3.7.4) holds under the LFC which is again given by $\mu_{[1]} = \cdots = \mu_{[k-1]} = \mu_{[k]} - \delta^*$. The functions $f(y)$ and $g_0(y)$ are exactly the same as these were in sections 3.5 – 3.6. First, we summarize certain first–order asymptotic results parallel to those in Theorems 3.5.1 and 3.6.1.

Theorem 3.7.1. *For the three-stage procedure (3.7.2) – (3.7.3), we have as* $\delta^* \to 0$:

(i) $N/C \to 1$ *w.p.* 1;

(ii) $E(N/C) \to 1$;

(iii) $C^{-\frac{1}{2}}(N-C) \xrightarrow{\mathcal{L}} N(0, 2(k\rho)^{-1})$;

(iv) liminf P(CS) $\geq$ P* *for all* $\underset{\sim}{\mu}\in\Omega(\delta^*)$;

(v) lim P(CS) = P* *under the LFC, namely* $\mu_{[1]} = \cdots = \mu_{[k-1]} = \mu_{[k]} - \delta^*$;

where $C = h^2\sigma^2/\delta^{*2}$ *and* h *comes from (3.2.4).*

Proof. Along the lines of (3.5.6), note that U(n) can be written as $(n-1)^{-1}\Sigma_{i=1}^{n-1}Y_i$ where Y's are i.i.d. $\sigma^2 k^{-1}\chi_k^2$, and hence (3.7.2) – (3.7.3) agrees with the representations in (2.3.1) – (2.3.2) having $\lambda = h^2/\delta^{*2}$. Now, parts (i) – (ii) follow from Theorem 1 in Vik and Mukhopadhyay (1988). From the proof given for Theorem 2.3.3 (ii), part (iii) easily follows. Parts (iv) – (v) follow along the lines of Theorem 3.5.1 (iv) – (v). Further details are omitted. □

3.7.1 SECOND–ORDER ASYMPTOTICS

In this section, we strengthen the results discussed in Theorem 3.7.1 (ii) and (v).

Theorem 3.7.2. *For the three–stage procedure (3.7.2) – (3.7.3), we have as* $\delta^* \to 0$:

(i) $E(N) = C + \beta' + o(1)$ *for all* $\underset{\sim}{\mu}\in\Omega(\delta^*)$;

(ii) $Var(N) = 2(k\rho)^{-1}C + o(\delta^{*-2})$ *for all* $\underset{\sim}{\mu}\in\Omega(\delta^*)$;

(iii) $E[|N-E(N)|^3] = o(\delta^{*-4})$ *for all* $\underset{\sim}{\mu}\in\Omega(\delta^*)$;

where $C = h^2\sigma^2/\delta^{*2}$, $\beta' = \frac{1}{2} - 2(k\rho)^{-1}$ *and* h *comes from (3.2.4).*

Proof. For parts (i) – (ii), we exploit Theorem 2.3.2 with $\omega = 1, 2$, $n^* = C$,

$\theta = \sigma^2$ and $\mathrm{Var}(W_1) = \theta^{-2}\mathrm{Var}(Y_1) = 2/k$ in order to obtain

$$E(N) = C + [\tfrac{1}{2} - 2(k\rho)^{-1}] + o(1),$$

$$E(N^2) = C^2 + [1 - 2(k\rho)^{-1}]C + o(C),$$

which imply

$$\begin{aligned} \mathrm{Var}(N) &= E(N^2) - \{E(N)\}^2 \\ &= C^2 + [1 - 2(k\rho)^{-1}]\, C + o(C) \\ &\quad - C^2 - [\tfrac{1}{2} - 2(k\rho)^{-1}]^2 - 2[\tfrac{1}{2} - 2\,(k\rho)^{-1}]\, C + o(C) \\ &= 2(k\rho)^{-1}C - [\tfrac{1}{2} - 2(k\rho)^{-1}]^2 + o(C) \\ &= 2(k\rho)^{-1}C + o(C). \end{aligned}$$

Part (iii) follows from Hall (1981). □

As will be seen shortly in the proof of Theorem 3.7.3, the above selection procedure (3.7.2) – (3.7.3) has the property that $P(CS) = P^* + c_1\delta^{*2}/\sigma^2 + o(\delta^{*2})$ under the LFC, as $\delta^* \to 0$, where c_1 is a computable constant. We now modify the above selection procedure slightly so as to be able to conclude that $P(CS) = P^* + o(\delta^{*2})$ under the LFC as $\delta^* \to 0$. Here, as in the above selection procedure, m_1 still must satisfy (3.7.1). The modified selection procedure is defined as follows: With $\beta' = \frac{1}{2} - 2(k\rho)^{-1}$, let

$$T = T(\delta^*) = \max\{[\rho h^2 \mathrm{U}(m_1)\delta^{*-2}]^* + 1, m_1\},$$

$$m_1^* = -\beta' - h^2(k\rho)^{-1}\{f''(h^2)/f'(h^2)\},$$

$$N = N(\delta^*) = \max\{[h^2 \mathrm{U}(T)\delta^{*-2} + m_1^*]^* + 1, T\}. \qquad (3.7.5)$$

Note that

$$E(N) = C + \beta' + m_1^* + o(1) \text{ as } \delta^* \to 0, \tag{3.7.6}$$

and that the moment expansions of Theorem 3.7.2 (ii) – (iii) still hold for the stopping time N of this modified selection procedure (3.7.5). Once we determine N, we implement $\mathscr{P}_N$ given by (3.2.1). Before proceeding with the derivation of the second–order expansion of P(CS) for the modified procedure (3.7.5) in Theorem 3.7.3, we state and prove the following lemma.

Lemma 3.7.1. *Let the stopping time* N *be defined as in (3.7.5). Then as* δ^* $\to 0$, *we have:*

(i) For any arbitrary, but fixed $\epsilon \in (0,1)$, $P\{N \le \epsilon E(N)\} = o(\delta^{*2t/v})$, *for every positive integer* t;

(ii) Suppose that (a) $v^*(x)$ *is a real valued function of* $x(>0)$ *such that there exists* $a_i > 0$, $\alpha_i > 0$, $i = 1,\ldots,r$ *for which* $|v^*(x)| \le \Sigma_{i=1}^r a_i/x^{\alpha_i}$;

(b) W^* *is a positive random variable such that* $\min\{\ell^2 N, \ell^2 E(N)\} \le W^* \le \max\{\ell^2 N, \ell^2 E(N)\}$ *w.p.* 1, *where* $\ell = \delta^*/\sigma$. *Then, we have* $E|v^*(W^*)\delta^{*6}\{N - E(N)\}^3| = o(\delta^{*2})$.

Proof. First we will prove part (i). Recall that $\beta' = \frac{1}{2} - 2(k\rho)^{-1}$. Thus,

$$P\{N \le \epsilon E(N)\} \le P\{h^2 U(T)/\delta^{*2} \le \epsilon(C + |\beta'| + |m_1^*| + o(1))\}$$

$$= P\{U(T) \le \epsilon\sigma^2 + (\epsilon/h^2)(|\beta'| + |m_1^*|)\delta^{*2} + o(\delta^{*2})\}.$$

Choose ϵ' such that $0 < \epsilon' < 1 - \epsilon$, and let $\epsilon'' = \epsilon' + \epsilon$. There exists a δ such that $\delta^* < \delta$ implies that $(\epsilon/h^2)(|\beta'| + |m_1^*|)\,\delta^{*2} + o(\delta^{*2}) < \epsilon'\sigma^2$, and thus for $\delta^* < \delta$ we have

$$P\{N \le \epsilon E(N)\} \le P\{U(T) \le \sigma^2(\epsilon + \epsilon')\}$$

$$\leq P\{|U(T) - \sigma^2| \geq \epsilon''\sigma^2\}$$

$$\leq P\{\max_{m_1 \leq n \leq n_1} |U(n) - \sigma^2| \geq \epsilon''\sigma^2\},$$

where $n_1 = [\epsilon E(N)]^*$, since $T \leq N \leq \epsilon E(N)$ on the set of interest here. Applying the Hajek–Renyi inequality [see Chow and Teicher (1978), p. 243], we have

$$P\{N \leq \epsilon E(N)\} \leq \{k\epsilon''\sigma^2\}^{-2} \Sigma_{j=a'}^{b'} E\{W_j - k\}^t j^{-t}$$

where $a' = m_1 - 1$, $b' = n_1 - 1$, the W_i's are i.i.d. χ_k^2 random variables, and t is any positive integer. This sum is $0(m_1^{-t})$. Now $m_1 = 0(\delta^{*-2/v})$, and thus we obtain $P\{N \leq \epsilon E(N)\} = 0(\delta^{*2t/v})$.

For part (ii), note that $E|v^*(W^*)\delta^{*6}(N - E(N))^3| \leq E(Q^*) + E(R^*)$, where

$$Q^* = |\{v^*(W^*)\delta^{*6}(N - E(N))^3\}| I(N > E(N)/2),$$

$$R^* = |\{v^*(W^*)\delta^{*6}(N - E(N))^3\}| I(N \leq E(N)/2).$$

Now on the set $\{N > E(N)/2\}$ we have $W^* \geq \ell^2 E(N)/2 = w_0$, say. Thus we have $v^*(W^*) \leq \Sigma_{i=1}^r a_i / w_0^{\alpha_i}$. Therefore,

$$E(Q^*) \leq (\Sigma_{i=1}^r a_i / w_0^{\alpha_i}) \delta^{*6} E|N - E(N)|^3 = o(\delta^{*2}) \text{ as } \delta^* \to 0,$$

by Theorem 3.7.2 (iii), since $w_0 = 0(1)$.

Similarly, on the set $\{N \leq E(N)/2\}$ we have $W^* \geq \ell^2 N \geq \ell^2 m_1$, and thus $v^*(W^*) \leq \Sigma_{i=1}^r a_i / (\ell^2 m_1)^{\alpha_i}$. Therefore,

$$E(R^*) \le \{\Sigma_{i=1}^{r} a_i/(\ell^2 m_1)^{\alpha_i}\}\delta^{*6} E\{|N - E(N)|^3 I(N \le E(N)/2)\}.$$

Also note that on the set $\{N \le E(N)/2\}$ we have

$$\{N-E(N)\}^3 \le N^3 + 3N\{E(N)\}^2 \le \{E(N)/2\}^3$$

$$+3\{E(N)/2\}\{E(N)\}^2 = (13/8)\{E(N)\}^3.$$

Thus,

$$E(R^*) \le \{\Sigma_{i=1}^{r} a_i/(\ell^2 m_1)^{\alpha_i}\}\delta^{*6}(13/8)(E(N))^3 P\{N \le E(N)/2\}.$$

Now $m_1 = 0(\delta^{*-2/v})$ implies that $\Sigma_{i=1}^{r} a_i/(\ell^2 m_1)^{\alpha_i} = 0(\delta^{*d})$, where $d = \{-2(v-1)/v\} \max \alpha_i$. Also, one notices that $\{E(N)\}^3 = 0(\delta^{*-6}\}$ as well as $P\{N \le E(N)/2\} = 0(\delta^{*2t/v})$. We have therefore,

$$E(R^*) \le \{\Sigma_{i=1}^{r} a_i/(\ell^2 m_1)^{\alpha_i}\}\delta^{*6}(13/8)(E(N))^3 P\{N \le E(N)/2\} = 0(\delta^{*z_0}),$$

where $z_0 = \{-2(v-1)/v\}\max \alpha_i + 2t/v$. For this to be $o(\delta^{*2})$ as $\delta^* \to 0$, we need $z_0 > 2$. This condition is satisfied if $t > v + (v-1) \max \alpha_i$. Since t can be chosen to be any positive integer, this condition can always be satisfied. Thus we have $0(\delta^{*z_0}) = o(\delta^{*2})$ as $\delta^* \to 0$. Therefore,

$$E|v^*(W^*)\delta^{*6}\{N - E(N)\}^3| \le E(Q^* + R^*) = o(\delta^{*2}) \text{ as } \delta^* \to 0. \qquad \square$$

Theorem 3.7.3. *Let the three-stage stopping rule be defined as in (3.7.5). Then, under the* LFC, *we have as* $\delta^* \to 0$:

$$P(CS) = P^* + o(\delta^{*2}).$$

Proof. In this proof, we will define $N = \max\{[h^2 U(T)\delta^{*-2} + m_1^*]^* + 1, T\}$, and show that we must have $m_1^* = -\beta' - h^2(k\rho)^{-1}\{f''(h^2)/f'(h^2)\}$. We know from (3.7.4) that under the LFC, we have

$$P(CS) = E\{\int_{-\infty}^{\infty} [\Phi(z + \{N^{\frac{1}{2}}\delta^* / \sigma\})]^{k-1} \phi(z)dz\} = E\{f(\ell^2 N)\}, \tag{3.7.7}$$

where $\ell = \delta^* / \sigma$. Expanding $f(\ell^2 N)$ in a Taylor series around $\ell^2 E(N)$ and taking expectations gives us

$$E\{f(\ell^2 N)\} = f(\ell^2 E(N)) + \tfrac{1}{2} f''(\ell^2 E(N)) \ell^4 E\{N - E(N)\}^2$$

$$+ E\{f'''(W_1^*) \ell^6 (N - E(N))^3 / 6\}, \tag{3.7.8}$$

where $\min(\ell^2 N, \ell^2 E(N)) \le W_1^* \le \max(\ell^2 N, \ell^2 E(N))$.

We will first evaluate $E|\{f'''(W_1^*) \ell^6 (N - E(N))^3/6\}|$ using Theorem 3.7.2 (ii). Letting $a^* = \int_{-\infty}^{\infty} |x| \phi(x)dx$ and $b^* = \int_{-\infty}^{\infty} x^2 \phi(x)dx$, one can easily verify that $|f'''(x)| \le \Sigma_{i=1}^{5} a_i / x^{\alpha_i}$, where $a_1 = (1/8)(k-1)(2\pi)^{-\frac{1}{2}}$, $a_2 = (1/8)(k-1)\{3(k-2)(2\pi)^{-1} + (2\pi)^{-\frac{1}{2}}(2a^*)\}$, $a_3 = (1/8)(k-1)\{(k-2) \times (k-3)(2\pi)^{-\frac{3}{2}} + 3(k-2)(2\pi)^{-1} + (2\pi)^{-\frac{1}{2}}(4 + b^*)\}$, $a_4 = (3/8)(k-1) \times \{(k-2)(2\pi)^{-1} + a^*(2\pi)^{-\frac{1}{2}}\}$, $a_5 = (3/8)(k-1)(2\pi)^{-\frac{1}{2}}$, and $\alpha_1 = 1/2$, $\alpha_2 = 1$, $\alpha_3 = 3/2$, $\alpha_4 = 2$, and $\alpha_5 = 5/2$. Thus the conditions of Lemma 3.7.1 are satisfied, and we have

$$E|\{f'''(W_1^*) \ell^6 (N - E(N))^3/6\}| = o(\delta^{*2}). \tag{3.7.9}$$

Also, note that

$$\ell^4 E\{N - E(N)\}^2/2 = (\delta^{*4}/2\sigma^4)\{2h^2\sigma^2(\delta^{*2}\rho k)^{-1} + o(\delta^{*-2})\}$$

$$= h^2 \delta^{*2} (\sigma^2 \rho k)^{-1} + o(\delta^{*2}), \tag{3.7.10}$$

as $\delta^* \to 0$, by Theorem 3.7.2 (ii).

Next we need to expand $f(\ell^2 E(N))$ in a Taylor series around h^2:

$$f(\ell^2 E(N)) = f(h^2) + f'(h^2)(\ell^2 E(N) - h^2) + \tfrac{1}{2} f''(W_2^*)\{\ell^2 E(N) - h^2\}^2,$$

where $$\min(\ell^2 E(N), h^2) \le W_2^* \le \max(\ell^2 E(N), h^2).$$

Now $\ell^2 E(N) - h^2 = (\delta^{*2}/\sigma^2)(\beta' + m_1^*) + o(\delta^{*2})$ as $\delta^* \to 0$. Hence, $\{\ell^2 E(N) - h^2\}^2 = o(\delta^{*2})$ as $\delta^* \to 0$. Using the fact that $\ell^2 E(N) \to h^2$ as $\delta^* \to 0$, we get $W_2^* \to h^2$ as $\delta^* \to 0$. Hence, this immediately implies that $f''(W_2^*)\{\ell^2 E(N) - h^2\}^2/2 = o(\delta^{*2})$ as $\delta^* \to 0$. Noting that $f(h^2) = P^*$, we have

$$f(\ell^2 E(N)) = P^* + f'(h^2)(\delta^{*2}\sigma^2)(\beta' + m_1^*) + o(\delta^{*2}) \text{ as } \delta^* \to 0. \tag{3.7.11}$$

Finally, we expand $f''(\ell^2 E(N))$ in a Taylor series around h^2:

$$f''(\ell^2 E(N)) = f''(h^2) + \tfrac{1}{2} f'''(W_3^*)\{\ell^2 E(N) - h^2\},$$

where $$\min(\ell^2 E(N), h^2) \le W_3^* \le \max(\ell^2 E(N), h^2).$$

Since $W_3^* \to h^2$ as $\delta^* \to 0$ and $\ell^2 E(N) - h^2 = o(\delta^*)$, we have

$$f''(\ell^2 E(N)) = f''(h^2) + o(\delta^*) \text{ as } \delta^* \to 0. \tag{3.7.12}$$

Combining (3.7.7) – (3.7.12), we have, under the LFC,

$$P(CS) = P^* + (\delta^{*2}/\sigma^2)\{f'(h^2)(\beta' + m_1^*) + f''(h^2)h^2(\rho k)^{-1}\} + o(\delta^{*2}).$$

In order to have $P(CS) = P^* + o(\delta^{*2})$ we need $f'(h^2)(\beta' + m_1^*) + f''(h^2)h^2(\rho k)^{-1} = 0$, and thus $m_1^* = -\beta' - \{f''(h^2)/f'(h^2)\}h^2(k\rho)^{-1}$. □

3.8 Moderate Sample Performances and Comments

In the case of the two–stage procedure (3.3.1) and the modified two–stage procedure (3.4.1) – (3.4.2), one has exact results regarding the probability of correct selection, namely that $P(CS) \geq P^*$ for all $\underset{\sim}{\mu} \in \Omega(\delta^*)$, the preference zone. However, there are several other additional aspects that need particular attention as well. One will note that most theoretical results obtained for the two–stage and other multistage selection procedures are indeed asymptotic in nature. It will be interesting to address a question such as "how asymptotic is asymptotic?" In other words, how large C has to be in practice so that the asymptotic expansions obtained for E(N) and P(CS) can also be assumed to provide reasonable approximations as practical guidelines. There should be other concerns as well. For example, how does one choose the starting sample size m_1 in all these methodologies? In this vein, we should add that there is no available optimality criteria in order to pick the best m_1, say. Hence, we tackle this issue by taking a couple of m_1 values and examining any visible effects on the moderate sample performances of our selection methodologies. By the same token, the accelerated sequential procedure (3.6.4) – (3.6.6) as well as the three–stage procedure (3.7.5) involve the choice of a proper fraction ρ. In the accelerated sequential case, we estimate ρC by means of purely sequential sampling followed by a batch sample, while in the case of the three–stage procedure, we estimate ρC by means of two–stage sampling followed by a batch sample. Naturally, one would expect that the choice of ρ value will very likely impact the performances of these methodologies for moderate values of C. But, again, there is no available optimality criteria yet in order to pick the best ρ, say. Hence, we try various values of $\rho \in (0, 1)$ and examine any visible effects or marks on such procedures. We accomplish such investigations and comparisons via extensive sets of computer simulations which we explain and summarize here.

Theorems 3.5.2, 3.6.2, 3.7.3 and (3.7.6) provide second–order expansions of $E(N) - C$ and $P(CS) - P^*$ for the purely sequential, accelerated sequential and three–stage selection methodologies. For example, in Theorem 3.5.2, one has $E(N) - C \approx \beta$ as well as $P(CS) - P^* \approx h^2C^{-1}\{\beta f'(h^2) + h^2k^{-1}f''(h^2)\}$, under the LFC. So, now if one estimates $E(N) - C$ or $P(CS) - P^*$ from a large scale simulation of the purely sequential procedure (3.5.1), a natural query will be whether such estimated values are "close" to the corresponding theoretical asymptotic values. For example, in the purely sequential case, one may like to know whether the simulated estimate of E(N) is "close" to $C + \beta$ for moderate values of C. In this sense, the quantity β may be viewed as the "correction term" for the estimated value of $E(N) - C$.

In this section we first provide Tables 3.8.1 – 3.8.4, to help evaluate the correction terms for $E(N - C)$ and $P(CS) - P^*$ for the purely sequential procedure (3.5.1), accelerated sequential procedure (3.6.4) – (3.6.6) and the three–stage procedure (3.7.5). These tables also help in computing the final sample size for the three–stage procedure (3.7.5) and the term q that is needed in the implementation of the accelerated sequential procedure (3.6.4) – (3.6.6). After this, we include Tables 3.8.5 – 3.8.7 which provide the simulated results of all the multistage selection procedures considered in this chapter. These tables also include the asymptotic values of $E(N - C)$ and $P(CS) - P^*$, whenever appropriate.

In Table 3.8.1, we provide values of $\nu(k)$ given by (3.5.7) and $\beta(k) = k^{-1}(\nu(k) - 2)$ for $k = 1(1)30$. In order to obtain these, we write our own program in Fortran and use IMSL subroutine DGAMIC to compute the incomplete gamma function. Any term bigger than 10^{-15} in magintude, was included in the infinite sum in (3.5.7). It is intuitively clear that as $k \to \infty$, $\beta(k) \to \frac{1}{2}$. However, one can see by looking at the column of $\beta(k)$ values, that the convergence rate is quite slow. Recall, from Theorem 3.5.2 (i), that the asymptotic value of $E(N - C)$ is $\beta(k)$ for the purely sequential procedure (3.5.1). Hence, we expect the average oversampling in comparison with C to be close to $\beta(k)$ and we will later see in Table 3.8.7a that such a feature is noticeable for the simulation studies carried out for the purely sequential procedure (3.5.1), when C is moderate in size.

In Table 3.8.2, we provide values of $f'(h^2)$, $f''(h^2)$ and $A = h^2\{\beta f'(h^2) + h^2k^{-1}f''(h^2)\}$ for $P^* = .90$, .95 and .99, for $k = 2(1)20$, where $\beta = \beta(k) = k^{-1}(\nu(k) - 2)$ and $\nu(k)$ comes from (3.5.7). Recall that the values of h, given k and P^*, can be obtained from Bechhofer (1954) and Table A.1 of Gibbons et al. (1977). Look at (3.2.4) one more time and observe that $h = h(k, P^*)$ also satisfies the following equation:

$$P\{Y_i - Y_k \le h,\ i = 1,\ldots,k-1\} = P^*, \tag{3.8.1}$$

where $Y_1,\ldots,Y_k$ are i.i.d. standard normal. In other words, let $(Y_1^*,\ldots,Y_{k-1}^*)$ denote a $(k-1)$–dimensional normal variable having zero means, unit variances, equicorrelation $\frac{1}{2}$ and then (3.8.1) can be rewritten as

$$P\{Y_i^* \le 2^{-\frac{1}{2}}h,\ i = 1,\ldots,k-1\} = P^*. \tag{3.8.2}$$

For $k = 3(1)21$ and $P^* = .90, .95, .99$, h values can also be obtained from Table A of Tong (1990). The values of $f'(h^2)$ and $f''(h^2)$ are needed in order to evaluate q (Theorem 3.6.2(iii)) in the case of accelerated sequential procedure (3.6.4) – (3.6.6) and also in computing m_1^* for the three–stage procedure (3.7.5). One observes that in Theorem 3.5.2 (ii), for the purely sequential procedure (3.5.1), the asymptotic value of $P(CS) - P^*$ equals $C^{-1}A$. In Table 3.8.7a, we use the asymptotic value of $P(CS) - P^*$ to compare with the estimated values of $P(CS) - P^*$, for the moderate sample sizes. To evaluate $f'(h^2)$, $f''(h^2)$ and A, we wrote programs in Mathematica (Version 2.0). A few selected values of $f'(h^2)$ and $f''(h^2)$ are also available in Table 2 of Mukhopadhyay and Judge (1991), however the values of $f''(h^2)$ are somewhat inaccurate. The values of $f'(h^2)$ reported in Mukhopadhyay and Judge (1991) match exactly with the corresponding values given in Table 3.8.2. It seems like their computer program had some subtle errors. In order to double check, we computed the numerical values of $f''(h^2)$ using IMSL also and these agreed with our Mathematica answers exactly.

In Table 3.8.3, we provide the values of q given in Theorem 3.6.2(iii) for $\rho = .25, .50, .75$; $P^* = .90, .95, .99$ and $k = 2(1)20$. We wrote the programs in Mathematica for evaluating these q values. Such values of q are needed in the implementation of the accelerated sequential procedure (3.6.4) – (3.6.6). Also, we use these to evaluate the lower and upper bounds for the asymptotic values of $E(N - C)$ in the accelerated sequential procedure as seen in Theorem 3.6.2 (i).

In Table 3.8.4, we present the values of m_1^* for $\rho = .25, .50, .75$; $P^* = .90, .95, .99$ and $k = 2(1)20$. Again, in order to compute m_1^*, we wrote programs in Mathematica. The values of m_1^* will be needed for implementing the three–stage procedure (3.7.5) and also to evaluate the asymptotic value of $E(N - C)$ for the procedure (3.7.5).

We now elaborate on the moderate sample size performances of the two–stage procedure (3.3.1), modified two–stage procedure (3.4.1) – (3.4.2), purely sequential procedure (3.5.1), accelerated sequential procedure (3.6.4) – (3.6.6) and the three–stage procedure (3.7.5). The VAX 11/780 mainframe computer at the University of New Orleans and programs written in Fortran are used to simulate all the selection procedures one by one. We consider $C = 25, 50, 100, 200, 400, 800$. Recall that C is the optimal fixed sample size, if the population variance σ^2 is known, one would need from each of the k populations to select the best population via $\mathscr{P}_n$ given by (3.2.1) where $n = [C]^* + 1$. When we simulate any of the selection procedures, we put in a value of σ in order to generate the populations but pretend that σ, and hence C, is unknown and then we compare the simulated estimates of $E(N)$ with the input values of C. We repeat this for each procedure, for each value of C, 5000 (= J, say) times. Each time we stop, we also check to see if the correct population was selected or not. To begin with, we used IMSL (1989, Edition 10) subroutine RNNOR to generate $(k - 1)$ pseudo normal populations each with mean = 5 and variance = 1, and another pseudo normal population with mean = $5 + \delta^*$, where $\delta^* = (h^2C^{-1})^{\frac{1}{2}}$. We consider $k = 5, 10$ and $P^* = .95$ only for simulation purposes. For a fixed set of values of k, m_1 and C, we independently implement each sampling procedure by repeating it independently J times. Let n_j be the value of N and $p_j = 1(0)$ if the correct population, that is the one with mean $5 + \delta^*$, is (is not) selected at the end of the jth iteration, $j = 1, 2,\ldots,J$. Let $\bar{n} = J^{-1} \Sigma_{j=1}^{J} n_j$ be the sample mean of n_j's and $s(\bar{n}) = \{(J^2 - J)^{-1} \Sigma_{j=1}^{J}(n_j - \bar{n})^2\}^{\frac{1}{2}}$ be the standard error of $\bar{n}$. We use $\bar{n}$ to estimate the value of C. Also, let $\bar{p} = J^{-1} \Sigma_{j=1}^{J} p_j$ be the relative frequency of the correct population being selected and $s(\bar{p}) =$

$\{\bar{p}(1-\bar{p})/J\}^{\frac{1}{2}}$ be the standard error of $\bar{p}$. Note that $\bar{p}$ will be used as an estimator for P(CS).

In Tables 3.8.5 – 3.8.7, we provide various columns showing the values of k, δ^*, C, $\bar{n}-C$ and $\bar{p}$. Also, underneath each value of $\bar{n}-C$ and $\bar{p}$, we provide their respective standard errors $s(\bar{n})$ and $s(\bar{p})$ within parenthesis. Again, wherever available, we provide the asymptotic values of $\bar{n}-C$ and $\bar{p}$ within triangular brackets $< >$ at the end of appropriate block of results.

Table 3.8.5 provides simulation results for the two–stage procedure (3.3.1). We include the values $m_1 = 10$, 20 and 30 for the starting sample size, and C = 25, 50, 100, 200, 400 and 800. Recall that $\bar{n}$ is the average of J(= 5000) simulated estimates of C, while pretending that σ is unknown. In columns 3 and 6 we provide the values of $\bar{n}-C$, the amount by which the two–stage procedure (3.3.1) "oversamples" compared with C. One will note that this oversampling is quite significant in terms of magnitude. For example in Table 3.8.5a, with $m_1 = 10$, k = 5 and C = 200 this oversampling amounts to about 16.4% of C. One notes that, the amount of oversampling keeps increasing for fixed k, however the magnitudes of oversampling come down as k increases and m_1 is held fixed. This is expected, since for fixed m_1 one has the degrees of freedom $k(m_1-1)$ for $U(m_1)$ in (3.3.1) while estimating σ^2 in C and naturally $k(m_1-1)$ increases as k increases (m_1 held fixed), thereby leading to a more precise estimation of C. This is reflected in having smaller oversampling as k increases when m_1 is held fixed. Underneath the columns 3 and 6, we provide the asymptotic value τ^2/h^2 for $\bar{n}/C$ and one will observe that the values of $\bar{n}/C$ are approximated fairly well by means of τ^2/h^2, even for moderate values of C, for example, when C = 200. We should add however that the rate of convergence of $\bar{n}/C$ to τ^2/h^2 appears to be slow, nonetheless one gets some feeling with regard to where one is in comparison with the expectation (asymptotic). One will note a few other facts. If k is held fixed, but the starting sample size m_1 increases, then this "oversampling" goes down steadily as can be seen from Tables 3.8.5a,b,c. However, very large values of m_1 may be quite impractical, in other words, one should use moderate values of m_1, for example 10 or 20 so that one may expect good results for many values of C, *across the board.* On the other hand, for fixed k, whatever be m_1, the values of $(\overline{n} - C)$ do increase steadily as C increases, that is as δ^* decreases, and such an observation is quite in harmony with equation (3.3.6). Our procedure is indeed consistent (Theorem 3.3.1(i)) and we do notice that $\overline{p}$ values are little larger than the target value P^*, *across the board,* for all values of C.

In Tables 3.8.6a,b,c, we present a summary of our findings in connection with moderate sample performances of the modified two–stage procedure (3.4.1) – (3.4.2). We first fix $m_0^* = 10$, $P^* = .95$. Next, we need to specify $\gamma \epsilon R^+$ in a certain suitable fashion. Note that $C = h^2\sigma^2/\delta^{*2} \geq h^2/\delta^{*2}$ when $\sigma \geq 1$ and also m_1 given by (3.4.1) will be approximated (for small δ^*) by $(h/\delta^*)^{2/(\gamma+1)}$. In other words, C is expected to be larger than m_1, as long as $\sigma \geq 1$, whatever $\gamma \epsilon R^+$. That is, the stopping variable N given by (3.4.2) will be essentially $\tau^2 \mathrm{U}(m_1)/\delta^{*2}$, whatever m_0^* and $\gamma \epsilon R^+$. So, how one should choose γ will particularly depend on the moderate sample performances observed when σ is smaller than unity. We tried $\sigma = .25$, .50, .75 and 1.0. There is another point to be noted, namely, that m_1 should be as large as practically possible without getting over C, having σ unknown. For small δ^*, this translates into picking small values of γ. In our simulations, we consider $\gamma = .1(.1).9$ and fix $k = 5$, 10 and a particular value of C, which in turn gives $\delta^* = (h^2\sigma^2/C)^{\frac{1}{2}}$. Now, we generate k normal populations as we did in the case of the two–stage procedure, having the common standard deviation σ. Very small γ will lead to a very large value of m_1 (perhaps, larger than C when $\sigma < 1$), if δ^* is also small. In each situation, we look at the values of $\overline{n} - C$, $s(\overline{n})$ and $\overline{p}$, $s(\overline{p})$. For $\sigma = .25$, .50 and .75, the choice of γ that gives rise to "small" magnitudes of $\overline{n} - C$ as well as close proximity between $\overline{p}$ and P^* turned out to be either .2, .3 or .4. These were also associated with fairly stable and reasonable estimated standard error values. The point is that we wish to give some useful recommendations about γ for a wide range of values of C, not just values which are too small or too large. In Tables 3.8.6a,b,c, we include summary findings when $\sigma = .5, .75$ and 1.0 with $\gamma = .2, .3$ and $.4$. The various entries are self explanatory.

From Tables 3.8.6a,b,c, we find that *across the board*, $(\bar{n} - C)$ values are fairly small when $\gamma = .3$ or $.4$. Considering both $(\bar{n} - C)$ and $\bar{p}$ values, $\gamma = .3$ or $.4$ appears to be a good choice in evaluating m_1 given by (3.4.1). Such a choice for γ also works well when $\sigma \geq 1$. To see this, one only needs to compare the $(\bar{n} - C)$ values in Table 3.8.6c with those in Table 3.8.5a in conjunction with a close scrutiny of the $\bar{p}$ values in relation with the target P^*. If one holds k, σ, P^* fixed, from Tables 3.8.6a,b,c, it becomes clear that $\bar{n} - C$ increases as C increases beyond certain magnitude. This is an undesirable feature of the modified two–stage methodology that empirically verifies the point made by equation (3.4.10). Also, $\bar{n}/C$ is very close to the asymptotic value, that is unity, which empirically justifies Theorem 3.4.1(ii).

Now, we simulate the other multistage procedures, one by one, and hope to see empirically that the associated $(\bar{n} - C)$ values are bounded, for moderate values of C. Such investigations are expected to add practical relevance to otherwise asymptotic concepts, such as second–order efficiency and expansions, etc. In Table 3.8.7a, we consider all the different sampling schemes for which second order results are available and, in particular, one observes that $E(N - C)$ is bounded. The methodologies consist of the purely sequential procedure (3.5.1), accelerated sequential procedure (3.6.4) – (3.6.6) and the three–stage procedure (3.7.5).

In the case of the two–stage and modified two–stage selection methodologies, the choice of m_1, the starting sample size, was a crucial factor in achieving "good" performances of these techniques in non–asymptotic situations. This was the case because the common variance σ^2 was to be estimated only once, based on first–stage or pilot samples, and there was no possibility of revision or updating such estimators in the followup. The purely sequential, accelerated sequential and the three–stage sampling techniques are very different on this count. In the purely sequential methodology, the estimators of σ^2 are updated with every new observation from each population arriving sequentially. So, as long as m_1 is not very large, one will essentially do just fine because even a wrongfully chosen small starting sample size m_1 is going to be "checked" and updated again and again, throughout the sampling process. In other words, it is quite alright to be conservative here and pick a small m_1. Similar arguments or their variants are also valid in the contexts of both the accelerated sequential and three–stage selection methodologies. In summary, we add that these three multistage selection procedures are fairly robust with respect to the choices of m_1, the starting sample size. On the other hand, we wish to empirically validate various second–order results given by a whole array of sophisticated Theorems, and hence m_1 will have to be chosen so as to satisfy all the sufficient conditions for the relevant Theorems. The choice "$m_1 = 10$" satisfies all such requirements in every single case under consideration. One should note another point. Unlike what we had seen in the case of the modified two–stage selection methodology, none of the design constants involved in the three multistage procedures given by (3.5.1), (3.6.4) – (3.6.6), and (3.7.5) depends on the actual magnitude of σ^2. In other words, we are not too concerned here about the situations where $\sigma < 1$ as opposed to $\sigma \geq 1$ as such. The primary reason being that σ^2 is indeed estimated and updated several times and this feature is embedded in our multistage sampling processes. Hence, without any loss of generality, we fix one particular value of σ^2 for simulation purposes, and throughout we consider $\sigma^2 = 1$. We fix $k = 5, 10$, $P^* = .95$, $C = 25$, 50, 100, 200, 400, 800 and determine $\delta^* = (h^2/C)^{\frac{1}{2}}$. Now, along the lines of the two–stage selection procedure's generation scheme, we build k pseudo normal populations and for each multistage procedure and k, C and δ^*, we estimate $E(N)$, the probability of correct selection, and their associated standard errors, via simulations using $J(= 5000)$ independent replications. The results are summarized in Table 3.8.7a. The notations are self explanatory. In each block, at the end of the column of values of $\bar{n} - C$ and $\bar{p}$, we provide the corresponding asymptotic values within < > for immediate comparisons. Further details follow.

In Table 3.8.7a, columns 2 and 3 show values of $\bar{n} - C$ and $\bar{p}$, along with their standard errors for the purely sequential procedure (3.5.1), while columns 4 – 5 and columns 6 – 7 show these figures respectively for the accelerated sequential procedure (3.6.4) – (3.6.6) and the three–stage procedure (3.7.5). For the accelerated sequential procedure, we take $\rho = .6$, the value recommended by Mukhopadhyay and Solanky (1992a), and for the three–stage procedure we take $\rho = .75$ for $k = 5$ and $\rho = .80$ for $k = 10$, the choices of ρ as recommended by Mukhopadhyay and Judge (1991). Let us briefly explain what these choices mean in practice. For the basic understanding, *pretend* that the approximate magnitude of C is known. The starting sample size m_1 is supposed to be "small" compared with C anyway. Now, suppose the *number of sampling operations* ($= S$) is defined as the number of steps taken by a particular sampling process to terminate. Let us consider an example. In the case $k = 3$, say, if one takes five samples at a time from each population and repeats this same process once more and quits, then we define $S = 2$. On the other hand, for $k = 3$, if one first takes a sample each from all these populations, and repeats this process nine more times, and then quits sampling, then we define $S = 10$. Note that in either situation, the experimenter ends up with ten random samples from each of three populations, but the first sampling scheme can be operationally a lot more convenient than the second one. For moderate and large values of C, the number of sampling operations S approximately equals C, ρC and 3, respectively for the purely sequential, accelerated sequential and three–stage selection methodologies. In other words, the purely sequential selection rule is operationally the most inconvenient one. For $k = 5, 10$, the accelerated sequential selection rule with $\rho = .6$ cuts such sampling operations by 40%, and the three–stage procedure cuts the amount of sampling operations to bare bones. There is another type of interpretation also. The accelerated sequential procedure with $\rho = .6$ starts out purely sequentially and estimates 60% of $\hat{C}$ through such a process, and then takes approximately 40% of $\hat{C}$ in one single batch. In contrast, the three–stage procedure, for example in the case $k = 5$, estimates 75% of $\hat{C}$ in the second stage, and estimates the remaining 25% of $\hat{C}$ in the third stage exclusively via batch sampling alone in just three steps. Now, how these three multistage selection techniques stack up against each other with reference to second–order properties can be explained through the entries given in Table 3.8.7a, and in what follows, we attempt to do just that.

First we address columns $2 - 3$ in Table 3.8.7a. From Theorem 3.5.2(i) and Table 3.8.1, we expect $(\overline{n} - C)$ to be close to $\beta(k)$, that is .25 and .39 respectively when $k = 5$ and 10, while $s(\overline{n})$ may be expected to be close to $(2C/Jk)^{\frac{1}{2}}$ for large C, in the purely sequential case. The second assertion follows from a combination of Theorem 3.5.1(iii) and equation (3.5.8). For $C \geq 200$, $s(\overline{n})$ values are seen to be very close to the asymptotic answers; for example, in the case $k = 5$, the expression $(2C/Jk)^{\frac{1}{2}}$ gives the values .089, .126, .178 and .253 respectively when $C = 100$, 200, 400 and 800. Also, when $C \geq 100$, one finds that $\beta(k)$ usually belongs to the 2–standard deviation interval, $(\overline{n} - C) \pm 2s(\overline{n})$. In other words, even though $(\overline{n} - C)$ values are not by themselves close to $\beta(k)$, by taking the variations into account, we may use $\beta(k)$ as a practical guideline for the expected magnitudes of $(\overline{n} - C)$ for $C \geq 100$. Also, one will note that $(\overline{n} - C)$ stays fairly constant for fixed k, when $C \geq 100$. This is an empirical justification of second–order efficiency property for moderate sample sizes.

In column 3, one finds the values of $\overline{p}$ for the purely sequential procedure. For both $k = 5$ and 10, $\overline{p}$ values appear to be near the set target P^*. In almost every case, P^* is not very far from $\overline{p} \pm 2s(\overline{p})$. The asymptotic second–order order analysis in Theorem 3.5.2(ii) reveals that the values of $(\overline{p} - P^*)$ may be expected to be close to A/C for large C, where $A = A(k, P^*)$ is a constant that is evaluated in Table 3.8.2. Also, one will note that the values of A/C are not very far from the 2–standard deviation interval $(\overline{p} - P^*) \pm 2s(\overline{p})$ for almost all values of $C \geq 100$.

Overall, by looking at columns 2 – 3 in Table 3.8.7a, we can safely say two things then. "Asymptotics" kick in quite early and also *second–order correction terms* lead to truly useful practical guidelines in terms of what one should expect as the magnitudes of $(\overline{n} - C)$ and $(\overline{p} - P^*)$, for C–values as small as 100 or so, in the case of the purely sequential methodology (3.5.1).

Now, we turn to columns 4 – 5 in Table 3.8.7a which give similar entries for $(\overline{n} - C)$, $\overline{p}$ and their associated estimated standard errors for the accelerated sequential procedure (3.6.4) – (3.6.6) with $k = 5, 10$, $\rho = .6$ and $q = \max\{0,\ 2(k\rho)^{-1} - h^2 f''(h^2)[k\rho f'(h^2)]^{-1}\}$, given by Theorem 3.6.2(iii). Various values of q are provided in Table 3.8.3 for different choices of ρ values. As before, the number in parentheses below each $\overline{n} - C$ and $\overline{p}$ value is the standard error of the number above it. The last row in the $k = 5$ block and also in the $k = 10$ block provides us with the asymptotic second order guidelines, wherever available. These numbers are written as earlier in triangular brackets. One notes that we do not quite have an asymptotic second–order expansion of $E(N - C)$ in the case of the accelerated sequential methodology. Instead, Theorem 3.6.2 (i) only provides the asymptotic *second–order upper* and *lower bounds* for $E(N - C)$. Thus, in order to compare the simulated results with the corresponding asymptotic answers, we can only provide an interval instead of one number. For the accelerated sequential procedure (3.6.4) – (3.6.6) at the end of each block, we thus provide the interval $< q - 2(k\rho)^{-1}, q + 1 - 2(k\rho)^{-1} >$. From Theorem 3.6.2 (i), one may observe that $E(N - C)$ asymptotically falls within this interval. By computing $\bar{n} - C \pm 2$ (standard error), once again the simulated values for the small and moderate sample sizes show remarkable agreement. The $\bar{p}$ values also exceed the target, namely $P^* = .95$. For the three–stage procedure (3.7.5), at the end of each block, we provide the values of $\beta' + m_1^*$, which is the asymptotic value of $E(N - C)$ by Theorem 3.7.2. Note that the values of $\bar{n} - C$ are much larger than the suggested second order asymptotic guidelines. For example, for $k = 5$, $C = 800$, $\bar{n} - C$ is 5.97 with standard error of .31, whereas its asymptotic value is .67. The explanation for such a behavior is that m_1 in our simulation of procedure (3.7.5) is a fixed number (=10), although by virtue of (3.7.1) it should be a function of δ^*. In fact, see Table 3.8.7b and observe that $\bar{n} - C$ starts becoming smaller and asymptotic agreement is approached as m_1 increases. The $\bar{p}$ values for procedure (3.7.5) are again above the target value of .95.

Let us now compare the procedures in the Table 3.8.7a with one another. One can easily see that the purely sequential procedure (3.5.1) looks better than the other two procedures. This should not be very surprising, because for the purely sequential procedure we update our estimators after each sample, and after each sample we decide whether to take another sample or not. However, this very property of bookkeeping after every sample makes the purely sequential procedure (3.5.1) operationally inconvenient and increases the cost of operation and implementation. So, to cut down the cost of operation and to make the procedure more easily implementable and yet to preserve the second order asymptotic properties, the accelerated sequential procedure (3.6.4) – (3.6.6) and the three–stage procedure (3.7.5) come into the picture as viable competitors. For $k = 5$, the values of $\bar{n} - C$ for the accelerated sequential procedure are around 2, and such entries are close to something around 5 for the three–stage procedures, as provided by Table 3.8.7a. These differences remain fairly stable as k increases from 5 to 10. Hence, the accelerated sequential procedure (3.6.4) – (3.6.6) seems to perform better than the three–stage procedure (3.7.5) in terms of smaller oversampling when $m_1 = 10$ or 20.

3.9 A Few generalizations

In this section, we first consider a general situation where the aim is to select the normal populations associated with the s largest means, $1 \le s < k$. Among other characteristics, we also address the behavior of $P(CS)$ under a few parameter configurations which are different from the associated LFC. Following Bechhofer's (1954) indifference zone formulation, given $\delta^* > 0$, we consider the *preference zone* $\Omega^{(s)}(\delta^*)$ for the problem of selecting the s–best populations with

$$\Omega^{(s)}(\delta^*) = \{\underset{\sim}{\mu} = (\mu_1,\ldots,\mu_k) : \mu_{[k-s+1]} - \mu_{[k-s]} \ge \delta^*\}. \tag{3.9.1}$$

Note that for $s = 1$, $\Omega^{(1)}(\delta^*)$ was earlier written as $\Omega(\delta^*)$ in (3.1.3). For a given $P^* \in (\binom{k}{s}^{-1}, 1)$, we are interested in obtaining selection procedures such that $P(CS)$ is *at least* P^* or *asymptotically* (as $\delta^* \to 0$) *at least* P^* whenever the true parameters $\underset{\sim}{\mu} \in \Omega^{(s)}(\delta^*)$.

3.9.1 PRELIMINARIES

For the moment, let us pretend that σ is known. Having recorded $X_{i1},\ldots,X_{in}$ from π_i, let us write $\overline{X}_{[1n]} \le \cdots \le \overline{X}_{[sn]} \le \ldots \le \overline{X}_{[kn]}$ for the ordered statistics obtained from the sample means $\overline{X}_{1n},\ldots,\overline{X}_{sn},\ldots,\overline{X}_{kn}$. The immediate generalization of the selection rule $\mathscr{P}_n$ given in (3.2.1) will lead us

Table 3.8.1. $\nu(k)$ and $\beta(k)$ values in Theorem 3.5.2

k	$\beta(k)$	$\nu(k)$
1	−1.18277	0.81723
2	−0.25497	1.49007
3	0.03483	2.10449
4	0.17160	2.68640
5	0.24954	3.24771
6	0.29916	3.79499
7	0.33316	4.33209
8	0.35770	4.86159
9	0.37616	5.38544
10	0.39048	5.90477
11	0.40188	6.42068
12	0.41114	6.93367
13	0.41880	7.44442
14	0.42525	7.95350
15	0.43073	8.46088
16	0.43544	8.96705
17	0.43954	9.47222
18	0.44314	9.97656
19	0.44633	10.48020
20	0.44916	10.98327
21	0.45172	11.48608
22	0.45401	11.98819
23	0.45609	12.48998
24	0.45798	12.99150
25	0.45971	13.49279
26	0.46130	13.99388
27	0.46277	14.49480
28	0.46413	14.99559
29	0.46539	15.49625
30	0.46656	15.99682

$\nu(k)$: defined in (3.5.7)

$\beta(k) = k^{-1}\{\nu(k) - 2\}$

Table 3.8.2. Values of $f'(h^2)$, $f''(h^2)$ and A

k	$P^* = .90$			$P^* = .95$			$P^* = .99$		
	$f'(h^2)$	$f''(h^2)$	A	$f'(h^2)$	$f''(h^2)$	A	$f'(h^2)$	$f''(h^2)$	A
2	.03424	−.01377	−.1030	.01568	−.00537	−.1002	.00286	−.00085	−.0576
3	.02987	−.00959	−.0739	.01437	−.00430	−.0736	.00275	−.00077	−.0428
4	.02808	−.00818	−.0449	.01378	−.00387	−.0499	.00270	−.00073	−.0315
5	.02703	−.00742	−.0222	.01341	−.00363	−.0319	.00266	−.00071	−.0233
6	.02630	−.00693	−.0045	.01315	−.00346	−.0181	.00263	−.00069	−.0172
7	.02575	−.00657	.0096	.01295	−.00333	−.0072	.00261	−.00068	−.0125
8	.02532	−.00630	.0212	.01279	−.00323	.0017	.00259	−.00067	−.0088
9	.02496	−.00609	.0308	.01265	−.00315	.0090	.00258	−.00066	−.0057
10	.02466	−.00591	.0389	.01254	−.00308	.0151	.00256	−.00065	−.0031
11	.02440	−.00576	.0459	.01244	−.00302	.0204	.00255	−.00064	−.0009
12	.02418	−.00563	.0520	.01235	−.00297	.0250	.00254	−.00064	.0010
13	.02397	−.00551	.0574	.01227	−.00293	.0290	.00253	−.00063	.0026
14	.02379	−.00541	.0622	.01220	−.00289	.0326	.00252	−.00063	.0041
15	.02363	−.00532	.0665	.01213	−.00285	.0358	.00251	−.00062	.0054
16	.02348	−.00524	.0704	.01208	−.00282	.0387	.00251	−.00062	.0065
17	.02335	−.00517	.0739	.01202	−.00279	.0413	.00250	−.00061	.0075
18	.02322	−.00510	.0772	.01197	−.00276	.0437	.00249	−.00061	.0085
19	.02311	−.00504	.0801	.01192	−.00274	.0459	.00249	−.00061	.0094
20	.02299	−.00499	.0829	.01188	−.00271	.0479	.00248	−.00060	.0102

$f(\cdot)$: $f(y) = g_0(y^{\frac{1}{2}})$, $g_0(y) = \int_{-\infty}^{\infty} \{\Phi(x+y)\}^{k-1}\phi(x)dx$, $y > 0$

$A = h^2\{\beta f'(h^2) + h^2k^{-1}f''(h^2)\}$, $\beta = \beta(k)$: Table 3.8.1

Table 3.8.3. Values of q from Theorem 3.6.2(iii)

k	P* = .90			P* = .95			P* = .99		
	ρ=.25	ρ=.50	ρ=.75	ρ=.25	ρ=.50	ρ=.75	ρ=.25	ρ=.50	ρ=.75
2	6.6424	3.3212	2.2141	7.7056	3.8528	2.5685	10.4121	5.2060	3.4707
3	4.7959	2.3979	1.5986	5.5972	2.7986	1.8658	7.5530	3.7765	2.5177
4	3.7509	1.8755	1.2503	4.3910	2.1955	1.4638	5.9255	2.9628	1.9752
5	3.0847	1.5424	1.0283	3.6186	1.8093	1.2062	4.8817	2.4408	1.6272
6	2.6232	1.3116	.8744	3.0817	1.5409	1.0272	4.1581	2.0791	1.3860
7	2.2843	1.1421	.7614	2.6868	1.3434	.8956	3.6258	1.8129	1.2086
8	2.0246	1.0123	.6749	2.3837	1.1919	.7946	3.2176	1.6088	1.0725
9	1.1892	.9096	.6064	2.1437	1.0718	.7146	2.8941	1.4471	.9647
10	1.6526	.8263	.5509	1.9487	.9743	.6496	2.6314	1.3157	.8772
11	1.5146	.7573	.5049	1.7870	.8935	.5957	2.4136	1.2068	.8045
12	1.3983	.6992	.4661	1.6050	.8254	.5503	2.2301	1.1150	.7434
13	1.2991	.6496	.4330	1.5344	.7672	.5115	2.0732	1.0366	.6911
14	1.2133	.6067	.4044	1.4337	.7168	.4779	1.9375	.9687	.6458
15	1.1384	.5692	.3795	1.3458	.6729	.4486	1.8190	.9095	.6063
16	1.0725	.5362	.3575	1.2682	.6341	.4427	1.7144	.8572	.5715
17	1.0139	.5069	.3380	1.1994	.5997	.3998	1.6217	.8108	.5406
18	.9615	.4808	.3205	1.1379	.5689	.3793	1.5386	.7693	.5129
19	.9144	.4572	.3048	1.0825	.5412	.3608	1.4639	.7320	.4880
20	.8719	.4359	.2906	1.0324	.5162	.3441	1.3964	.6982	.4655

$q = \max\{0, 2(k\rho)^{-1} - h^2 f''[k\rho f'(h^2)]^{-1}\}$

$f'(h^2)$, $f''(h^2)$: Table 3.8.2

Table 3.8.4. Values of m_1^* from (3.7.5)

	P* = .90			P* = .95			P* = .99		
k	ρ=.25	ρ=.50	ρ=.75	ρ=.25	ρ=.50	ρ=.75	ρ=.25	ρ=.50	ρ=.75
2	6.1424	2.8212	1.7141	7.2056	3.3528	2.0685	9.9121	4.7060	2.9707
3	4.2959	1.8979	1.0986	5.0972	2.2986	1.3658	7.0530	3.2765	2.0177
4	3.2509	1.3755	.7503	3.8910	1.6955	.9637	5.4237	2.4619	1.4746
5	2.5847	1.0424	.5282	3.1186	1.3093	.7062	4.3817	1.9408	1.1272
6	2.1232	.8116	.3744	2.5817	1.0409	.5272	3.6581	1.5791	.8860
7	1.7843	.6421	.2614	2.1868	.8434	.3956	3.1260	1.3130	.7087
8	1.5246	.5123	.1749	1.8837	.6919	.2946	2.7176	1.1088	.5725
9	1.3192	.4096	.1064	1.6437	.5718	.2146	2.3941	.9471	.4647
10	1.1526	.3263	.0509	1.4487	.4743	.1496	2.1314	.8157	.3771
11	1.0146	.2573	.0049	1.2870	.3935	.0957	1.9136	.7068	.3045
12	.8983	.1992	−.0339	1.1508	.3254	.0503	1.7301	.6150	.2434
13	.7991	.1496	−.0670	1.0344	.2672	.0115	1.5732	.5366	.1911
14	.7133	.1067	−.0956	.9337	.2168	−.0221	1.4375	.4687	.1458
15	.6384	.0692	−.1205	.8457	.1729	−.0514	1.3190	.4095	.1063
16	.5725	.0362	−.1425	.7682	.1341	−.0773	1.2144	.3572	.0715
17	.5139	.0069	−.1620	.6994	.0997	−.1002	1.1217	.3108	.0406
18	.4615	−.0193	−.1795	.6379	.0689	−.1207	1.0386	.2693	.0129
19	.4144	−.0428	−.1952	.5825	.0412	−.1392	.9639	.2320	−.0120
20	.3719	−.0641	−.2094	.5323	.0162	−.1559	.8963	.1982	−.0346

$\beta' = \frac{1}{2} - 2(k\rho)^{-1}$

$m_1^* = -\beta' - h^2(k\rho)^{-1}\,\{f''(h^2)/f'(h^2)\}$

$f'(h^2)$, $f''(h^2)$: Table 3.8.2

Table 3.8.5a. Two–Stage Procedure (3.3.1) Under LFC: $m_1 = 10$, $P^* = .95$

	k = 5			k = 10		
C	δ^*	$\bar{n}$–C $(s(\bar{n}))$ < >[1]	$\bar{p}$ $(s(\bar{p}))$	δ^*	$\bar{n}$–C $(s(\bar{n}))$ < >[1]	$\bar{p}$ $(s(\bar{p}))$
25	.61	4.53 (.03)	.972 (.002)	.68	3.52 (.03)	.969 (.002)
50	.43	9.38 (.18)	.968 (.002)	.48	7.81 (.17)	.966 (.003)
100	.31	17.80 (.36)	.963 (.003)	.34	15.35 (.36)	.965 (.003)
200	.22	32.77 (.72)	.966 (.003)	.24	26.31 (.71)	.961 (.003)
400	.15	69.82 (1.43)	.961 (.003)	.17	66.07 (1.39)	.963 (.003)
800	.11	96.72 (2.84)	.961 (.003)	.12	88.64 (2.69)	.960 (.003)
		<1.059>			<1.036>	

1: The values of τ^2/h^2; Theorem 3.3.1(iii)

Note: $\bar{n}/C$ is to be compared with the asymptotic values in < >

Table 3.8.5b. Two–Stage Procedure (3.3.1) Under LFC: $m_1 = 20$, $P^* = .95$

	k = 5			k = 10		
C	δ^*	$\bar{n}$–C $(s(\bar{n}))$ < >[1]	$\bar{p}$ $(s(\bar{p}))$	δ^*	$\bar{n}$–C $(s(\bar{n}))$ < >[1]	$\bar{p}$ $(s(\bar{p}))$
25	.61	5.15 (.05)	.970 (.002)	.68	4.63 (.07)	.976 (.002)
50	.43	6.03 (.20)	.963 (.003)	.48	6.07 (.21)	.969 (.002)
100	.31	14.31 (.36)	.964 (.003)	.34	13.26 (.33)	.965 (.003)
200	.22	28.42 (.71)	.966 (.003)	.24	25.70 (.73)	.961 (.003)
400	.15	60.72 (1.43)	.961 (.003)	.17	58.84 (1.32)	.960 (.003)
800	.11	88.61 (2.73)	.961 (.003)	.12	82.68 (2.73)	.960 (.003)
		<1.026>			<1.018>	

1: The values of τ^2/h^2; Theorem 3.3.1(iii)

Note: $\bar{n}/C$ is to be compared with the asymptotic values in < >

Table 3.8.5c. Two–Stage Procedure (3.3.1) Under LFC: $m_1 = 30$, $P^* = .95$

		k = 5			k = 10	
C	δ^*	$\bar{n}$–C $(s(\bar{n}))$ $\langle\ \rangle^1$	$\bar{p}$ $(s(\bar{p}))$	δ^*	$\bar{n}$–C $(s(\bar{n}))$ $\langle\ \rangle^1$	$\bar{p}$ $(s(\bar{p}))$
25	.61	6.23 (.07)	.972 (.002)	.68	6.12 (.07)	.972 (.002)
50	.43	6.11 (.19)	.966 (.003)	.48	5.85 (.19)	.970 (.002)
100	.31	13.97 (.33)	.961 (.003)	.34	12.65 (.31)	.963 (.003)
200	.22	27.12 (.77)	.964 (.003)	.24	25.71 (.76)	.961 (.003)
400	.15	59.45 (1.39)	.961 (.003)	.17	55.77 (1.29)	.961 (.003)
800	.11	84.63 (2.71)	.961 (.003)	.12	81.66 (2.71)	.960 (.003)
		<1.015>			<1.013>	

1: The values of τ^2/h^2; Theorem 3.3.1(iii)

Note: $\bar{n}/C$ is to be compared with the asymptotic values in $\langle\ \rangle$

Table 3.8.6a. Modified Two–Stage Procedure (3.4.1)–(3.4.2) Under LFC
$\sigma = .50$, $m_0^* = 10$, $P^* = .95$

k	δ^*	C	$\gamma = .2$ $\bar{n}$–C $(s(\bar{n}))$ < >[1]	$\gamma = .2$ $\bar{p}$ $(s(\bar{p}))$	$\gamma = .3$ $\bar{n}$–C $(s(\bar{n}))$ < >[1]	$\gamma = .3$ $\bar{p}$ $(s(\bar{p}))$	$\gamma = .4$ $\bar{n}$–C $(s(\bar{n}))$ < >[1]	$\gamma = .4$ $\bar{p}$ $(s(\bar{p}))$
5	.31	25	22.00 (.00)	1.000 (.000)	10.00 (.00)	.997 (.001)	2.41 (.01)	.983 (.002)
	.22	50	33.00 (.00)	1.000 (.000)	9.00 (.00)	.989 (.002)	3.15 (.01)	.961 (.003)
	.15	100	48.00 (.00)	.997 (.001)	3.44 (.24)	.967 (.003)	4.57 (.29)	.952 (.003)
	.11	200	63.75 (.01)	.976 (.002)	6.11 (.59)	.955 (.003)	5.01 (.59)	.944 (.003)
	.08	400	68.19 (.01)	.992 (.001)	7.53 (.91)	.948 (.003)	7.13 (1.01)	.951 (.003)
	.05	800	34.53 (.02)	.980 (.002)	9.42 (1.43)	.951 (.003)	8.93 (1.39)	.950 (.003)
	.04	1600	5.15 (2.04)	.948 (.003)	11.21 (2.22)	.953 (.003)	12.21 (2.31)	.951 (.003)
10	.34	25	22.00 (.00)	1.000 (.000)	10.00 (.00)	.993 (.001)	2.69 (.14)	.961 (.003)
	.24	50	33.00 (.00)	1.000 (.000)	9.00 (.00)	.991 (.001)	3.11 (.35)	.963 (.003)
	.17	100	48.73 (.01)	.976 (.002)	3.62 (.28)	.957 (.003)	4.98 (.60)	.949 (.003)
	.12	200	63.21 (.01)	.993 (.001)	6.05 (.65)	.961 (.003)	5.53 (.83)	.953 (.003)
	.09	400	68.19 (.01)	.991 (.001)	8.83 (.98)	.948 (.003)	8.62 (1.23)	.947 (.003)
	.06	800	34.02 (.02)	.958 (.003)	10.36 (1.58)	.949 (.003)	9.07 (1.72)	.949 (.003)
	.04	1600	6.11 (2.12)	.951 (.003)	11.31 (2.37)	.950 (.003)	13.21 (2.84)	.951 (.003)
			<1>		<1>		<1>	

1 : $\bar{n}/C$ is to be compared with the asymptotic values in < >

Table 3.8.6b. Modified Two–Stage Procedure (3.4.1)–(3.4.2) Under LFC
$\sigma = .75$, $m_0^* = 10$, $P^* = .95$

			$\gamma = .2$		$\gamma = .3$		$\gamma = .4$	
k	δ^*	C	$\bar{n}$–C (s($\bar{n}$)) < >[1]	$\bar{p}$ (s($\bar{p}$))	$\bar{n}$–C (s($\bar{n}$)) < >[1]	$\bar{p}$ (s($\bar{p}$))	$\bar{n}$–C (s($\bar{n}$)) < >[1]	$\bar{p}$ (s($\bar{p}$))
5	.31	56.25	1.77	.961	1.27	.970	1.66	.963
			(.38)	(.003)	(.31)	(.002)	(.33)	(.003)
	.22	112.5	2.97	.939	2.73	.963	2.80	.944
			(.52)	(.003)	(.46)	(.003)	(.49)	(.003)
	.15	225	2.99	.951	2.62	.947	3.81	.977
			(.74)	(.003)	(.73)	(.003)	(.79)	(.002)
	.11	450	3.41	.980	3.39	.952	4.02	.971
			(1.23)	(.002)	(1.22)	(.003)	(1.31)	(.002)
	.08	900	5.69	.963	5.53	.957	5.48	.946
			(1.97)	(.003)	(2.14)	(.003)	(2.09)	(.003)
	.05	1800	9.98	.949	8.49	.950	9.56	.939
			(2.91)	(.003)	(2.66)	(.003)	(2.63)	(.003)
	.04	3600	10.97	.953	9.51	.946	10.06	.951
			(3.94)	(.003)	(3.51)	(.003)	(3.42)	(.003)
10	.34	56.25	1.99	.963	1.49	.969	1.72	.964
			(.39)	(.003)	(.33)	(.003)	(.37)	(.003)
	.24	112.5	2.96	.951	2.88	.961	2.86	.949
			(.55)	(.003)	(.56)	(.003)	(.47)	(.003)
	.17	225	3.05	.962	3.04	.944	3.34	.952
			(.69)	(.003)	(.73)	(.003)	(.86)	(.003)
	.12	450	3.61	.966	3.83	.950	4.11	.939
			(1.44)	(.003)	(1.51)	(.003)	(1.53)	(.003)
	.09	900	6.01	.951	6.11	.955	6.15	.944
			(2.01)	(.003)	(2.14)	(.003)	(2.31)	(.003)
	.06	1800	10.44	.950	8.91	.947	9.14	.952
			(2.98)	(.003)	(2.84)	(.003)	(2.61)	(.003)
	.04	3600	10.91	.952	9.38	.951	10.33	.949
			(3.71)	(.003)	(3.41)	(.003)	(3.53)	(.003)
			<1>		<1>		<1>	

1 : $\bar{n}/C$ is to be compared with the asymptotic values in < >

Table 3.8.6c. Modified Two–Stage Procedure (3.4.1)–(3.4.2) Under LFC
$\sigma = 1.0$, $m_0^* = 10$, $P^* = .95$

k	C	δ^*	$\gamma = .2$ $\bar{n}$–C $(s(\bar{n}))$ $< >^1$	$\gamma = .2$ $\bar{p}$ $(s(\bar{p}))$	$\gamma = .3$ $\bar{n}$–C $(s(\bar{n}))$ $< >^1$	$\gamma = .3$ $\bar{p}$ $(s(\bar{p}))$	$\gamma = .4$ $\bar{n}$–C $(s(\bar{n}))$ $< >^1$	$\gamma = .4$ $\bar{p}$ $(s(\bar{p}))$
5	25	.61	3.22	.963	4.99	.957	4.69	.962
			(.05)	(.003)	(.09)	(.003)	(.07)	(.003)
	50	.43	3.53	.966	5.90	.963	6.03	.964
			(.67)	(.003)	(.91)	(.003)	(1.02)	(.003)
	100	.31	2.56	.968	2.86	.957	5.30	.966
			(.98)	(.003)	(1.20)	(.003)	(1.25)	(.003)
	200	.22	4.08	.952	4.05	.961	5.52	.960
			(1.51)	(.003)	(1.58)	(.003)	(2.14)	(.003)
	400	.15	7.53	.960	5.46	.954	6.44	.951
			(2.29)	(.003)	(2.51)	(.003)	(2.67)	(.003)
	800	.11	8.33	.953	6.11	.961	8.23	.957
			(2.67)	(.003)	(2.63)	(.003)	(3.04)	(.003)
10	25	.68	2.38	.961	3.22	.956	3.99	.951
			(.06)	(.003)	(.08)	(.003)	(.09)	(.003)
	50	.48	4.01	.952	3.97	.951	4.87	.954
			(.66)	(.003)	(.87)	(.003)	(.89)	(.003)
	100	.34	2.81	.955	3.18	.962	5.15	.960
			(.88)	(.003)	(1.23)	(.003)	(1.27)	(.003)
	200	.24	3.99	.960	3.85	.966	5.89	.957
			(1.27)	(.003)	(1.49)	(.003)	(2.01)	(.003)
	400	.17	6.52	.958	5.21	.959	6.02	.961
			(1.89)	(.003)	(2.47)	(.003)	(2.51)	(.003)
	800	.12	7.37	.961	6.03	.953	7.81	.955
			(2.22)	(.003)	(2.70)	(.003)	(2.83)	(.003)
			<1>		<1>		<1>	

1 : $\bar{n}/C$ is to be compared with the asymptotic values in $< >$

Table 3.8.7a. Comparison of Procedures Under LFC; $m_1=10, P^*=.95, \sigma^2=1$

	Purely Sequential Procedure (3.5.1)		Accelerated Sequential Procedure (3.6.4)–(3.6.6) $\rho=.6$		Three–Stage Procedure (3.7.5) k=5 $\rho=.75$; k=10, $\rho=.80$	
C	$\bar{n}-C$	$\bar{p}$	$\bar{n}-C$	$\bar{p}$	$\bar{n}-C$	$\bar{p}$
δ^*	$(s(\bar{n}))$ < >[1]	$(s(\bar{p}))$ < >[2]	$(s(\bar{n}))$ < >[3]	$(s(\bar{p}))$	$(s(\bar{n}))$ < >[4]	$(s(\bar{p}))$
			k = 5			
25	1.35	.955	2.61	.976	1.99	.956
.61	(.07)	(.003)	(.06)	(.002)	(.06)	(.003)
50	1.20	.946	2.55	.957	2.54	.957
.43	(.06)	(.003)	(.09)	(.003)	(.07)	(.003)
100	1.30	.952	2.14	.952	4.61	.950
.31	(.09)	(.003)	(.13)	(.003)	(.11)	(.003)
200	.92	.945	2.07	.953	4.87	.952
.22	(.13)	(.003)	(.18)	(.003)	(.16)	(.003)
400	.81	.947	2.01	.957	5.86	.958
.15	(.18)	(.003)	(.23)	(.003)	(.21)	(.003)
800	.79	.947	1.96	.951	5.97	.951
.11	(.26)	(.003)	(.34)	(.003)	(.31)	(.003)
	<.25>	<–.032>	<.83,1.83>		<.67>	
			k = 10			
25	.96	.961	2.84	.977	1.93	.968
.68	(.07)	(.003)	(.06)	(.002)	(.06)	(.003)
50	1.14	.950	2.63	.961	2.40	.953
.48	(.05)	(.003)	(.07)	(.003)	(.06)	(.003)
100	1.26	.947	2.07	.954	4.28	.954
.34	(.07)	(.003)	(.10)	(.003)	(.07)	(.003)
200	.93	.951	2.00	.958	4.63	.951
.24	(.09)	(.003)	(.12)	(.003)	(.11)	(.003)
400	.85	.947	1.76	.951	5.13	.953
.17	(.13)	(.003)	(.18)	(.003)	(.16)	(.003)
800	.68	.946	1.64	.951	5.51	.952
.12	(.18)	(.003)	(.24)	(.003)	(.21)	(.003)
	<.39>	<.015>	<.48,1.48>		<.36>	

1 : Values of $\beta(k)$ from Table 3.8.1; $(\bar{n}-C)$ is to be compared with $\beta(k)$

2 : Values of A from Table 3.8.2; $(\bar{p}-P^*)$ is to be compared with A/C

3 : The interval is $(q-2(k\rho)^{-1}, q+1-2(k\rho)^{-1})$, where q comes from Table 3.8.3

4 : Values of $\beta' + m_1^*$; Look at (3.7.6)

Table 3.8.7b. Three–Stage Procedure (3.7.5) Under LFC; P* = .95, $\sigma^2 = 1$

			$m_1 = 20$		$m_1 = 40$		$m_1 = 60$	
k	C	δ^*	$\bar{n}$–C $(s(\bar{n}))$ $< >^1$	$\bar{p}$ $(s(\bar{p}))$	$\bar{n}$–C $(s(\bar{n}))$ $< >^1$	$\bar{p}$ $(s(\bar{p}))$	$\bar{n}$–C $(s(\bar{n}))$ $< >^1$	$\bar{p}$ $(s(\bar{p}))$
5	100	.31	4.03 (.19)	.957 (.003)	3.80 (.17)	.956 (.003)	2.15 (.15)	.956 (.003)
	200	.22	4.81 (.26) <.67>	.964 (.003)	4.43 (.25) <.67>	.963 (.003)	3.47 (.23) <.67>	.961 (.003)
10	100	.34	3.62 (.27)	.958 (.003)	2.51 (.24)	.957 (.003)	1.99 (.23)	.957 (.003)
	200	.24	4.79 (.33) <.36>	.952 (.003)	3.81 (.33) <.36>	.951 (.003)	2.28 (.31) <.36>	.953 (.003)

1 : Values of $\beta' + m_1^*$; Look at (3.7.6)

to propose the following selection rule in order to suit the present situation.

$$\mathscr{P}_n^{(s)}: \text{ Select } \pi_{i_j} \text{ in the set of s–best}$$

$$\text{population if } \bar{X}_{i_j n} \geq \bar{X}_{[k-sn]},\ j = 1,\ldots,s. \tag{3.9.2}$$

In other words, the selection rule $\mathscr{P}_n^{(s)}$ picks those s populations which are associated with the s largest sample means, that is $\bar{X}_{[k-s+1n]},\ldots,\bar{X}_{[kn]}$. Note that $\mathscr{P}_n$ from (3.2.1) coincides with $\mathscr{P}_n^{(1)}$.

Let us continue to write $\pi_{(i)}$ for the population associated with $\mu_{[i]}$ and let us denote $\bar{X}_{(in)}$ for the corresponding sample mean arising from $\pi_{(i)}$, $i = 1,\ldots,k$. Suppose $\delta_{i1} = \mu_{[k-s+1]} - \mu_{[i]}$, $i = 1,\ldots,k-s$ and $\delta_{j2} = \mu_{[k-s+1]} - \mu_{[k-j]}$, $j = 0, 1,\ldots,s-2$. Now, under the preference zone $\Omega^{(s)}(\delta^*)$ we have $\delta_{i1} \geq \delta^*$, $\delta_{j2} \leq 0$ for $i = 1,\ldots,k-s$, $j = 0, 1,\ldots,s-2$. One can easily verify that

$$P_n(CS)$$
$$= s\int_{-\infty}^{\infty} \prod_{i=1}^{k-s} \Phi(y + n^{\frac{1}{2}}\delta_{i1}\sigma^{-1}) \prod_{j=0}^{s-2} \{1 - \Phi(y + n^{\frac{1}{2}}\delta_{j2}\sigma^{-1})\}\phi(y)\, dy. \tag{3.9.3}$$

From (3.9.3), it is easy to see that

$$\operatorname*{Inf}_{\underset{\sim}{\mu}\in\Omega^{(s)}(\delta^*)\cup\Omega^{(s)c}(\delta^*)} P_n(CS)$$

$$= s\int_{-\infty}^{\infty} \Phi^{k-s}(y)\{1 - \Phi(y)\}^{s-1}\, \phi(y)dy = \binom{k}{s}^{-1}.$$

However, for all $\underset{\sim}{\mu}\in\Omega^{(s)}(\delta^*)$, that is when $\delta_{i1} \geq \delta^*$, $\delta_{j2} \leq 0$, $i = 1,\ldots,k-s$, and $j = 0, 1, \ldots, s-2$, one has

$$P_n(CS) \geq s\int_{-\infty}^{\infty} \prod_{i=1}^{k-s} \Phi\left(y + n^{\frac{1}{2}}\delta^*\sigma^{-1}\right) \prod_{j=0}^{s-2} \{1 - \Phi(y)\}\ \phi(y)dy$$

and the equality holds if $\delta_{i1} = \delta^*$ and $\delta_{j2} = 0$ for all $i = 1, \ldots, k-s$, and $j = 0, 1, \ldots, s-2$. Hence,

$$\operatorname*{Inf}_{\underset{\sim}{\mu}\in\Omega^{(s)}(\delta^*)} P_n(CS)$$

$$= s\int_{-\infty}^{\infty} \Phi^{k-s}\left(y + n^{\frac{1}{2}}\delta^*\sigma^{-1}\right)\{1 - \Phi(y)\}^{s-1}\ \phi(y)dy, \tag{3.9.4}$$

the infimum being attained when $\mu_{[1]} = \cdots = \mu_{[k-s]} = \mu_{[k-s+1]} - \delta^* = \cdots = \mu_{[k]} - \delta^*$, a parameter configuration that is referred as the LFC. We define $h^{(s)} = h^{(s)}(k, s, P^*)$ such that

$$s\int_{-\infty}^{\infty} \Phi^{k-s}\left(y + h^{(s)}\right) \{1 - \Phi(y)\}^{s-1}\ \phi(y)dy = P^*. \tag{3.9.5}$$

Recall that for $s = 1$, we wrote h instead of $h^{(1)}$ in section 3.2, given by (3.2.4). Now it follows from (3.9.4) that

$$P_n(CS) \geq P^* \text{ for all } \underset{\sim}{\mu}\in\Omega^{(s)}(\delta^*),$$

if n is the smallest integer $\geq h^{(s)2}\sigma^2/\delta^{*2} = C^{(s)}$, say. In other words, had σ been known, one would take $[C^{(s)}]^* + 1$ samples from each π and implement the corresponding selection rule (3.9.2). The values of $h^{(s)}$ had been tabulated in Bechhofer (1954). However σ is unknown and hence we again opt for various multistage selection procedures.

Suppose in general, one devises a statistical methodology giving rise to a positive integer valued random variable N estimating $C^{(s)}$ and selects s populations via (3.9.2).

Theorem 3.9.1. *Suppose that* N *satisfies the conditions of Theorem 3.2.1. Then,*

$$\inf_{\underset{\sim}{\mu} \in \Omega^{(s)}(\delta^*)} P(CS)$$

$$= s\, E\left[\int_{-\infty}^{\infty} \Phi^{k-s}(y + N^{\frac{1}{2}}\delta^*\sigma^{-1})\, \{1 - \Phi(y)\}^{s-1}\, \phi(y)dy\right]$$

and this infimum is attained when $\mu_{[1]} = \cdots = \mu_{[k-s]} = \mu_{[k-s+1]} - \delta^* = \cdots = \mu_{[k]} - \delta^*$.

Its proof is omitted in view of the proof of Theorem 3.2.1 and (3.9.3) – (3.9.4).

3.9.2 PURELY SEQUENTIAL PROCEDURE

Along the lines of section 3.5, we start with $m_1(\geq 2)$ samples from each π. Then, we proceed by taking one more sample at a time from each π according to the following stopping rule:

$$N = N(\delta^*) = \inf\{n \geq m_1 : n \geq h^{(s)2} U(n) \delta^{*-2}\}, \tag{3.9.6}$$

where $U(n)$, defined in (3.1.1), is an estimator of σ^2. It is easy to see that for fixed $\underset{\sim}{\mu}$, σ^2, m_1 and δ^*, $P(N < \infty) = 1$. Based on $X_{i1},...,X_{iN}$ from π_i, $i = 1,...,k$, we now implement the selection rule $\mathcal{R}_N^{(s)}$ defined in (3.9.2). Throughout $\underset{\sim}{\mu}$, σ are kept fixed.

Theorem 3.9.2. *For the purely sequential procedure (3.9.6), we have as* $\delta^* \to 0$:

(i) $N/C^{(s)} \to 1$ *w.p.* 1;

(ii) $E(N/C^{(s)}) \to 1$;

(iii) $C^{(s)-\frac{1}{2}}(N - C^{(s)}) \xrightarrow{\mathscr{L}} N(0, 2/k)$;

(iv) $\liminf P(CS) \geq P^*$ *for all* $\underset{\sim}{\mu} \in \Omega^{(s)}(\delta^*)$;

(v) $\lim P(CS) = P^*$ *under the LFC*;

where $C^{(s)} = h^{(s)2}\, \sigma^2/\delta^{*2}$ *and* $h^{(s)}$ *comes from (3.9.5).*

The proof is similar to that of Theorem 3.5.1 and hence it is omitted.

3.9.2.1 Second–Order Asymptotics

In this section, we strengthen the results given in Theorem 3.9.2 (ii) and (v) by providing expansions of (a) E(N) for all $\underset{\sim}{\mu}\in\Omega^{(s)}(\delta^*)$ and (b) P(CS) under both the LFC as well as certain non LFC. Let us define

$$\delta_{i3} = a_i\delta^*,\ \delta_{j4} = b_j\delta^* \text{ for } a_i \geq 1,\ b_j \leq 0;$$

$$g_{01}^{(s)}(x) = s\int_{-\infty}^{\infty} \prod_{i=1}^{k-s} \Phi(y+a_i x) \prod_{j=0}^{s-1} \{1 - \Phi(y + b_j x)\}\ \phi(y)dy;$$

$$f_{01}^{(s)}(x) = g_{01}^{(s)}(x^{\frac{1}{2}});\ x > 0,\ i = 1,\ldots,k-s,\ j = 0, 1,\ldots,s-2.$$

Note that under the LFC, one has $\delta_{i1} = \delta_{i3}$ with $a_1 = \cdots = a_{k-s} = 1$ and $\delta_{j2} = \delta_{j4}$ with $b_o = b_1 = \cdots = b_{s-2} = 0$. Under the LFC for $s = 1$, $g_{01}^{(1)}(x)$ and $f_{01}^{(1)}(x)$ were written as $g_o(x)$ and $f(x)$ respectively in section 3.5. Now, we have the following results.

Theorem 3.9.3. *For the purely sequential procedure (3.9.6), we have as* $\delta^* \to 0$:

(i) $E(N) = C^{(s)} + \beta + o(1)$ *for all* $\underset{\sim}{\mu} \in \Omega^{(s)}(\delta^*)$ *if* $m_1 \geq 3$ *when* $k = 2$, (b) $m_1 \geq 2$ *when* $k \geq 3$;

(ii) $P(CS) = f_{o1}^{(s)}(h^{(s)2}) + h^{(s)2}C^{(s)-1}\{\beta f_{o1}^{(s)\prime}(h^{(s)2}) + h^{(s)2}k^{-1} \times f_{o1}^{(s)\prime\prime}(h^{(s)2})\} + o(\delta^{*2})$ *under* $\Omega^{(s)}(\delta^*)$ *with* $\delta_{i3} = a_i\delta^*$, $\delta_{j4} = b_j\delta^*$, *for fixed* $a_i \geq 1$, $b_j \leq 0$, $i = 1,\ldots,k-s$ *and* $j = 0, 1,\ldots,s-2$ *if* (a) $m_1 \geq 4$ *when* $k = 2$, (b) $m_1 \geq 3$ *when* $k = 3, 4, 5$, (c) $m_1 \geq 2$ *when* $k \geq 6$;

where $\beta = \beta(k) = k^{-1}\{\nu(k) - 2\}$ *with* $\nu(k)$ *defined in (3.5.7).*

Proof. The proof of part (i) is similar to that of Theorem 3.5.2 (i). For part (ii), first observe that one has $f_{o1}^{(s)\prime}(x) = \frac{1}{2}x^{-\frac{1}{2}}g_{o1}^{(s)\prime}(x^{\frac{1}{2}})$, $f_{o1}^{(s)\prime\prime}(x) = \frac{1}{4}\{x^{-1}g_{o1}^{(s)\prime\prime}(x^{\frac{1}{2}}) - x^{-\frac{3}{2}}g_{o1}^{(s)\prime}(x^{\frac{1}{2}})\}$ for $x > 0$. Also,

$$|g_{o1}^{(s)\prime}(x)| \leq s\,\Sigma_{i=1}^{k-1}(2\pi)^{-\frac{1}{2}}a_i - s\,\Sigma_{j=0}^{s-2}(2\pi)^{-\frac{1}{2}}b_j, \tag{3.9.7}$$

$$\begin{aligned}|g_{o1}^{(s)\prime\prime}(x)| \leq{}& s\,\Sigma_{i=1}^{k-s}\,\Sigma_{\substack{j=1\\ j\neq i}}^{k-s}(2\pi)^{-1}a_ia_j + s\Sigma_{i=1}^{k-s}(2\pi)^{-\frac{1}{2}}a_i^2x \\ &- 2s\,\Sigma_{i=1}^{k-s}\,\Sigma_{j=0}^{s-2}(2\pi)^{-1}a_ib_j + s\Sigma_{i=0}^{s-2}\,\Sigma_{\substack{j=0\\ j\neq i}}^{s-2}(2\pi)^{-1}b_ib_j \\ &+ s\,\Sigma_{j=0}^{s-2}(2\pi)^{-\frac{1}{2}}b_j^2\,x.\end{aligned} \tag{3.9.8}$$

Now, from Theorem 3.9.1, one obtains

$$P(CS) = E[f_{o1}^{(s)}(Nh^{(s)2}/C^{(s)})].$$

One can easily verify that $|f_{01}^{(s)\prime\prime}(x)| \le \Sigma_{i=1}^{3} d_i\, x^{-e_i}$ where $e_1 = 1$, $e_2 = \frac{1}{2}$, $e_3 = \frac{3}{2}$ and

$$d_1 = s\, \Sigma_{i=1}^{k-s} \Sigma_{\substack{j=1\\ j\neq i}}^{k-s} (2\pi)^{-1} a_i a_j - s\Sigma_{i=1}^{k-s} \Sigma_{j=0}^{s-2} \pi^{-1} a_i b_j + s\, \Sigma_{i=0}^{s-2} \Sigma_{\substack{j=0\\ j\neq i}}^{s-2} (2\pi)^{-1}\, b_i b_j,$$

$$d_2 = s\, \Sigma_{i=1}^{k-s} (2\pi)^{-\frac{1}{2}}\, a_i^2 + s\, \Sigma_{j=0}^{s-2} (2\pi)^{-\frac{1}{2}}\, b_j^2,$$

$$d_3 = s\, \Sigma_{i=1}^{k-s} (2\pi)^{-\frac{1}{2}}\, a_i - s\, \Sigma_{j=0}^{s-2} (2\pi)^{-\frac{1}{2}}\, b_j.$$

Now, one utilizes the techniques given in the proofs of Lemma 3.5.1 and Theorem 3.5.2 (ii). □

Remark 3.9.1. In the case $\underset{\sim}{\mu}\in\Omega^{(s)}(\delta^*)$, we can claim that $f_{01}^{(s)}(h^{(s)2})$, that is the first term in the expansion of P(CS), is at least P^* in the non LFC situations. However, $f_{01}^{(s)}(h^{(s)2}) = P^*$ in the LFC case.

3.9.3 ACCELERATED SEQUENTIAL PROCEDURE

Along the lines of section 3.6, for technical reasons, we consider estimating σ^2 by means of $U^*(n)$ defined in (3.1.1). We first choose and fix $0 < \rho < 1$ and $q^{(s)} \ge 0$. Along the lines of (2.5.1) – (2.5.3), we define

$$t = t(\delta^*) = \inf\{n \ge m_1 : n \ge \rho h^{(s)2} U^*(n)/\delta^{*2}\}, \tag{3.9.9}$$

$$N_1 = N_1(\delta^*) = [h^{(s)2} U^*(t)\delta^{*-2} + q^{(s)}]^* + 1, \tag{3.9.10}$$

$$N = N(\delta^*) = \max\{t(\delta^*), N_1(\delta^*)\}, \tag{3.9.11}$$

where $m_1(\ge 2)$ is the starting sample size from each π. The accelerated sequential procedure (3.9.9) – (3.9.11) is implemented in the same way as

(3.6.4) – (3.6.6). Based on $X_{i1},...,X_{iN}$ from π_i, $i = 1,...,k$, we then select the s–best populations via $\mathscr{R}_N^{(s)}$ defined in (3.9.2). Let us define

$$g_{o2}^{(s)}(x) = s\int_{-\infty}^{\infty} \Phi^{k-s}(y + x)\{1 - \Phi(y)\}^{s-1}\phi(y)dy,$$

$$f_{o2}^{(s)}(x) = g_{o2}^{(s)}(x^{\frac{1}{2}}),\ x > 0.$$

Note that $g_{o2}^{(s)}(x)$ and $f_{o2}^{(s)}(x)$ coincides with $g_{o1}^{(s)}(x)$ and $f_{o1}^{(s)}(x)$ respectively in the LFC case. The asymptotic second–order properties are now summarized in the following theorem while we omit its proof for brevity.

Theorem 3.9.4 *For the accelerated sequential procedure (3.9.9) – (3.9.11), we have as* $\delta^* \to 0$:

(i) $q^{(s)} - 2(k\rho)^{-1} \le \liminf E(N - C^{(s)}) \le \limsup E(N - C^{(s)}) \le q^{(s)} + 1 - 2(k\rho)^{-1}$ *for all* $\underset{\sim}{\mu}\epsilon\Omega^{(s)}(\delta^*)$ *if* $m_1 \ge 4$;

(ii) $\mathrm{Var}(N) = 2(k\rho)^{-1} C^{(s)} + o(\delta^{*-2})$ *for all* $\underset{\sim}{\mu}\epsilon\Omega^{(s)}(\delta^*)$ *if* (a) $m_1 \ge 3$ *when* $k = 2$, (b) $m_1 \ge 2$ *when* $k \ge 3$;

(iii) $P(CS) \ge P^* + o(\delta^{*2})$ *under* $\Omega^{(s)}(\delta^*)$ *if* (a) $m_1 \ge 5$ *when* $k = 2$, (b) $m_1 \ge 4$ *when* $k \ge 3$, *with* $q^{(s)} = \max\{0,\ 2(k\rho)^{-1} - h^{(s)2} f_{o2}^{(s)''}(h^{(s)2})\,[k\rho f_{o2}^{(s)'}(h^{(s)2})]^{-1}\}$;

where $C^{(s)} = h^{(s)2}\sigma^2/\delta^{*2}$ *and* $h^{(s)}$ *comes from (3.9.5).*

3.9.4 THREE–STAGE PROCEDURE

In this section, a three–stage selection methodology is discussed along the lines of section 3.7. Now, we go back to estimating σ^2 by means of $U(n)$.

One starts with $m_1(\geq 2)$ samples from each π where it is assumed that

$$m_1 = m_1(\delta^*) = o(C^{(s)1/r}),$$

for some $r > 1$, as in (3.7.1). We also choose and fix $0 < \rho < 1$. Then, one defines

$$T = T(\delta^*) = \max\{m_1, [\rho h^{(s)2} U(m_1)\delta^{*-2}]^* + 1\}, \tag{3.9.13}$$

$$m_1^{*(s)} = -\beta' - h^{(s)2}(k\rho)^{-1}\{f_{o2}^{(s)''}(h^{(s)2})/f_{o2}^{(s)'}(h^{(s)2})\},$$

$$N = N(\delta^*) = \max\{T, [h^{(s)2} U(T)\delta^{*-2} + m_1^{*(s)}]^* + 1\}, \tag{3.9.14}$$

where $\beta' = \frac{1}{2} - 2(k\rho)^{-1}$ and $h^{(s)}$ comes from (3.9.5). The three-stage procedure (3.9.13) – (3.9.14) is implemented in the same way as in section 3.7. Based on $X_{i1},\ldots,X_{iN}$ from π_i, $i = 1,\ldots,k$, one then selects the s-best populations via $\mathscr{R}_N^{(s)}$ defined in (3.9.2). The asymptotic second-order properties are now summarized in the following theorem while we omit its proof.

Theorem 3.9.5. *For the three-stage procedure (3.9.14), we have as* $\delta^* \to 0$:

(i) $E(N) = C^{(s)} + \beta' + m_1^* + o(1)$ *for all* $\underset{\sim}{\mu} \in \Omega^{(s)}(\delta^*)$;

(ii) $Var(N) = 2(k\rho)^{-1} C^{(s)} + o(\delta^{*-2})$ *for all* $\underset{\sim}{\mu} \in \Omega^{(s)}(\delta^*)$;

(iii) $P(CS) \geq P^* + o(\delta^{*2})$ *for all* $\underset{\sim}{\mu} \in \Omega^{(s)}(\delta^*)$;

where $C^{(s)} = h^{(s)2}\sigma^2/\delta^{*2}$, $\beta' = \frac{1}{2} - 2(k\rho)^{-1}$ *and* $h^{(s)}$ *comes from (3.9.5).*

3.9.5 MODERATE SAMPLE PERFORMANCES AND COMMENTS

In this section we present discussions on some moderate sample performances of the purely sequential procedure (3.9.6), accelerated sequential procedure (3.9.9) – (3.9.11) and the three stage procedure (3.9.14) for $1 \leq s < k$. We will consider the LFC as well as some non LFC situations for the simulation studies in case of the purely sequential procedure. As in section 3.8, for LFC case we used IMSL (1989, Edition 10) subroutine RNNOR to generate $(k - s)$ pseudo normal populations each with mean = 5 and variance = 1, the remaining s pseudo normal populations are generated with mean = $5 + \delta^*$ and variance = 1, where $\delta^* = (h^{(s)2}/C^{(s)})^{\frac{1}{2}}$.

In Table 3.9.1, we provide our findings under LFC for the purely sequential procedure (3.9.6), accelerated sequential procedure (3.9.9) – (3.9.11) and the three–stage procedure (3.9.14). As in Table 3.8.7a, we provide the corresponding standard errors under each column of $\bar{n} - C$ and $\bar{p}$ values. One should note that the purely sequential procedure (3.9.6) performs better than the other two procedures. This is along the lines of our findings presented in section 3.8. Also, the accelerated sequential procedure (3.9.9) – (3.9.11) appears to be better than the three–stage procedure (3.9.14) in terms of smaller values of $\bar{n} - C$. For $k = 5$ and $s = 2$, $f_{o2}^{(s)\prime}(h^{(s)2}) = .01317$, $f_{o2}^{(s)\prime\prime}(h^{(s)2}) = -.00346$, $q^{(s)} = 1.60728$ and $m_1^{*(s)} = 1.10728$.

In Table 3.9.2, we compare the purely sequential procedure (3.9.6) under various configurations different from LFC in $\Omega^{(s)}(\delta^*)$ given by (3.9.1). The two configurations we consider are "$\delta^*/8$ apart" and "$\delta^*/20$ apart". In "$\delta^*/20$ apart" the population means $\mu_{[k-s+1]} - \mu_{[k-s]} = \delta^*$ and rest all other populations are stacked $\frac{\delta^*}{20}$ above one another. In other words, $a_i = 1 + (\frac{k-s-i}{20})$ and $b_j = \frac{(s-j-1)}{20}$, $i = 1,\dots,k-s$, $j = 0,\dots,s-2$. Note that the $\bar{n} - C$ values in Table 3.9.2 are quite similar to the ones given in Table 3.9.1 for $k = 10$, whereas the $\bar{p}$ values are considerably larger than .95. Considering the small magnitude of $\delta^*/20$, it is interesting to note that $\bar{p}$ values are quite sensitive to small departures from LFC and tend to approach 1 very fast. As an exercise to the reader, we leave various aspects of comparisons of these $\bar{p}$

Table 3.9.1. Comparison of Procedures Under LFC
$\sigma = 1.0$, $m_0^* = 10$, $P^* = .95$, $s = 2$ for $k = 5$ and $s = 3$ for $k = 10$

	Purely Sequential Procedure (3.9.6)		Accelerated Sequential Procedure (3.9.9)–(3.9.11) ρ=.6		Three–Stage Procedure (3.9.14) k=5 ρ=.75; k=10, ρ=.80	
$C^{(s)}$	$\bar{n}-C^{(s)}$	$\bar{p}$	$\bar{n}-C^{(s)}$	$\bar{p}$	$\bar{n}-C^{(s)}$	$\bar{p}$
δ^*	$(s(\bar{n}))$	$(s(\bar{p}))$	$(s(\bar{n}))$	$(s(\bar{p}))$	$(s(\bar{n}))$	$(s(\bar{p}))$
			k = 5			
25	1.26	.952	2.93	.968	1.89	.957
.66	(.05)	(.003)	(.06)	(.003)	(.06)	(.003)
50	1.39	.951	2.64	.963	2.83	.952
.46	(.06)	(.003)	(.09)	(.003)	(.09)	(.003)
100	1.26	.955	2.51	.947	4.73	.951
.33	(.09)	(.003)	(.13)	(.003)	(.12)	(.003)
200	1.53	.950	2.26	.951	4.91	.949
.23	(.13)	(.003)	(.17)	(.003)	(.17)	(.003)
400	1.50	.950	2.11	.950	4.97	.951
.16	(.18)	(.003)	(.23)	(.003)	(.24)	(.003)
800	1.46	.955	2.01	.951	5.23	.951
.12	(.25)	(.003)	(.27)	(.003)	(.29)	(.003)
			k = 10			
25	1.64	.955	2.93	.962	1.96	.956
.77	(.04)	(.003)	(.06)	(.003)	(.06)	(.003)
50	1.40	.952	2.61	.960	2.81	.950
.55	(.06)	(.003)	(.08)	(.003)	(.09)	(.003)
100	1.37	.952	2.48	.951	4.63	.952
.39	(.08)	(.003)	(.11)	(.003)	(.13)	(.003)
200	1.41	.954	2.20	.951	4.72	.951
.27	(.10)	(.003)	(.14)	(.003)	(.17)	(.003)
400	1.35	.952	2.07	.950	4.81	.951
.19	(.13)	(.003)	(.19)	(.003)	(.23)	(.003)
800	1.29	.945	1.97	.951	4.88	.951
.14	(.16)	(.003)	(.22)	(.003)	(.26)	(.003)

Table 3.9.2. Purely Sequential Procedure Under Non LFC
$\sigma = 1.0$, $m_0^* = 10$, $P^* = .95$, $k = 10$, $s = 3$

$c^{(s)}$	δ^*	$\delta^*/8$ apart $\bar{n}-C$ $(s(\bar{n}))$	$\bar{p}$ $(s(\bar{p}))$	$\delta^*/20$ apart $\bar{n}-C$ $(s(\bar{n}))$	$\bar{p}$ $(s(\bar{p}))$
25	.77	1.48 (.03)	.995 (.001)	1.45 (.03)	.984 (.002)
50	.55	1.35 (.05)	.994 (.001)	1.36 (.05)	.984 (.002)
100	.39	1.39 (.06)	.995 (.001)	1.34 (.06)	.982 (.002)
200	.27	1.44 (.09)	.993 (.001)	1.33 (.09)	.983 (.002)
400	.19	1.37 (.13)	.995 (.001)	1.31 (.13)	.983 (.002)
800	.14	1.27 (.15)	.994 (.001)	1.25 (.15)	.983 (.002)

values in Table 3.9.2 with $f_{01}^{(s)}(h^{(s)2})$. Also, it is easy to verify that the standard errors of $\bar{n}$ for all the procedures in Table 3.9.1 and 3.9.2 compare quite favorably with their asymptotic values.

3.10 NOTES AND OVERVIEW

Only the case of unknown and equal variance has been given attention in this chapter. The two–stage procedure of Dudewicz and Dalal (1975) was designed to handle the challenging situation when the variances were unknown and unequal, and the sample sizes from the populations were also taken to be unequal. The results of Rinott (1978) and Mukhopadhyay (1979) are particularly relevant in this area. These are briefly addressed in section 8.4.1.

Going back to section 3.2 and equation (3.2.2), one notes that $P_n(CS) \geq \int_{-\infty}^{\infty} \prod_{i=1}^{k-1} \Phi[y + n^{\frac{1}{2}}\delta_i\sigma_i^{-1}]\phi(y)dy$, if it is assumed that π_i has the population variance σ_i^2. Thus, even when σ_i's are unknown and unequal, if one can *assume that* $\sigma_i^2 \leq \sigma^2$ *for all* i *and* σ^2 *is known, then*

$$\operatorname*{Inf}_{\underset{\sim}{\mu}\in\Omega(\delta^*)} P_n(CS) \geq P^*, \tag{3.10.1}$$

under $\mathscr{P}_n$ *where* $n = [C]^* + 1$, *that is a single stage procedure will "suffice".* In this situation, how does an appropriate modification of Dudewicz and Dalal's (1975) procedure compare with the above single stage procedure? Investigations are needed in this area.

Further investigations of ways to properly design purely sequential, accelerated sequential and three–stage procedures would be very worthwhile in the unequal and unknown variances case. In this situation, as well as the common and unknown variance case, it will be interesting and challenging to incorporate Paulson (1964)–type elimination of "inferior" populations in the selection process in the light of second–order analysis. See Swanepoel and Geertsema (1973) for some ideas. The likelihood based selection rules of Bechhofer et al. (1968) needs attention in this light too. See section 8.3.

Other types of problems such as selection of "good" populations, comparisons of treatments with a control or standard, ordering in r–way classification, etc. also hold much promise. See Gupta and Panchapakesan (1979, 1991) and Hochberg and Tamhane (1987) for a review of the types of results which are already available. It will be truly challenging to obtain asymptotic *second–order properties* of the associated existing and new multistage procedures in each of these problems and other closely related ones.

Complements and Problems

3.1 Show that the sufficient conditions given in Theorem 3.2.1 are satisfied by the stopping variables N arising from the two–stage, modified two–stage, purely sequential, accelerated sequential and the three–stage selection methodologies.

[*Hints*: Use Problems 2.1 – 2.2.]

3.2 Give an alternative proof of Theorem 3.5.1 (ii) by showing first $E(N) \le C + o(1)$.

[*Hints*: Look at Problem 2.6 (c) for ideas.]

3.3 In the sequential selection procedure (3.5.1), suppose one uses $U^*(n)$ instead of $U(n)$ in order to estimate σ^2. Prove Theorem 3.5.1 for this new selection methodology. Verify "Theorem 3.5.2" with appropriate modifications. Give particular attention to sufficient conditions involving m_1.

3.4 For the selection procedure (3.5.5), show that $N \ge (h/\delta^*)^{2/(\gamma+1)}$ w.p. 1. In other words, this sequential process is going to take at least $[(h/\delta^*)^{2/(\gamma+1)}]^* + 1$ many samples anyway. Does the choice for m_1 given in (3.4.1) look very reasonable now?

3.5 Prove Theorem 3.7.2 (ii) – (iii) for N defined in (3.7.5).

3.6 In the proof of Theorem 3.5.2, we repeatedly used the fact that $P\{N \le \frac{1}{2}C\} = O(C^{-\frac{1}{2}k(m_1-1)})$. Give a direct proof of this result, without appealing to Theorem 2.4.8 (i).

[*Hints*: Look at Problem 2.11.]

3.7 You are told that an experimenter had implemented the purely sequential selection methodology (3.5.1) and $\mathscr{R}_N$. Then, it is realized that the variances are not all really the same; instead the variance of the population π_i is σ_i^2 and $\sigma_i = b_i\sigma$ where $b_1,...,b_k$ are known positive numbers, however, $\sigma \in R^+$ is considered unknown. In a simple situation, consider $k = 2$ and without any loss of generality, let $b_1 = 1$, $b_2 = b^2$, $b > 0$. Show that, for all $\underset{\sim}{\mu} \in \Omega(\delta^*)$, one has $P(CS) \ge E[\min\{\int_{-\infty}^{\infty} \Phi(xb^{-1} + N^{\frac{1}{2}}\delta^*(\sigma b)^{-1})\phi(x)dx\}, \{\int_{-\infty}^{\infty} \Phi(xb + N^{\frac{1}{2}}\delta^*\sigma^{-1})\phi(x)dx\}]$.

[*Hints*: Recall the derivation of (3.2.2) for fixed n. Let $X_{(1)}, X_{(2)}$ be i.i.d. $N(0, 1)$. Then, observe that $P_n(CS) = P\{b_{(2)}X_{(2)} + n^{\frac{1}{2}}\delta_2\sigma^{-1} > b_{(1)}X_{(1)}\}$ $= \int_{-\infty}^{\infty}\Phi(\frac{b_{(2)}x+n^{\frac{1}{2}}\delta_2\sigma^{-1}}{b_{(1)}})\ \phi(x)dx$, that is, one has $\underset{\underset{\sim}{\mu}\in\Omega(\delta^*)}{\text{Inf}} P_n(CS) \ge$ $\min[\int_{-\infty}^{\infty}\Phi(\frac{x+n^{\frac{1}{2}}\delta^*\sigma^{-1}}{b})\ \phi(x)dx, \int_{-\infty}^{\infty}\Phi(xb + n^{\frac{1}{2}}\delta^*\sigma^{-1})\ \phi(x)dx]$. Now, proceed as in Theorem 3.2.1.]

3.8 In the accelerated sequential procedure (3.6.4) – (3.6.6), suppose one uses $U(n)$ instead of $U^*(n)$ in order to estimate σ^2. Prove Theorem 3.6.1 for this new selection methodology.

3.9 In the three–stage selection procedure (3.7.2) – (3.7.3), suppose one uses $U^*(n)$ instead of $U(n)$ in order to estimate σ^2. Prove Theorem 3.7.1 for this new selection methodology.

3.10 Prove Theorem 3.7.3 via Theorem 2.3.4.

[*Hints*: Look at (3.7.7) and note that this is of the form (2.3.27) with $\zeta = 1$, $a = h^2$, $L = \delta^{*2}$, $\theta^* = \sigma^2$, $\lambda = h^2/\delta^{*2}$ and $G^*(x) = f(x)$. Hence, Theorem 2.3.4 will imply that $m_1^* = \epsilon^* = (k\rho)^{-1}\{2 - \frac{h^2 f''(h^2)}{f'(h^2)}\} - \frac{1}{2} = -\beta' - (k\rho)^{-1}h^2 f''(h^2)/f'(h^2)$, since $\mathrm{Var}(W_1) = 2/k$.]

3.11 In the two–stage procedure (3.3.1), suppose that one uses $U^*(n)$, instead of $U(n)$, as the estimator of σ^2. For such a two–stage procedure, verify the properties given in Theorem 3.3.1 (i) and (iv). What happens to part (ii)?

3.12 In the modified two–stage procedure (3.4.1) – (3.4.2), suppose that one uses $U^*(n)$, instead of $U(n)$, as the estimator of σ^2. For such a modified two–stage procedure, verify the properties given in Theorem 3.4.1.

3.13 (a) Construct a two–stage procedure of the type (3.3.1) for selecting the s–best, $1 \le s < k$, normal populations and derive results similar to the Theorem 3.3.1.

(b) Revise the two–stage procedure proposed in part (a) above using the motivations provided in section 3.4 to construct a modified two–stage procedure and derive results similar to the Theorem 3.4.1.

3.14 (a) Prove Theorem 3.9.2 utilizing the proof of the Theorem 3.5.1.

(b) Prove Theorem 3.9.3 (i) utilizing the proof of the Theorem 3.5.2 (i).

3.15 (a) For the accelerated sequential procedure (3.9.9) – (3.9.11) obtain the first–order properties along the lines of Theorem 3.6.1.

(b) Prove Theorem 3.9.4.

3.16 (a) For the three–stage procedure (3.9.13) – (3.9.14) obtain the first order properties along the lines of Theorem 3.7.1.

(b) Prove Theorem 3.9.5.

3.17 Numerically evaluate $f_{01}^{(s)}(h^{(s)2})$ for $k = 10$, $s = 3$ and $P^* = .95$ and empirically verify Theorem 3.9.3 (ii) with the help of simulated values given in Table 3.9.2 under the specific non LFC situations considered there.

REFERENCES

Abramowitz and Stegun (1965), Anscombe (1952), Bechhofer (1954), Bechhofer, Dunnett and Sobel (1954), Bechhofer, Kiefer and Sobel (1968), Chow and Robbins (1965), Chow and Teicher (1978), Dudewicz (1971), Dudewicz and Dalal (1975), Dunnett and Sobel (1955), Ghosh and Mukhopadhyay (1976, 1981), Gibbons, Olkin and Sobel (1977), Gupta (1963), Gupta and Miescke (1984), Gupta, Nagel and Panchapakesan (1973), Gupta and Panchapakesan (1979, 1991), Gupta and Sobel (1957), Hall (1981), Hochberg and Tamhane (1987), Johnson and Kotz (1972), Krishnaiah and Armitage (1965), Mukhopadhyay (1979, 1980, 1981, 1983a,b), Mukhopadhyay and Judge (1989, 1991), Mukhopadhyay and Solanky (1992a), Paulson (1964), Rinott (1978), Robbins, Sobel and Starr (1968), Simons (1968), Stein (1945), Swanepoel and Geertsema (1973), Tong (1990), Vik and Mukhopadhyay (1988), Wiener (1939).

4

SELECTING THE BEST NEGATIVE EXPONENTIAL POPULATION

4.1 INTRODUCTION

Suppose that we have k (≥ 2) *independent* populations where the ith population π_i is assumed to have a negative exponential distribution with the p.d.f. $f(x;\ \theta_i,\ \sigma)$ given by (2.2.5), $i = 1,\ldots,k$. It is also assumed that the location parameters $\theta_i \in \mathbb{R}$ and the *common scale parameter* $\sigma \in \mathbb{R}^+$, $i = 1,\ldots,k$, are all *unknown*. Let $\theta_{[1]} \leq \cdots \leq \theta_{[k]}$ be the ordered θ–values. Since the scale parameter is the same in all populations, we sample equally from each population and at any point of time, whenever we need "new" samples from π's, we take a certain equal number, to be specified later, of samples from each π at that step. Having recorded i.i.d. random variables $X_{i1},\ldots,X_{in}$ from π_i, we write

$$\tilde{X}_{in} = \min\{X_{i1},\ldots,X_{in}\},\ V_{in} = (n-1)^{-1}\,\Sigma_{j=1}^{n}(X_{ij} - \tilde{X}_{in}),$$

$$V(n) = k^{-1}\Sigma_{i=1}^{k}\,V_{in},\ V^*(n) = n(n-1)^{-1}\,V(n) \qquad (4.1.1)$$

for $i = 1,\ldots,k$ and $n \geq 2$. One will note that θ_i is to be estimated by $\tilde{X}_{in}$. The common unknown scale parameter σ is often estimated by the pooled unbiased estimator $V(n)$. For technical reasons, in section 4.6, σ will be estimated by $V^*(n)$. It should be added that both $V(n)$ and $V^*(n)$ are consistent estimators of σ, that is, $V(n) \xrightarrow{P} \sigma$ as well as $V^*(n) \xrightarrow{P} \sigma$, as $n \to \infty$.

The distribution (2.2.5) has been used extensively in many reliability and life testing experiments for describing the failure rates of complex equipment, vacuum tubes, etc. It has also been recommended as a statistical model in clinical trials, such as the studies of behavior of tumor systems in animals and analysis of survival data in cancer research. One is referred to Zelen (1966). Other applications can be found in Johnson and Kotz (1970), Bain (1978), Lawless and Singhal (1980), Grubbs (1971). The population π_i can be thought to be characterized by the "life distribution" for the ith brand of a particular type of equipment or part, say, and the location parameter θ_i will then relate to the *minimum guarantee time* of "survival" in some sense. In such contexts, one will of course assume that $\theta_i \in R^+$ for $i = 1,...,k$. In general, however, let us keep assuming that $\theta_i \in R$ for $i = 1,...,k$. Obviously, $\max_{1 \le i \le k} P(X_{i1} \ge a^*)$ is attained, for all $a^* \in R$, in the population that is associated with $\theta_{[k]}$. In other words, the population or the brand associated with $\theta_{[k]}$ will have the maximum "survival" probability beyond any specified minimal requirements, such as a^*. Hence, our aim is to *select the population associated with the largest location parameter*, that is $\theta_{[k]}$, and such a population is called the *best population* in the present context. We assume that there is just one population associated with $\theta_{[k]}$. We do not, however assume any knowledge about the association of the θ_i's with $\theta_{[i]}$'s.

In what follows, we again adopt Bechhofer's (1954) *indifference zone* approach, and given $\delta^* > 0$, we let

$$\Omega(\delta^*) = \{\underset{\sim}{\theta} = (\theta_1,...,\theta_k) : \theta_{[k]} - \theta_{[k-1]} \ge \delta^*\}. \tag{4.1.2}$$

As in Chapter 3, for a given $P^* \in (k^{-1}, 1)$, we are interested in selection procedures such that the P(CS) of the population associated with $\theta_{[k]}$ is at least P^* or *asymptotically* at least P^* whenever the true parameter $\underset{\sim}{\theta} \in \Omega(\delta^*)$. We reacll that $\Omega(\delta^*)$ is called the *preference zone*, while $\Omega^c(\delta^*)$ is known as the *indifference zone* and the experimenter wishes to select the best population with the nominal level of P(CS) only when the parameter $\underset{\sim}{\theta}$ comes from the preference zone.

4.2 Preliminaries

For a moment let us pretend that σ is known and we have recorded $X_{i1},\ldots,X_{in}$ from π_i, $i = 1,\ldots,k$. The natural selection rule is simply to pick the population associated with the largest $\tilde{X}_{in}$ value, $i = 1,\ldots,k$, that is

$$\mathscr{P}_n^* : \text{Select } \pi_j \text{ as the best population if } \tilde{X}_{jn} = \underset{1 \le i \le k}{\text{Max}} \tilde{X}_{in}. \tag{4.2.1}$$

Throughout this chapter, we work with the selection procedure $\mathscr{P}_n^*$ given by (4.2.1).

Let $\pi_{(i)}$ stand for the population associated with the location parameter $\theta_{[i]}$ and let us write $\tilde{X}_{(in)}$ for the sample minimum arising from the population $\pi_{(i)}$, $i = 1,\ldots,k$. Also, write $\delta_i = \theta_{[k]} - \theta_{[i]}$ and note that

$$\delta_i \ge 0 \text{ for arbitrary } \underset{\sim}{\theta} \in \Omega(\delta^*) \cup \Omega^c(\delta^*)$$

and

$$\delta_i \ge \delta^* \text{ for arbitrary } \underset{\sim}{\theta} \in \Omega(\delta^*).$$

Then, the probability of correct selection is given by

$$P_n(CS) = P\{\tilde{X}_{(kn)} > \tilde{X}_{(in)},\ i = 1,\ldots,k-1\}$$

$$= P\Big[n\{\tilde{X}_{(kn)} - \mu_{[k]}\}\,\sigma^{-1} + n\sigma^{-1}\{\theta_{[k]} - \theta_{[i]}\}$$

$$> n\{\tilde{X}_{(in)} - \mu_{[i]}\}\,\sigma^{-1},\ i = 1,\ldots,k-1\Big]$$

$$= P\{Z_{kn} + n\delta_i\sigma^{-1} > Z_{in},\ i = 1,\ldots,k-1\}, \qquad (4.2.2)$$

where $Z_{in} = n\{\tilde{X}_{(in)} - \mu_{[i]}\}/\sigma$, $i = 1,\ldots,k$. From (4.2.2), for all $\underset{\sim}{\theta} \in \Omega(\delta^*)$,

$$P_n(CS) = \int_0^{\infty} \prod_{i=1}^{k-1} [1 - \exp(-z-n\delta_i\sigma^{-1})]\exp(-z)dz$$

$$\geq \int_0^{\infty} [1 - \exp(-z-n\delta^*\sigma^{-1})]^{k-1}\exp(-z)dz,$$

where the equality holds when $\delta_i = \delta^*$ for all $i = 1,\ldots,k-1$, that is when $\theta_{[1]} = \cdots = \theta_{[k-1]} = \theta_{[k]} - \delta^*$, a parameter configuration that is referred to as the *least favorable configuration* (LFC). Hence,

$$\underset{\underset{\sim}{\theta}\in\Omega(\delta^*)}{\text{Inf}}\ P_n(CS) = \int_0^{\infty} [1 - \exp(-z-n\delta^*\sigma^{-1})]^{k-1}\exp(-z)dz, \qquad (4.2.3)$$

while this infimum is attained at the LFC. Let $a = a(k, P^*)$ be such that

$$\int_0^{\infty} \{1 - \exp(-z-a)\}^{k-1}\exp(-z)dz = P^*, \qquad (4.2.4)$$

and then it follows from (4.2.3) that

$$P_n(CS) \geq P^* \text{ for all } \underset{\sim}{\theta} \in \Omega(\delta^*)$$

when $n \geq a\sigma/\delta^* = C$, say. The values of "a" had been tabulated in Raghavachari and Starr (1970). See also Table 4.8.2 in this context. In other words, had σ been known, one would take $[C]^* + 1$ samples from each π and implement the corresponding selection rule (4.2.1). Thus, C is known as the *optimal fixed sample size* needed from each π, in conjunction with (4.2.1), in order to meet the P*–requirement for P(CS) in $\Omega(\delta^*)$, had σ been known. In fact, however, C is unknown. Moreover, one can appropriately modify the results of Dudewicz (1971) to show that there exists no fixed–sample size procedure that will meet our P*–requirement for P(CS) in $\Omega(\delta^*)$, uniformly in $\sigma \in R^+$. In what follows, we develop two–stage, purely sequential, accelerated sequential and three–stage selection methodologies for this problem.

Suppose, in general, one devises a statistical methodology giving rise to a positive integer valued random variable N estimating C and selects the population associated with the largest sample minimum, that is to implement the selection procedure $\mathscr{P}_N^*$ given by (4.2.1).

Theorem 4.2.1. *Suppose that* N *satisfies the following conditions. The distribution of* N *does not depend on* $\underset{\sim}{\theta}$; $I(N = n)$ *is independent of* $(\tilde{X}_{1n},\ldots,\tilde{X}_{kn})$ *for every fixed* $n \geq 2$; *and* $P(N < \infty) = 1$. *Then,*

$$\underset{\underset{\sim}{\theta}\in\Omega(\delta^*)}{\text{Inf}} P(CS) = E[\int_0^\infty \{1 - \exp(-z-N\delta^*\sigma^{-1})\}^{k-1}\exp(-z)dz] \quad (4.2.5)$$

and this infimum is attained when $\theta_{[1]} = \cdots = \theta_{[k-1]} = \theta_{[k]} - \delta^*$.

Proof. As before, write $\delta_i = \theta_{[k]} - \theta_{[i]}$ and $\delta_i \geq \delta^*$ whenever $\underset{\sim}{\theta} \in \Omega(\delta^*)$. Suppose that $m_1(\geq 2)$ is the starting sample size from each π that ultimately gives rise to N. The selection rule $\mathscr{P}_N^*$ is well–defined since $P(N < \infty) = 1$. Now,

$$P(CS) = \Sigma_{n=m_1}^\infty P(CS \mid N = n)P(N = n)$$

$$= \Sigma_{n=m_1}^\infty P_n(CS)P(N = n)$$

along the lines of (3.2.5). Hence, utilizing (4.2.3) will complete the proof. □

4.3 TWO–STAGE PROCEDURE

Desu et al. (1977) had proposed a two–stage procedure for this problem along the lines of Stein (1945). That procedure looks like (2.2.1). Let $m_1(\geq 2)$ be the starting sample size from each π and $\tau_0 = \tau_0(m_1, k, P^*)$ be a positive real number, to be determined specifically a little later. Define

$$N = N(\delta^*) = \max\{m_1, [b^* V(m_1)\delta^{*-1}]^* + 1\} \qquad (4.3.1)$$

where $b^* = b^*(m_1) = 2\tau_0 k(m_1 - 1)$. If $N = m_1$, we do not take any more samples from any population. However, if $N > m_1$, then we sample the difference in the second stage, that is, we take $(N - m_1)$ more samples from each π in the second stage. Based on the totality of all samples, that is having $X_{i1},\ldots,X_{iN}$ from π_i, $i = 1,\ldots,k$, one now implements the usual selection procedure $\mathscr{P}_N^*$ given by (4.2.1). We keep $\underset{\sim}{\theta}$ and σ fixed.

Theorem 4.3.1. *For the two–stage procedure (4.3.1), with* τ_0 *defined in (4.3.3), we have:*

(i) $P(CS) \geq P^*$ *for all* $\underset{\sim}{\theta} \in \Omega(\delta^*)$;

(ii) $b^*\sigma\delta^{*-1} \leq E(N) \leq m_1 + b^*\sigma\delta^{*-1}$;

(iii) $E(N/C) \to b^* a^{-1}$ *as* $\delta^* \to 0$;

(iv) $\liminf P(CS) \geq P^*$, *for all* $\underset{\sim}{\theta} \in \Omega(\delta^*)$ *as* $\delta^* \to 0$;

where $C = a\sigma/\delta^*$, a *comes from (4.2.4) and* $b^* = b^*(m_1) = 2\tau_0 k(m_1 - 1)$.

Proof. The basic inequality (2.2.3) turns out to be

$$b^*V(m_1)\delta^{*-1} \le N \le b^*V(m_1)\delta^{*-1} + m_1 \tag{4.3.2}$$

and now taking expectations throughout leads to part (ii), since $E[V(m_1)] = \sigma$. Next, dividing throughout the expression in part (ii) by C and then taking limits as $\delta^* \to 0$ leads to part (iii).

In order to prove part (i), first observe that Theorem 4.2.1 holds for the two–stage selection procedure (4.3.1) and thus $\inf_{\underset{\sim}{\theta}\in\Omega(\delta^*)} P(CS)$ is given by the expression in (4.2.5) and the LFC turns out to be $\theta_{[1]} = \cdots = \theta_{[k-1]} = \theta_{[k]} - \delta^*$. Along the lines of Desu et al. (1977), one can write,

$$\inf_{\underset{\sim}{\theta}\in\Omega(\delta^*)} P(CS) \ge E[\int_0^\infty \{1 - \exp(-z - b^*V(m_1)\sigma^{-1})\}^{k-1}\exp(-z)dz]$$

$$= E\{P[Y_i < Y_k + b^*V(m_1)\sigma^{-1}, i = 1,\ldots,k-1 \mid V(m_1)]\}$$

where $Y_1,\ldots,Y_k$ are i.i.d. standard exponentials, that is the common p.d.f. is $f(y;\ 0,\ 1)$ given by (2.2.5), and these are chosen independently of $V(m_1)$, since $N\delta^* \ge b^*V(m_1)$ w.p. 1. Note that $T^* = 2k(m_1 - 1)V(m_1)\sigma^{-1} \sim \chi^2_{2k(m_1-1)}$ and $b^*V(m_1) = \tau_0 T^*$. Hence, we can write

$$\inf_{\underset{\sim}{\theta}\in\Omega(\delta^*)} P(CS) \ge P[Y_i < Y_k + \tau_0 T^*, i = 1,\ldots,k-1]$$

$$= \int_0^\infty\int_0^\infty \{1 - \exp(-z - \tau_0 t)\}^{k-1}\exp(-z)g^*(t)dzdt$$

where $g^*(t)$ is the p.d.f. of T^*. Thus, the P^* requirement is met if we determine $\tau_0 = \tau_0(m_1, k, P^*)$ in such a way that

$$\int_0^{\infty}\{\int_0^{\infty}[1-\exp(-z-\tau_0 t)]^{k-1}\exp(-z)dz\}g^*(t)dt = P^*. \tag{4.3.3}$$

Expanding $\{1-\exp(-z-\tau_0 t)\}^{k-1}$ as a finite series and carrying out the term by term integration, we obtain

$$\Sigma_{j=1}^{k}(-1)^{j+1}(k-1)!/\{[j!(k-j)!][1+2\tau_0(j-1)]^{k(m_1-1)}\}=P^*. \tag{4.3.4}$$

That is, alternatively, one can determine $\tau_0 = \tau_0(m_1, k, P^*)$ such that it satisfies (4.3.4). Hence, part (i) now follows. In order to prove part (iv), first notice from (4.3.2) that $N\delta^* \to b^*V(m_1)$ w.p. 1 as $\delta^* \to 0$. From (4.2.5) and the dominated convergence theorem one obtains

$$\liminf_{\delta^*\to 0} P(CS) \geq E[\int_0^{\infty}\{1-\exp(-z-b^*V(m_1)\sigma^{-1})\}^{k-1}\exp(-z)dz] \tag{4.3.5}$$

for all $\underset{\sim}{\theta} \in \Omega(\delta^*)$. Hence, part (iv) follows in view of (4.3.3). □

Remark 4.3.1. The table in Desu et al. (1977) provides the values of τ_0 for $k = 2(1)6$, $P^* = .95$ and $m_1 = 2(1)30$. See also Table 4.8.1 in this context.

We have already noted in Chapter 3 that in a two–stage procedure, the choice of m_1 is very crucial. Both "very large" as well as "very small", but fixed, choices of m_1 values will tend to make N overestimate C by a very wide margin. In Theorem 2 of Mukhopadhyay (1986), it was shown that $b^* > a$ only for $k = 2$. From the Table of Desu et al. (1977) and our Table 4.8.1, it *appears* as though $b^* > a$ for all $k \geq 3$. Hence, it is obvious that

$$\liminf_{\delta^*\to 0} E(N - C) = \infty, \tag{4.3.6}$$

for $k = 2$ and *we conjecture that (4.3.6) holds for all* $k \geq 3$ *as well.* That is, as long as m_1 is held fixed, N is expected to *overestimate* C *even asymptotically*.

4.4 MODIFIED TWO–STAGE PROCEDURE

In this procedure, proposed along the lines of sections 2.2.2 and 3.4, we determine m_1 in such a way that m_1 grows as $\delta^* \to 0$, but m_1 still stays fairly small compared with C. In other words, the idea, as before, is to choose m_1 larger when C is large, while picking a smaller m_1 when one suspects that C is small. But, then the problem is that C is unknown. Following Mukhopadhyay (1986), we pick $m_0^*(\geq 2)$, $\gamma \in R^+$ and define

$$m_1 = m_1(\delta^*) = \max\{m_0^*, [(a\delta^{*-1})^{1/(\gamma+1)}]^* + 1\} \tag{4.4.1}$$

where recall that "a" comes from (4.2.4). This m_1 is then the experimenter's starting sample size from each population. Now, let

$$N = N(\delta^*) = \max\{m_1, [b^*V(m_1)\delta^{*-1}]^* + 1\} \tag{4.4.2}$$

where $b^* = b^*(m_1) = 2\tau_0 k(m_1 - 1)$, and from a practical point of view, we implement the modified two–stage procedure (4.4.1) – (4.4.2) along the lines of the original two–stage procedure (4.3.1) in conjunction with the customary selection procedure $\mathscr{P}_N^*$ given by (4.2.1). Again, we hold $\underset{\sim}{\theta}$ and σ fixed.

Theorem 4.4.1. *For the modified two–stage procedure (4.4.1) – (4.4.2), with* τ_0 *defined in (4.3.3), we have:*

(i) $P(CS) \geq P^*$ *for all* $\underset{\sim}{\theta} \in \Omega(\delta^*)$;

(ii) $N/C \to 1$ *w.p.* 1 *and* $E(N/C) \to 1$ *as* $\delta^* \to 0$;

(iii) $\liminf P(CS) \geq P^*$, *for all* $\underset{\sim}{\theta} \in \Omega(\delta^*)$, *as* $\delta^* \to 0$;

where $C = a\sigma/\delta^*$, $0 < \gamma < \infty$ *is arbitrary, but fixed and* a *comes from (4.2.4).*

Proof. Part (i) is nonasymptotic in nature and it follows immediately along the lines of Theorem 4.3.1 (i). Now, we turn to part (ii). Note that $m_1(\delta^*) \to \infty$ and $m_1(\delta^*)/C \to 0$ as $\delta^* \to 0$, whatever $\gamma \in R^+$. The basic inequality is given by

$$b^*V(m_1)\delta^{*-1} \leq N \leq m_0^* + \{a\delta^{*-1}\}^{1/(\gamma+1)} + b^*V(m_1)\delta^{*-1} \qquad (4.4.3)$$

where one notes that $\tau_0 = \tau_0(m_1, k, P^*)$ is actually a function of δ^*, since k and P^* are held fixed. Using Theorem 2 of Mukhopadhyay (1986), one can see that $b^* \to a$ as $\delta^* \to 0$. Now, from (4.4.3) one gets

$$(b^*/a)(V(m_1)/\sigma) \leq N/C \leq m_0^*C^{-1} + 0(\delta^{*1-(\gamma+1)^{-1}})$$
$$+ (b^*/a)(V(m_1)/\sigma) \qquad (4.4.4)$$

which implies $N/C \to 1$ w.p. 1 as $\delta^* \to 0$, since $V(n) \to \sigma$ w.p. 1 as $n \to \infty$. Taking expectations throughout in (4.4.4), one gets

$$b^*a^{-1} \leq E(N/C) \leq m_1^*C^{-1} + 0(\delta^{*\gamma/(\gamma+1)}) + b^*a^{-1}$$

and hence $\lim_{\delta^* \to 0} E(N/C) = 1$.

Now in view of part (ii) and the dominated convergence theorem, one immediately obtains from (4.2.5)

$$\liminf_{\delta^* \to 0} P(CS) \geq E[\int_0^\infty \{1 - \exp(-z-a)\}^{k-1}\exp(-a)dz] = P^*,$$

for all $\underset{\sim}{\theta} \in \Omega(\delta^*)$. □

Following the terminologies from Chapter 3, in particular, referring to (3.4.5) – (3.4.8), some remarks are in order.

Remark 4.4.1. Both the two–stage procedure (4.3.1) and the modified two–stage procedure (4.4.1) – (4.4.2) are *consistent* and *asymptotically consistent*. However, the two–stage procedure is *not asymptotically first–order efficient* while the modified version is.

Remark 4.4.2. It is *conjectured* that $E(N - C)$ is *unbounded* in the limit for the modified two–stage procedure (4.4.1) – (4.4.2).

4.5 Purely Sequential Procedure

The purely sequential procedure discussed here was originally proposed in Mukhopadhyay (1986). Recall that $C = a\sigma/\delta^*$ which is the optimal sample size required from each population, had σ been known. Since σ is unknown, the basic idea is to keep updating estimates of σ until we arrive at some kind of "stability", that is the sample size n and the corresponding estimator $V(n)$ of σ satisfies a certain relationship which is to be made precise very shortly.

We start with $m_1(\geq 2)$ samples from each π. Then, we keep taking one more sample at a time from each population according to the following stopping rule:

$$N = N(\delta^*) = \inf\{n \geq m_1 \ : \ n \geq aV(n)\delta^{*-1}\}, \qquad (4.5.1)$$

where a comes from (4.2.4). For fixed $\underset{\sim}{\theta}$, σ, m_1 and δ^*, we have $P[N < \infty] = 1 - \lim_{n\to\infty} P[N > n] \geq 1 - \lim_{n\to\infty} P[n < aV(n)\delta^{*-1}] = 1$, since $V(n) \to \sigma$ w.p. 1 as $n \to \infty$. That is, $P(N < \infty) = 1$, in other words the purely sequential procedure (4.5.1) terminates with probability one. Based on the totality of all samples, that is having $X_{i1},\ldots,X_{iN}$ from π_i, $i = 1,\ldots,k$, one now implements the customary selection procedure $\mathcal{P}_N^*$ given by (4.2.1). Throughout, $\underset{\sim}{\theta}$ and σ are kept fixed.

Theorem 4.5.1. *For the purely sequential procedure (4.5.1), we have as* $\delta^* \to 0$:

(i) $N/C \to 1$ *w.p.* 1;

(ii) $E(N/C) \to 1$;

(iii) $C^{-\frac{1}{2}}(N - C) \xrightarrow{\mathcal{L}} N(0, k^{-1})$;

(iv) $\liminf P(CS) \geq P^*$ *for all* $\underset{\sim}{\theta} \in \Omega(\delta^*)$;

(v) $\lim P(CS) = P^*$ *under the LFC, namely* $\theta_{[1]} = \cdots = \theta_{[k-1]} = \theta_{[k]} - \delta^*$;

where $C = a\sigma/\delta^*$ *and* a *comes from (4.2.4).*

Proof. Utilizing Lemma 1 of Chow and Robbins (1965), it follows that as $\delta^* \to 0$, we have

$$N \to \infty, \quad V(N) \to \sigma \quad \text{and} \quad V(N-1) \to \sigma \quad \text{w.p. 1.}$$

The sequential procedure (4.5.1) agrees with (2.4.1) where $\nu = [\delta^{*-1}]^* + 1$, $\psi_\nu = a/\delta^*$, $T_n = V(n)$ and $m = m_1$. The basic inequality (2.4.3) is then equivalent to

$$aV(N)\delta^{*-1} \leq N \leq m_1 + aV(N-1)\delta^{*-1}. \tag{4.5.2}$$

Now divide throughout (4.5.2) by C and take limits as $\delta^* \to 0$. This leads to part (i).

Observe that there is a basic difference between the sequences of U(n) and $V(n)$ statistics defined respectively in (3.1.1) and (4.1.1). By employing Helmert's orthogonal transformation (see Problem 2.2), $U(n)$ could be represented as the sample mean of i.i.d. random variables such as $(n-1)^{-1}\Sigma_{i=1}^{n-1} Z_i$. However, in this exponential case, weighted spacings between successive order statistics play the crucial role of such Z_i's, but then (see Problem 2.4) we can only write $V(n) = (n-1)^{-1}\Sigma_{i=1}^{n-1} Z_{in}$ where $Z_{1n},\ldots,Z_{n-1n}$ are i.i.d. for every fixed $n \geq 2$. However, the embedding techniques of Lombard and Swanepoel (1978) come to rescue in this setup since the two infinite sequences $\{V(n) : n \geq m_1\}$ and $\{(n-1)^{-1}\Sigma_{i=1}^{n-1} W_i' : n \geq m_1\}$ have identical probability distributions, that is, for all practical purposes, we can write

$$V(n) = (n-1)^{-1}\Sigma_{i=1}^{n-1} W_i', \tag{4.5.3}$$

where $W_1', W_2',\ldots$ are i.i.d. $\frac{1}{2}k^{-1}\sigma\chi_{2k}^2$ random variables. In fact, from Lombard and Swanepoel (1978) it does follow that N given by (4.5.1) has exactly the same distribution as that of

$$N = \inf\{n \geq m_1 : n \geq a\delta^{*-1}(n-1)^{-1}\Sigma_{i=1}^{n-1}W_i'\}, \tag{4.5.4}$$

where $W_1', W_2',\ldots$ are i.i.d. $(2k)^{-1}\sigma\,\chi_{2k}^2$ random variables. Use Anscombe's (1952) results to claim the sufficient conditions given in (2.4.2) with $a = \sigma$ and $b = k^{-\frac{1}{2}}\sigma$. Now, part (iii) follows from Theorem 2.4.1.

In order to prove parts (iv) – (v), first observe that Theorem 4.2.1 holds for the purely sequential selection procedure (4.5.1). Hence, for all $\underset{\sim}{\theta} \in \Omega(\delta^*)$,

$$P(CS) \geq E[\int_0^\infty \{1-\exp(-z-N\delta^*\sigma^{-1})\}^{k-1}\exp(-z)dz]. \tag{4.5.5}$$

Also, from part (i), one gets $N\delta^*\sigma^{-1} \to a$ w.p. 1 as $\delta^* \to 0$, and hence (4.5.5) together with the dominated convergence theorem will lead to part (iv). On the other hand, note that the LFC is given by $\theta_{[1]} = \cdots = \theta_{[k-1]} = \theta_{[k]} - \delta^*$ and under LFC, equality holds in (4.5.5). Thus, similar arguments then lead to part (v).

Now, we turn to prove part (ii). From the rhs of the basic inequality (4.5.2), it follows that

$$N \le m_1 + a\delta^{*-1} V^*,$$

that is, $NC^{-1} \le m_1 + \sigma^{-1}V^*$ for sufficiently small δ^*, where we write $V^* = \sup_{n\ge 2} \{(n-1)^{-1}\Sigma_{i=1}^{n-1} W_i'\}$. By Wiener's (1939) dominated ergodic theorem one can conclude that $E(V^*) < \infty$. Now, the dominated convergence theorem and part (i) together can easily be seen to imply part (ii). □

Remark 4.5.1. The purely sequential selection procedure (4.5.1) is *asymptotically* both *first-order efficient* and *consistent*, but, one should note that the exact consistency property [see (3.4.5)], which holds for the two-stage as well as the modified two-stage versions, no longer holds for the purely sequential selection procedure.

Remark 4.5.2. Suppose that the stopping rule (4.5.1) is modified slightly along the following lines:

$$N = N(\delta^*) = \inf\{n \ge m_1 \ : \ n \ge a\delta^{*-1}[V(n) + n^{-\gamma}]\} \tag{4.5.6}$$

for $0 < \gamma < \infty$. By taking (4.5.3) – (4.5.4) into account, one can easily claim that the stopping rule (4.5.6) is indeed of the form given in (2.4.6) with $\delta = 1$, $t = 0$, $\alpha = 1$, $m = m_1$ and $\psi_\nu = a/\delta^*$ with $\nu = [\delta^{*-1}]^* + 1$. Hence, part (i), (ii), (iv) and (v) of Theorem 4.5.1 holds for (4.5.6) for all $\gamma \in R^+$. Theorem 2.4.4 on the other hand implies that Theorem 4.5.1 (iii) holds when $\gamma > \frac{1}{2}$. Additionally, one can now obtain the rate of convergence also in Theorem 4.5.1 (iii), when the stopping variable N is taken from (4.5.6). For $\gamma > \frac{3}{4}$,

Theorem 2.4.7 gives the following:

$$\sup_{x} |P[(C/k)^{-\frac{1}{2}}(N - C) \le x] - \Phi(x)| = 0(\delta^{*\frac{1}{4} - \frac{\lambda}{2}})$$

for $0 < \lambda < \frac{1}{2}$, as $\delta^* \to 0$.

4.5.1 SECOND–ORDER ASYMPTOTICS

In this section, we strengthen the results given in Theorem 4.5.1 (ii) and (v) by providing expansions of (a) E(N) for all $\underset{\sim}{\theta} \in \Omega(\delta^*)$ and (b) P(CS) under the LFC, for the purely sequential selection methodology (4.5.1). Use the representation given in (4.5.4) to rewrite

$$N = \inf\{n \ge m_1 \; : \; 2kn(n-1)C^{-1} \ge \Sigma_{i=1}^{n-1} W_i\}$$

and hence $N = Q + 1$ where

$$Q = \inf\{n \ge m_1 - 1 : \; 2kn(n+1)C^{-1} \ge \Sigma_{i=1}^{n} W_i\}, \tag{4.5.7}$$

W's being i.i.d.. χ^2_{2k}. The stopping variable Q is of the form given in (2.4.7) with $\delta = 2$, $L_0 = 1$, $m = m_1 - 1$, $h^* = 2k/C$, $\theta = 2k$, $\tau^2 = 4k$, $\beta^* = 1$, $n_0^* = C$, $p = 1/k$, $b = k$ and

$$\nu = \nu(k) = (1+k) - \Sigma_{n=1}^{\infty} n^{-1} E[\max(0, \chi^2_{2nk} - 4nk)]. \tag{4.5.8}$$

See (2.4.8) – (2.4.9) in this context. The second–order theory summarized here was originally developed in Mukhopadhyay (1986) for general $k \ge 2$. The case when $k = 2$ was considered in Mukhopadhyay (1984). Let

$$q_1(x) = \exp(-x)I(x > 0), \; b(k, x) = \binom{k-1}{x}(-1)^x(x+1)^{-1}, \tag{4.5.9}$$

$$b_i = \Sigma_{x=0}^{k-1} x^i b(k, x)\, q_1(xa), \text{ for } i = 1, 2, \text{ and} \tag{4.5.10}$$

$$g_1(y) = \int_0^\infty \{1 - \exp(-z-y)\}^{k-1} \exp(-z)dz \text{ for } y > 0. \tag{4.5.11}$$

Now, we have the following theorem.

Theorem 4.5.2. *For the purely sequential procedure (4.5.1), we have as* $\delta^* \to 0$:

(i) $E(N) = C + \beta'' + o(1)$ *for all* $\underset{\sim}{\theta} \in \Omega(\delta^*)$ *if* (a) $m_1 \geq 3$ *when* $k = 2$, (b) $m_1 \geq 2$ *when* $k \geq 3$;

(ii) $P(CS) = P^* + \delta^* \sigma^{-1}\{(1-P^*)\beta'' - \frac{1}{4}ak^{-1}\} + o(\delta^*)$ *under the LFC for* $m_1 \geq 3$ *when* $k = 2$;

(iii) $P(CS) = P^* + \delta^* \sigma^{-1}\{\frac{1}{2}ab_2k^{-1} - b_1\beta''\} + o(\delta^*)$ *under the LFC for* $m_1 \geq 2$ *when* $k \geq 3$;

where $\beta'' = \beta''(k) = k^{-1}\{\frac{1}{2}\nu(k) - 1\}$ *with* $\nu(k)$ *defined in (4.5.8),* a *comes from (4.2.4), and* b_1, b_2 *comes from (4.5.10).*

Proof. Let us first prove part (i). Note that η in (2.4.10) corresponding to Q given in (4.5.7) simplifies to $k^{-1}(\frac{1}{2}\nu - 1) - 1$ and hence Theorem 2.4.8 (v) with $\omega = 1$ leads to

$$E(N) = 1 + E(Q) = 1 + C + \eta + o(1) = C + (\tfrac{1}{2}\nu - 1)k^{-1} + o(1) \tag{4.5.12}$$

if $m_1 - 1 > 2k^{-1}$, that is, if $m_1 > 1 + 2k^{-1}$. This is part (i). Let us turn now to prove part (ii). In the special case $k = 2$, we have under the LFC,

$$P(CS) = 1 - \tfrac{1}{2} E[\exp(-N\delta^*/\sigma)]$$
$$= 1 - \tfrac{1}{2} E[e^{-a} - (N\delta^*\sigma^{-1} - a)e^{-a} + \tfrac{1}{2}(N\delta^*\sigma^{-1} - a)^2 e^{-Y}] \quad (4.5.13)$$

for some suitable random variable Y between a and $N\delta^*/\sigma$. Hence, from (4.5.13),

$$P(CS) = (1 - \tfrac{1}{2}e^{-a}) + \tfrac{1}{2}e^{-a}\delta^*\sigma^{-1}E(N - C)$$
$$- \tfrac{1}{4}a\delta^*\sigma^{-1}E[U^{*2}e^{-Y}] \quad (4.5.14)$$

where repeated use of Theorem 2.4.8 implies that $U^* = C^{-\frac{1}{2}}(N - C) \xrightarrow{\mathscr{L}} N(0, k^{-1})$ and U^{*2} is uniformly integrable if $m_1 > 1 + k^{-1}$. Thus, $U^{*2}e^{-Y}$ is uniformly integrable if $m_1 > 1 + k^{-1}$. Also, $U^{*2}e^{-Y} \xrightarrow{\mathscr{L}} k^{-1}\chi_1^2$ as $\delta^* \to 0$ and $P^* = 1 - \tfrac{1}{2}e^{-a}$. Hence, combining with part (i), for $m_1 \geq 3$, one gets from (4.5.14)

$$P(CS) = P^* + \delta^*\sigma^{-1}\{(1-P^*)\beta'' - \tfrac{1}{4}ak^{-1}\} + o(\delta^*), \quad (4.5.15)$$

which is part (ii).
In order to prove part (iii), one writes

$$P(CS) = E\{g_1(N\delta^*/\sigma)\}$$
$$= \Sigma_{j=0}^{k-1} b(k, j)E[q_1(jN\delta^*/\sigma)]$$
$$= \Sigma_{j=0}^{k-1} b(k, j)q_1(ja) - \delta^*\sigma^{-1}E(N-C)\Sigma_{j=0}^{k-1} jb(k, j)q_1(ja)$$
$$+ \tfrac{1}{2}a\delta^*\sigma^{-1}\Sigma_{j=0}^{k-1} j^2 b(k, j)E\{U^{*2}q_1(W_j^*)\}, \quad (4.5.16)$$

for a suitable random variable W_j^* between ja and $jN\delta^*/\sigma$. Recall that $U^* = C^{-\frac{1}{2}}(N - C)$. Again, $U^{*2}q_1(W_j^*)$ is uniformly integrable if $m_1 > 1 + k^{-1}$. Also, $U^{*2}q_1(W_j^*) \xrightarrow{\mathscr{L}} k^{-1}q_1(ja)\chi_1^2$ as $\delta^* \to 0$. Now, from (4.5.16), one obtains for $m_1 > 1 + 2k^{-1}$, that is, for $m_1 \geq 2$,

$$P(CS) = P^* - \delta^*\sigma^{-1}(\beta'' + o(1))b_1 + \tfrac{1}{2}a\delta^*(k\sigma)^{-1}(b_2 + o(1))$$

$$= P^* - \delta^*\sigma^{-1}\{\beta''b_1 - \tfrac{1}{2}ab_2k^{-1}\} + o(\delta^*), \qquad (4.5.17)$$

in view of (4.5.9) – (4.5.11) and (4.5.12). This is part (iii). □

One should note that Theorem 4.5.1 (ii) and (iv) show that the purely sequential selection procedure (4.5.1) is *asymptotically* both *first–order efficient* and *consistent*. Theorem 4.5.2 (i), however, shows that the same methodology is also *asymptotically second–order efficient*. Recall (3.4.5) – (3.4.8) in this context. As we have already noted in section 3.5, the purely sequential rule can indeed be quite cumbersome for implementation, because here the experimenter takes one sample at a time from each population and the process may take a while to stop. It will be desirable to reduce such sampling operations and yet maintain the asymptotic second–order efficiency property. In the next two sections, the selection methodologies discussed aim at achieving this goal, that is to have more operational convenience in conjunction with asymptotic second–order properties.

Remark 4.5.3. Observe that $\nu(k)$ obtained from (4.5.8) is equivalent to $\nu(2k)$ obtained from (3.5.7). Table 3.8.1 will be useful here also. The case $k = 2$ was treated separately in Theorem 4.5.2 while evaluating $P(CS)$ because the expressions in the proof are fairly simple in that special situation.

4.6 Accelerated Sequential Procedure

Recall that $C = a\sigma/\delta^*$ and thus far σ has been estimated by $V(n)$. For technical reasons, however, we will now estimate σ by $V^*(n)$ given in (4.1.1). Asymptotically, that is, as $n \to \infty$, there is of course not much difference between $V(n)$ and $V^*(n)$. In fact, the sequential procedure (4.5.1) can *essentially* be replaced by

$$L = L(\delta^*) = \inf\{n \geq m_1 : n \geq aV^*(n)\delta^{*-1}\} \qquad (4.6.1)$$

where $m_1(\geq 2)$ is the starting sample size from each π. Once we stop sampling with the rule (4.6.1), we would use selection rule $\mathscr{P}_L^*$ given by (4.2.1). Note that $L = Q + 1$, where

$$Q = \inf\{n \geq m_1 - 1 \ : \ 2kn^2C^{-1} \geq \Sigma_{i=1}^n W_i\}, \tag{4.6.2}$$

W's being i.i.d. χ^2_{2k}. Thus, one can prove along the lines of Theorem 4.5.2 (i) that

$$E(L) = C + k^{-1}(\tfrac{1}{2}\nu - 1) + 1 + o(1) \tag{4.6.3}$$

if $m_1 > 1 + 2k^{-1}$, for all $\underset{\sim}{\theta} \in \Omega(\delta^*)$. Since $V^*(n)$ overestimates σ slightly for each $n \geq 2$, the fact that the purely sequential procedure (4.6.1) takes *on the average* one extra sample compared with the original purely sequential rule (4.5.1) is quite intuitive indeed. The second–order expansion of P(CS) given in Theorem 4.5.2 (ii) and (iii) also holds for the purely sequential procedure (4.6.1), having replaced β'' by the expression $(\tfrac{1}{2}\nu - 1)k^{-1} + 1$. In these senses, the two purely sequential rules (4.5.1) and (4.6.1) can be considered *essentially same*. Along the lines of the discussions in section 3.6, we motivate our accelerated version of the selection methodology via (4.6.1). This procedure was originally developed in Mukhopadhyay and Solanky (1992b).

We first choose and fix $0 < \rho < 1$ and $q \geq 0$. Let

$$t = t(\delta^*) = \inf\{n \geq m_1 : n \geq \rho a V^*(n)/\delta^*\}, \tag{4.6.4}$$

$$N = N_1(\delta^*) = [aV^*(t)\delta^{*-1} + q]^* + 1, \tag{4.6.5}$$

$$N = N(\delta^*) = \max\{t(\delta^*), N_1(\delta^*)\}. \tag{4.6.6}$$

We start the experiment with $m_1(\geq 2)$ samples from each π, and then sample one at a time from each π sequentially ending up with $X_{i1},\ldots,X_{it}$

from π_i, $i = 1,...,k$. Note that $t(\delta^*)$ estimates ρC, a fraction of the optimal fixed sample size. Now, $N_1(\delta^*)$ estimates C based on t samples observed from each π_i. If $N = t$, we do not take any more samples. If $N > t$, we sample the difference $(N - t)$ from each π in *one single batch* at this stage. Now, based on the totality of all samples, that is having $X_{i1},...,X_{iN}$ from π_i, $i = 1,...,k$, one now implements the customary selection procedure $\mathscr{P}_N^*$ given by (4.2.1). Throughout, $\underset{\sim}{\theta}$ and σ are kept fixed.

Observe that Theorem 4.2.1 again holds for the accelerated sequential selection procedure (4.6.4) – (4.6.6). Hence, for all $\underset{\sim}{\theta} \in \Omega(\delta^*)$,

$$P(CS) \geq E[\int_0^\infty \{1 - \exp(-z-N\delta^*\sigma^{-1})\}^{k-1}\exp(-z)dz] \tag{4.6.7}$$

as in (4.5.5), while the equality in (4.6.7) holds under LFC given by $\theta_{[1]} = \cdots = \theta_{[k-1]} = \theta_{[k]} - \delta^*$, as before. The constant $\nu = \nu(k)$ is the same as in (4.5.8). First, we discuss certain first–order asymptotic results parallel to those in Theorem 4.5.1.

Theorem 4.6.1. *For the accelerated sequential procedure (4.6.4) – (4.6.6), we have as* $\delta^* \to 0$:

(i) $N/C \to 1$ *w.p.* 1;

(ii) $E(N/C) \to 1$;

(iii) $C^{-\frac{1}{2}}(N - C) \xrightarrow{\mathscr{L}} N(0, (k\rho)^{-1})$;

(iv) $\liminf P(CS) \geq P^*$ *for all* $\underset{\sim}{\theta} \in \Omega(\delta^*)$;

(v) $\lim P(CS)=P^*$ *under the LFC, namely* $\theta_{[1]}=\cdots=\theta_{[k-1]}=\theta_{[k]}- \delta^*$;

where $C = a\sigma/\delta^*$ *and* a *comes from (4.2.4).*

Proof. Note that along the lines of (4.6.2), one can write

$$t = t(\delta^*) = 1 + \inf\{n \geq m_1 - 1 : 2knC^{-1} \geq \rho\overline{W}_n\}$$

where $\overline{W}_n = n^{-1}\Sigma_{i=1}^n W_i$, the W's being i.i.d. χ^2_{2k}, and it is easy to observe that the accelerated sequential procedure (4.6.4) – (4.6.6) agrees with (2.5.1) – (2.5.3) with $\delta = 1$, $h^* = 2k/C$, $m = m_1 - 1$, $\theta = 2k$, $\tau^2 = 4k$, $n_0 = \rho C$, $n_0^* = C$ and $p = 1/k$. Now, the rest of the proof is similar to the proof of Theorem 3.6.1. □

Remark 4.6.1. In Remark 3.6.1, we had mentioned that the accelerated sequential selection procedure was more variable than the purely sequential one in the normal problem. Once again, we note here that Var(N | Acc.seq.) is approximately $(k\rho)^{-1}$ while one has Var(N | Purely seq.) approximately k^{-1}, and of course $\rho \in (0, 1)$. In other words, a similar feature persists in the exponential problem as expected.

4.6.1 SECOND–ORDER ASYMPTOTICS

In this section, we strengthen the results given in Theorem 4.6.1 (ii) and (v) by providing expansions of (a) E(N) for all $\underset{\sim}{\theta} \in \Omega(\delta^*)$, and (b) P(CS) under the LFC, for the accelerated sequential methodology (4.6.4) – (4.6.6). Recall $q_1(\cdot)$, b_1, b_2 and $g_1(\cdot)$ from (4.5.9) – (4.5.11). Now, we have the following theorem.

Theorem 4.6.2. *For the accelerated sequential procedure (4.6.4) – (4.6.6), we have as* $\delta^* \to 0$:

(i) $q - (k\rho)^{-1} \leq \liminf E(N - C) \leq \limsup E(N - C) \leq q + 1 - (k\rho)^{-1}$ *for all* $\underset{\sim}{\theta} \in \Omega(\delta^*)$ *if* $m_1 \geq 4$;

(ii) $\mathrm{Var(N)} = (k\rho)^{-1}C + o(\delta^{*-1})$ *for all* $\underset{\sim}{\theta} \in \Omega(\delta^*)$ *if* $m_1 \geq 2$;

(iii) $P(CS) \geq P^* + o(\delta^*)$ *under the LFC if* $m_1 \geq 4$, *with* $q = \max\{0, (k\rho)^{-1}[1 + \frac{1}{2}ab_2b_1^{-1}]\}$;

where $C = a\sigma/\delta^*$ *and* a *comes from (4.2.4).*

Proof. Along the lines of Theorem 3.6.2, we again utilize the tools developed in section 2.5.1 extensively. With $\delta = 1$, $m_0(\delta)$ given in (2.5.7) indeed depends on k, and thus we write $m_0(k)$ instead and it is easy to check that (having $r' = 2$, $s = 1.01$) $m_0(k) = 2$ for all $k \geq 2$. Part (i) follows immediately from Corollary 2.5.1, for $m = m_1 - 1 > m_o(k)$. Part (ii) follows from Theorem 2.5.2 (ii) for $m = m_1 - 1 > 1/k$. In part (iii), we restrict our attention only to LFC. Now, for a suitable random variable W_j^* between ja and $jN\delta^*/\sigma$, with $U^* = C^{-\frac{1}{2}}(N - C)$ as in the proof of Theorem 4.5.2, we have

$$
\begin{aligned}
P(CS) &= E\{g_1(N\delta^*/\sigma)\} \\
&= \Sigma_{j=0}^{k-1} b(k, j)E\{q_1(jN\delta^*/\sigma)\} \\
&= \Sigma_{j=0}^{k-1} b(k, j)q_1(ja) - \delta^*\sigma^{-1}E\{N - C\}\Sigma_{j=0}^{k-1} jb(k, j)q_1(ja) \\
&+ \tfrac{1}{2} a\delta^*\sigma^{-1}\Sigma_{j=0}^{k-1} j^2 b(k, j)E\{U^{*2}q_1(W_j^*)\}. \qquad (4.6.8)
\end{aligned}
$$

Now, $U^{*2}q_1(W_j^*) \leq U^{*2}$ and U^{*2} is uniformly integrable if $m = m_1 - 1 > \max\{m_0(k), 2/k\}$, that is if $m_1 \geq 4$. See Remark 2.5.2. Since $U^* \xrightarrow{\mathscr{L}} N(0, (k\rho)^{-1})$ and $W_j^* \xrightarrow{P}$ ja as $\delta^* \to 0$, we can simplify (4.6.8) to obtain:

$$P(CS) \geq P^* + \delta^*\sigma^{-1}\{-b_1(q - [k\rho]^{-1}) + \tfrac{1}{2}ab_2(k\rho)^{-1}\} + o(\delta^*)$$

$$= P^* + o(\delta^*),$$

for $m_1 \geq 4$, if $q = \max\{0, (k\rho)^{-1}[1 + \frac{1}{2}ab_2b_1^{-1}]\}$ in view of part (i). Hence the proof is complete. □

4.7 Three–stage procedure

The three–stage selection methodology developed here reduces the sampling operations practically to the bare minimum and yet one maintains second–order asymptotic characteristics. This procedure was originally discussed in Mukhopadhyay (1987).

Again recall that $C = a\sigma/\delta^*$ and we go back to estimating σ by means of $V(n)$. In this three–stage selection procedure, we start with an initial sample of size $m_1 = m_1(\delta^*)$ from each of the k populations, where

$$m_1 = m_1(\delta^*) = o(C^{1/r}) \quad (4.7.1)$$

for some $r > 1$. This assumption is along the lines of (3.7.1). We choose and fix $0 < \rho < 1$. The idea is to first estimate ρC, a fraction of C, by means of two–stage sampling, and then estimate all of C in the third stage of sampling. In other words, this methodology consists of batch sampling all through. We now write

$$T = T(\delta^*) = \max\{m_1, [\rho a V(n)\delta^{*-1}]^* + 1\}, \quad (4.7.2)$$

$$N = N(\delta^*) = \max\{T, [aV(T)\delta^{*-1}]^* + 1\}. \quad (4.7.3)$$

In terms of implementation, we proceed as follows. We start with $X_{i1}, X_{i2},...,X_{im_1}$ from π_i, $i = 1,...,k$. If $T = m_1$, then we do not sample any more from any population at the second stage. However, if $T > m_1$, then we sample the difference at the second stage to extend $X_{i1}, X_{i2},...,X_{im_1}$ to X_{i1},

$X_{i2},\ldots,X_{iT}$ from π_i, $i = 1,\ldots,k$. Based on T samples from each population, we determine V(T) and thus we find N. If $N = T$, we do not sample any more from any population in the third stage. However, if $N > T$, then we sample the difference in the third stage to extend $X_{i1}, X_{i2},\ldots,X_{iT}$ to X_{i1}, $X_{i2},\ldots,X_{iN}$ from π_i, $i = 1,\ldots,k$. Once one has N samples from each population, one implements the customary selection procedure $\mathscr{P}^*_N$ given by (4.2.1). Throughout $\underset{\sim}{\theta}$ and σ are kept fixed.

Observe that Theorem 4.2.1 holds for the three-stage selection procedure (4.7.2) – (4.7.3). Hence, for all $\underset{\sim}{\theta} \in \Omega(\delta^*)$,

$$P(CS) \geq E[\int_0^\infty \{1 - \exp(-z-N\delta^*\sigma^{-1})\}^{k-1}\exp(-z)dz] \qquad (4.7.4)$$

as in (4.6.7), while the equality in (4.7.4) holds under the LFC which is again given by $\theta_{[1]} = \cdots = \theta_{[k-1]} = \theta_{[k]} - \delta^*$. The functions $q_1(x)$, $b(k, x)$, b_1 and b_2 are exactly same as these were in (4.5.9) – (4.5.11). First, we summarize certain first-order asymptotic results parallel to those in Theorems 4.5.1 and 4.6.1.

Theorem 4.7.1. *For the three-stage procedure (4.7.2) – (4.7.3), we have as* $\delta^* \to 0$:

(i) $N/C \to 1$ *w.p.* 1;

(ii) $E(N/C) \to 1$;

(iii) $C^{-\frac{1}{2}}(N - C) \xrightarrow{\mathscr{L}} N(0, (k\rho)^{-1})$;

(iv) $\liminf P(CS) \geq P^*$ *for all* $\underset{\sim}{\theta} \in \Omega(\delta^*)$;

(v) $\lim P(CS) = P^*$ *under the LFC, namely* $\theta_{[1]} = \cdots = \theta_{[k-1]} =$

$\theta_{[k]} - \delta^*$;

where $C = a\sigma/\delta^*$ *and* a *comes from (4.2.4).*

Proof. Along the lines of (4.5.3), note that $V(n)$ can be written as $(n-1)^{-1}\Sigma_{i=1}^{n-1} Y_i$, where Y's are i.i.d. $(2k)^{-1}\sigma\,\chi^2_{2k}$. The rest of the proof now can be given along the lines of the proof of Theorem 3.7.1. □

4.7.1 SECOND–ORDER ASYMPTOTICS

In this section, we strengthen the results discussed in Theorem 4.7.1 (ii) and (v). First, let us use the representation given in (4.5.3) to write $V(n) = (n-1)^{-1}\Sigma_{i=1}^{n-1} Y_i = \overline{Y}_n$, where Y's are i.i.d. $(2k)^{-1}\sigma\,\chi^2_{2k}$. Then, we can rewrite the three–stage procedure (4.7.2) – (4.7.3) as

$$T = T(\delta^*) = \max\{m_1, [\rho\, a\, \overline{Y}_{m_1} \delta^{*-1}]^* + 1\}, \tag{4.7.5}$$

$$N = N(\delta^*) = \max\{T, [a\, \overline{Y}_T\, \delta^{*-1}]^* + 1\}. \tag{4.7.6}$$

Theorem 4.7.2. *For the three–stage procedure (4.7.2) – (4.7.3), we have as* $\delta^* \to 0$:

(i) $E(N) = C + \beta_0' + o(1)$ *for all* $\underset{\sim}{\theta} \in \Omega(\delta^*)$;

(ii) $\mathrm{Var}(N) = (k\rho)^{-1}C + o(\delta^{*-1})$ *for all* $\underset{\sim}{\theta} \in \Omega(\delta^*)$;

(iii) $E[|N - E(N)|^3] = o(\delta^{*-2})$ *for all* $\underset{\sim}{\theta} \in \Omega(\delta^*)$;

where $C = a\sigma/\delta^*$, $\beta_0' = \frac{1}{2} - (k\rho)^{-1}$ *and* a *comes from (4.2.4).*

Proof. For parts (i) – (ii), we once again exploit Theorem 2.3.2 with $\omega = 1, 2$, $n^* = C$ and $Var(W_1) = Var(Y_1/\sigma) = 1/k$ to obtain

$$E(N) = C + [\tfrac{1}{2} - (k\rho)^{-1}] + o(1),\ E(N^2) = C^2 + [1 - (k\rho)^{-1}]\, C + o(C),$$

which imply

$$Var(N) = E(N^2) - \{E(N)\}^2 = (k\rho)^{-1}C + o(C).$$

Part (iii) follows from Hall (1981). □

We now modify the three–stage procedure (4.7.2) – (4.7.3) slightly so as to be able to conclude $P(CS) = P^* + o(\delta^*)$. So, along the lines of section 3.7.1, the modified three–stage selection procedure is defined as follows: Let

$$T = T(\delta^*) = \max\{m_1, [\rho a V(n)\, \delta^{*-1}]^* + 1\},$$

$$m_1^* = \{a\, b_2 - b_1\rho k + 2b_1\}/(2\rho k b_1),$$

$$N = N(\delta^*) = \max\{T, [aV(T)\delta^{*-1} + m_1^*]^* + 1\}. \tag{4.7.7}$$

Note that

$$E(N) = C + \beta_0' + m_1^* + o(1) \ \text{ as } \ \delta^* \to 0 \tag{4.7.8}$$

and that the moment expansions of Theorem 4.7.2 (ii) – (iii) still hold for the stopping time N of this modified selection procedure (4.7.7). Once we determine N, we implement $\mathscr{P}_N^*$ given by (4.2.1).

Theorem 4.7.3. *Let the three–stage stopping rule be defined as in (4.7.7). Then, under the LFC, we have as* $\delta^* \to 0$:

$$P(CS) = P^* + o(\delta^*).$$

Proof. From Theorem 4.2.1, we have

$$P(CS) = E[\int_0^\infty \{1 - \exp(-z-N\delta^*\sigma^{-1})\}^{k-1}\exp(-z)dz]$$

$$= \Sigma_{j=0}^{k-1} b(k, j)\, E\{q_1(jN\delta^*/\sigma)\}. \tag{4.7.9}$$

For a fixed integer $j \in \{0, 1, ..., k-1\}$, let us consider $E\{q_1(l\,N)\}$ where $l = j\delta^*/\sigma$. Now,

$$E\{q_1(l\,N)\} = q_1(l\,E(N)) + \tfrac{1}{2}l^2E\{(N-E(N))^2\}q_1''(l\,E(N)) + r_1(\delta^*), \tag{4.7.10}$$

where $|r_1(\delta^*)| \le l^3\, E\{|N - E(N)|^3\}/6 = o(\delta^*)$, by Theorem 4.7.2 (iii).

Now, again we have

$$q_1\{l\,E(N)\} = q_1(ja) + \{l\,E(N)-ja\}q_1'(ja) + r_2(\delta^*) \text{ say}, \tag{4.7.11}$$

where $r_2(\delta^*) = \frac{1}{2}\{l\,E(N)-ja\}^2\, q_1''(z)$ for a suitable positive number z.

Let us use $\lambda = \lambda(\delta^*) = a(1 + \epsilon)/\delta^*$, and with this choice, $|r_2(\delta^*)| = o(\delta^* + |\epsilon|)$. Also, we have

$$l\,E(N) - ja = j\delta^*\sigma^{-1}\{a(1+\epsilon)\sigma\delta^{*-1}-(\rho k)^{-1} + \tfrac{1}{2} + o(1)\} - ja$$

$$= ja\epsilon + j\delta^*\sigma^{-1}\{\tfrac{1}{2}-(qk)^{-1}\} + o(\delta^*). \tag{4.7.12}$$

Thus, combining (4.7.11) and (4.7.12), we get

$$q_1(l\,E(N)) = q_1(ja) - q_1(ja)\{ja\epsilon + j\delta^*\sigma^{-1}[\tfrac{1}{2}-(\rho k)^{-1}] + o(\delta^*)\}$$

$$+ o(\delta^* + |\epsilon|). \tag{4.7.13}$$

Also, we have

$$\tfrac{1}{2}\, l^2E\{(N - E(N))^2\} = \tfrac{1}{2}j^2\delta^{*2}\{a\sigma(\rho\delta^*k)^{-1} + o(\delta^{*-1})\}/\sigma^2$$

$$= a\delta^* j^2 (2\rho\sigma k)^{-1} + o(\delta^*), \tag{4.7.14}$$

and this follows from Theorem 4.7.2 (ii). Thus (4.7.13), (4.7.14) and (4.7.10) together lead to

$$E\{\exp(-jN\delta^*/\sigma)\} = q_1(ja)\{ja\epsilon + j\delta^*\sigma^{-1}[\tfrac{1}{2} - (\rho k)^{-1}] + o(\delta^*)\}$$

$$+ [a\delta^* j^2 (2\rho\sigma k)^{-1} + o(\delta^*)]q_1(ja) + o(\delta^* + |\epsilon|). \tag{4.7.15}$$

Now, combining (4.7.10) and (4.7.15), we obtain

$$P(CS) = \Sigma_{j=0}^{k-1} b(k,j)q_1(ja) - \Sigma_{j=0}^{k-1} b(k,j)q_1(ja)\{ja\epsilon + j\delta^*\sigma^{-1}[\tfrac{1}{2} - (\rho k)^{-1}]$$

$$- a\delta^* j^2 (2\rho\sigma k)^{-1}\} + o(\delta^*) + o(\delta^* + |\epsilon|)$$

$$= P^* + \{ab_1\epsilon + b_1\delta^*\sigma^{-1}[\tfrac{1}{2} - (\rho k)^{-1}] - ab_2\delta^*(2\rho\sigma k)^{-1}\}$$

$$+ o(\delta^*) + o(\delta^* + |\epsilon|). \tag{4.7.16}$$

In order to make the expression after P^* in (4.7.16) vanish, we choose $\epsilon = \delta^*\{ab_2(2\rho k)^{-1} - b_1[\tfrac{1}{2} - (\rho k)^{-1}]\}\,(ab_1\sigma)^{-1}$. With this ϵ, we have

$$\lambda\sigma = a(1 + \epsilon)\sigma/\delta^* = C + a\sigma\epsilon\delta^{*-1},$$

which is $C + m_1^*$. Combining (4.7.16), (4.7.7) and (4.7.3), we complete the proof of this theorem. □

4.8 Moderate sample performances and comments

Along the lines of section 3.8, here we address the question of "validity" of asymptotic results for moderate samples for the two–stage and other multistage selection procedures. We will tackle the problem of choosing a suitable m_1, by considering a couple of m_1 values and examining any visible effects on the moderate sample performances of our selection methodologies. We will also recommend choices for a value of ρ in the case of accelerated sequential procedure (4.6.4) – (4.6.6) and the three–stage procedure (4.7.7). But, again, one should note that there is no available optimality criteria as such in order to pick the best ρ, say. Hence, we tried working with various values of ρ and examined any visible effects or marks on these procedures. We accomplish such investigations and comparisons via extensive sets of computer simulations which we explain and summarize here.

Theorems 4.5.2, 4.6.2, 4.7.2 and 4.7.3 provide second–order expansions of $E(N) - C$ and $P(CS) - P^*$ for the purely sequential, accelerated sequential and three–stage selection methodologies. Once again, it is important to see the "closeness" of estimates of $E(N) - C$ or $P(CS) - P^*$ from large scale simulation studies to their respective asymptotic values provided by these theorems. In this sense, the asymptotic values provide some practical guidelines and correction terms.

We first provide Tables 4.8.1 – 4.8.6 in order to help evaluate the correction terms for $E(N) - C$ and $P(CS) - P^*$ for the purely sequential procedure (4.5.1), accelerated sequential procedure (4.6.4) – (4.6.6) and the three–stage procedure (4.7.7). These tables also help in computing the final sample size for the three–stage procedure and the term q that is needed in the implementation of the accelerated sequential procedure (4.6.4) – (4.6.6). After this, we include Tables 4.8.7 – 4.8.9 which provide simulated results of all the multistage selection procedures considered in this chapter. These tables also include the asymptotic values of $E(N - C)$ and $P(CS) - P^*$, whenever appropriate.

In Table 4.8.1a,b we provide the values of $\tau_0 = \tau_0(m_1, k, P^*)$ needed to implement the two–stage procedure (4.3.1) for $m_1 = 2(1)30, 40, 50$, $k = 2(1)10$ and $P^* = .90, .95$ respectively. In order to obtain these τ_0 values, we wrote programs in Fortran. The values of τ_0, correct up to five decimal places, are provided in these tables.

In Table 4.8.2, we provide the value of "a" given by (4.2.4) for $k = 2(1)30$ and $P^* = .90, .95$ and .99. To compute the values of "a" we use the equivalent form of (4.2.4) given by

$$\frac{e^a}{k}\{1-(1-e^{-a})^k\} = P^*. \tag{4.8.1}$$

In order to obtain these values, we again wrote programs in Fortran and we add here that these "a" values provide P^* correct up to six decimal places. Note that the values of "a" are also available in Raghavachari and Starr (1970) for $k = 1(1)15$ and $P^* = .90, .95, .975$ and $.99$. Although the values in Table 4.8.2 and in Raghavachari and Starr (1970) do not match exactly up to six decimal places, they both provide values of P^* correct up to six decimal places.

In Tables 4.8.3, we provide the values of $\nu(k)$ given by (4.5.8) and $\beta''(k) = k^{-1}\{\frac{1}{2}\nu(k) - 1\}$ for $k = 1(1)30$. In order to obtain these, we wrote a program in Fortran and used IMSL subroutine DGAMIC to compute the incomplete gamma function. Any term bigger than 10^{-15} in magnitude, was included in the infinite sum in (4.5.8). It is intuitively clear that as $k \to \infty$, $\beta''(k) \to \frac{1}{2}$. However, as noted in Table 3.8.1, one can see here also by looking at the column of $\beta''(k)$ values that the convergence rate is quite slow. Recall, from Theorem 4.5.2 (i), that the asymptotic value of $E(N - C)$ is $\beta''(k)$ for the purely sequential procedure (4.5.1). Hence, we expect the average oversampling in comparison with C to be close to $\beta''(k)$ and we will see later in Table 4.8.9 that such a feature is noticeable through the simulation studies carried out for the purely sequential procedure (4.5.1), when C is moderate in size.

In Table 4.8.4, we provide the values of b_1, b_2 and $A' = \frac{1}{2}ab_2k^{-1} - b_1\beta''$ for $P^* = .90, .95$ and $.99$, for $k = 2(1)30$, where $\beta'' = \beta''(k) = k^{-1}\{\frac{1}{2}\nu(k)-1\}$ and $\nu(k)$ comes from (4.5.8). The values of b_1 and b_2 are needed in order to evaluate q (Theorem 4.6.2 (iii)) in the case of accelerated sequential procedure (4.6.4) – (4.6.6) and also in computing m_1^* for the three–stage procedure (4.7.7). One observes that in Theorem 4.5.2 (ii), for the purely sequential procedure (4.5.1), the asymptotic value of $P(CS) - P^*$ equals $aC^{-1}A'$. In Table 4.8.9, we use the asymptotic value of $P(CS) - P^*$ to compare with the estimated values of $P(CS) - P^*$, for the moderate sample sizes. To compute b_1, b_2 and A', we wrote programs in Fortran and used IMSL subroutine DFAC to evaluate various factorials.

In Table 4.8.5, we provide the values of q given in Theorem 4.6.2 (iii) for $\rho = .25, .50, .75$; $P^* = .90, .95, .99$ and $k = 2(1)30$. We wrote the programs in Fortran using the IMSL subroutine DFAC to evaluate the factorials, for these q values. Such values of q are needed in the implementation of the accelerated sequential procedure (4.6.4) – (4.6.6). Also, we use these to evaluate the lower and upper bounds for the asymptotic values of $E(N - C)$ in the case of the accelerated sequential procedure as seen in Theorem 4.6.2 (i).

In Table 4.8.6, we present the values of m_1^* for $\rho = .25, .50, .75$; $P^* = .90, .95, .99$ and $k = 2(1)30$. Again, in order to compute m_1^*, we wrote programs in Fortran using IMSL subroutine DFAC. The values of m_1^* will be needed for implementing the three–stage procedure (4.7.7) and also to evaluate the asymptotic value of $E(N - C)$ for the procedure (4.7.7).

We now elaborate on the moderate sample size performances of the two–stage procedure (4.3.1), modified two–stage procedure (4.4.1) – (4.4.2), purely sequential procedure (4.5.1), accelerated sequential procedure (4.6.4) – (4.6.6) and the three–stage procedure (4.7.7). The VAX 11/780 mainframe computer at the University of New Orleans and programs written in Fortran are used to simulate all the selection procedures one by one. We consider $C = 25, 50, 100, 200, 400$ and 800. Recall that C is the optimal fixed sample size (if the common scale parameter σ were known) one would need from each of the k populations to select the best population via $\mathscr{P}_n^*$ given by (4.2.1). As in section 3.8 and 3.9, when we simulate any of the selection procedures, we put in a value of σ in order to generate the populations but pretend that σ, and hence C, is unknown and then we compare the simulated estimators of $E(N)$ and $P(CS)$ respectively with the input values of C and P^*. For each value of C we repeat this for each procedure J $(= 5000)$ times. Each time we stop sampling, we also check to see if the correct population was selected or not. To begin with, note that we used IMSL (1989, Edition 10) subroutine RNEXP to generate $(k - 1)$ pseudo negative exponential populations each with location parameter $= 5$ and scale parameter $\sigma = 1$, and another pseudo negative exponential population with location parameter $= 5 + \delta^*$ and $\sigma = 1$, where $\delta^* = a\,C^{-1}$. We consider $k = 5, 10$ and $P^* = .95$ for simulation purposes. For a fixed set of values of k, m_1 and C, we implement each sampling procedure by repeating it independently J times. Along the lines of our discussions in section 3.8, let n_j be the value of N and $p_j = 1(0)$ if the correct population, that is the one with location parameter $5 + \delta^*$, is (is not) selected at the end of the jth iteration, $j = 1, 2, \ldots, J$. Let $\bar{n} = J^{-1}\,\Sigma_{j=1}^{J} n_j$, $s(\bar{n}) = \{(J^2 - J)^{-1}\,\Sigma_{j=1}^{J}(n_j - \bar{n})^2\}^{\frac{1}{2}}$, $\bar{p} = J^{-1}\,\Sigma_{j=1}^{J} p_j$ and $s(\bar{p}) = \{\bar{p}(1 - \bar{p})/J\}^{\frac{1}{2}}$. Then $\bar{n}$ and $\bar{p}$ estimate C and $P(CS)$ respectively with $s(\bar{n})$ and $s(\bar{p})$ estimating their respective standard errors.

In Tables 4.8.7 – 4.8.9, we provide various columns showing the values of k, δ^*, C, $\bar{n} - C$ and $\bar{p}$. Also, underneath each value of $\bar{n} - C$ and $\bar{p}$, we provide their respective standard errors $s(\bar{n})$ and $s(\bar{p})$ within parenthesis. Again, whenever available, we provide the asymptotic values of $\bar{n} - C$ and $\bar{p}$ within triangular brackets $< >$ at the end of appropriate columns.

Tables 4.8.7a,b,c provide simulation results for the two–stage procedure (4.3.1). We include the values of $m_1 = 10, 20, 30$ for the starting sample size, and $C = 25, 50, 100, 200, 400$ and 800. Recall, that $\bar{n}$ is the average of $J(= 5000)$ simulated estimates of C, while pretending that σ is unknown. In columns 3 and 6 we provide the values of $\bar{n} - C$, the amount by which the two–stage procedure (4.3.1) "oversamples" compared with C. One will note that this oversampling is quite significant in terms of magnitude. For example, in Table 4.8.7a with $m_1 = 10$, $k = 5$ and $C = 200$ this oversampling amounts to about 15.7% of C. One notes that the amount of oversampling keeps increasing for fixed k; however, the magnitudes of oversampling come down as k increases and m_1 is held fixed. This can once again be explained via an increase in degrees of freedom $(= 2k(m_1 - 1))$ for $V(m_1)$ in (4.3.1) while estimating σ in the expression of C. Underneath the columns 3 and 6, we provide the asymptotic value b^*/a for $\bar{n}/C$ and one will observe that the values of $\bar{n}/C$ are approximated reasonably by means of

b^*/a, where $b^* = 2\tau_0 k(m_1 - 1)$, for moderate values of C. For example, when $C = 200$, $\bar{n}/C$ is 1.16 and b^*/a is 1.04, for $m_1 = 10$, $P^* = .95$, $k = 5$. One also observes from Tables 4.8.7a,b and c that as m_1 increases, this "oversampling" goes down. However, very large values of m_1 are quite impractical to use. Also, for fixed k, whatever be m_1, the values of $(\bar{n} - C)$ do increase steadily as C increases, that is as δ^* decreases, and such an observation is quite in harmony with equation (4.3.6). The procedure is indeed consistent (Theorem 4.3.1(i)) and we do notice that $\bar{p}$ values are a little larger than the target value P^*, *across the board*, for all values of C.

In Tables 4.8.8a,b,c, we present a summary of our findings in connection with moderate sample performances of the modified two–stage procedure (4.4.1) – (4.4.2). We first fix $m_0^* = 10$ and $P^* = .95$. Next, we need to specify $\gamma \in R^+$ in a certain suitable fashion. Note that $C = a\sigma/\delta^* \geq a/\delta^*$ when $\sigma \geq 1$ and also m_1 given by (4.4.1) will be approximated (for small δ^*) by $(a/\delta^*)^{\frac{1}{\gamma+1}}$. In other words, C is expected to be larger than m_1, as long as $\sigma \geq 1$, whatever $\gamma \in R^+$. That is, the stopping variable N given by (4.4.2) will be essentially $b^*V(m_1)/\delta^*$, whatever m_0^* and $\gamma \in R^+$. So, how one should choose γ will particularly depend on the moderate sample performances observed when σ is smaller than unity. We tried $\sigma = .25, .50, .75$ and 1.0. Also, as in the case of section 3.8, it is desirable to have m_1 as large as practically possible without going over C, having σ unknown. The experimenter would like to have $\bar{n}$ values close to C, that is the first–order efficiency property. For small δ^*, this translates into picking small values of γ. In our simulations, we considered $\gamma = .1(.1).9$ and kept fixed $k = 5, 10$ with a particular value of C, which in turn provided $\delta^* = a\sigma/C$. For $\sigma = .25, .50$ and .75, the choice of γ that gives rise to "small" magnitudes of $\bar{n} - C$ as well as close proximity between $\bar{p}$ and P^* turned out to be either .2, .3 or .4. These were also associated with fairly stable and reasonable estimated

Table 4.8.1a. τ_0 Values from (4.3.3), $P^* = 0.90$

m_1	k = 2	3	4	5	6	7	8	9	10
2	.61831	.54219	.45054	.38233	.33195	.29363	.26360	.23945	.21960
3	.24777	.22570	.19341	.16774	.14797	.13249	.12009	.10994	.10148
4	.15388	.14174	.12261	.10706	.09493	.08534	.07759	.07121	.06587
5	.11146	.10320	.08967	.07857	.06984	.06290	.05728	.05264	.04875
6	.08734	.08111	.07067	.06204	.05523	.04980	.04539	.04175	.03868
7	.07179	.06680	.05830	.05125	.04567	.04121	.03759	.03459	.03206
8	.06093	.05678	.04962	.04366	.03893	.03515	.03207	.02952	.02738
9	.05293	.04937	.04318	.03802	.03392	.03064	.02797	.02575	.02389
10	.04678	.04367	.03822	.03367	.03006	.02716	.02480	.02284	.02118
11	.04191	.03915	.03429	.03022	.02698	.02439	.02227	.02051	.01903
12	.03796	.03548	.03108	.02740	.02448	.02213	.02021	.01862	.01728
13	.03469	.03243	.02843	.02507	.02240	.02025	.01850	.01704	.01582
14	.03194	.02987	.02619	.02310	.02064	.01867	.01705	.01572	.01458
15	.02959	.02768	.02428	.02142	.01914	.01731	.01582	.01458	.01353
16	.02756	.02579	.02263	.01997	.01785	.01614	.01475	.01359	.01262
17	.02580	.02415	.02119	.01870	.01672	.01512	.01382	.01274	.01182
18	.02424	.02270	.01992	.01758	.01572	.01422	.01300	.01198	.01112
19	.02287	.02141	.01879	.01659	.01483	.01342	.01227	.01131	.01050
20	.02164	.02026	.01779	.01571	.01404	.01271	.01161	.01071	.00994
21	.02053	.01923	.01689	.01491	.01333	.01207	.01103	.01017	.00944
22	.01954	.01830	.01607	.01419	.01269	.01149	.01050	.00968	.00899
23	.01863	.01746	.01533	.01354	.01211	.01096	.01002	.00923	.00857
24	.01781	.01669	.01466	.01294	.01158	.01048	.00958	.00883	.00820
25	.01705	.01598	.01404	.01240	.01109	.01004	.00918	.00846	.00785
26	.01636	.01533	.01347	.01190	.01064	.00963	.00881	.00812	.00754
27	.01572	.01473	.01295	.01144	.01023	.00926	.00846	.00780	.00725
28	.01513	.01418	.01246	.01101	.00985	.00891	.00815	.00751	.00698
29	.01458	.01367	.01201	.01061	.00949	.00859	.00786	.00724	.00673
30	.01407	.01319	.01159	.01024	.00916	.00829	.00758	.00699	.00649
40	.01043	.00978	.00860	.00760	.00680	.00616	.00563	.00519	.00482
50	.00823	.00777	.00683	.00604	.00541	.00489	.00448	.00413	.00383

Table 4.8.1b. τ_0 Values from (4.3.3), $P^* = 0.95$

m_1	k = 2	3	4	5	6	7	8	9	10
2	1.08113	.82087	.63845	.52098	.44032	.38190	.33768	.30306	.27519
3	.38936	.31739	.25900	.21819	.18872	.16654	.14926	.13541	.12406
4	.23402	.19471	.16123	.13720	.11954	.10608	.09549	.08694	.07989
5	.16684	.14016	.11687	.09994	.08739	.07777	.07016	.06399	.05888
6	.12953	.10941	.09161	.07857	.06885	.06137	.05543	.05061	.04662
7	.10581	.08971	.07531	.06472	.05679	.05067	.04581	.04186	.03858
8	.08943	.07601	.06393	.05501	.04832	.04315	.03904	.03569	.03290
9	.07743	.06593	.05554	.04784	.04205	.03757	.03401	.03110	.02868
10	.06826	.05821	.04909	.04232	.03722	.03327	.03013	.02756	.02542
11	.06104	.05211	.04398	.03794	.03339	.02985	.02704	.02474	.02283
12	.05519	.04717	.03984	.03438	.03027	.02707	.02453	.02244	.02071
13	.05037	.04308	.03641	.03143	.02768	.02477	.02244	.02054	.01895
14	.04632	.03964	.03352	.02895	.02550	.02282	.02068	.01893	.01747
15	.04287	.03671	.03106	.02683	.02364	.02116	.01918	.01756	.01621
16	.03990	.03418	.02893	.02500	.02203	.01972	.01788	.01637	.01511
17	.03732	.03198	.02708	.02340	.02063	.01847	.01674	.01533	.01415
18	.03505	.03005	.02545	.02200	.01939	.01736	.01574	.01442	.01331
19	.03304	.02834	.02400	.02075	.01830	.01639	.01486	.01361	.01256
20	.03125	.02681	.02271	.01964	.01732	.01551	.01406	.01288	.01189
21	.02964	.02543	.02155	.01864	.01644	.01472	.01335	.01223	.01129
22	.02819	.02419	.02050	.01774	.01565	.01401	.01271	.01164	.01075
23	.02687	.02307	.01956	.01692	.01492	.01337	.01213	.01111	.01026
24	.02568	.02205	.01869	.01617	.01427	.01278	.01159	.01062	.00981
25	.02458	.02111	.01790	.01549	.01366	.01224	.01110	.01017	.00940
26	.02357	.02025	.01717	.01486	.01311	.01175	.01066	.00976	.00902
27	.02265	.01946	.01650	.01428	.01260	.01129	.01024	.00938	.00867
28	.02179	.01872	.01588	.01375	.01213	.01087	.00986	.00903	.00834
29	.02100	.01804	.01531	.01325	.01169	.01048	.00950	.00871	.00804
30	.02026	.01741	.01477	.01279	.01128	.01011	.00917	.00840	.00776
40	.01498	.01289	.01094	.00948	.00837	.00750	.00681	.00624	.00576
50	.01189	.01023	.00869	.00753	.00665	.00596	.00541	.00496	.00458

Table 4.8.2. Values of a Given by (4.2.4)

	P*		
k	.90	.95	.99
2	1.609429	2.302578	3.911920
3	2.267439	2.978621	4.601503
4	2.661198	3.378388	5.005646
5	2.943026	3.663203	5.292547
6	3.162655	3.884670	5.515884
7	3.342637	4.065839	5.697729
8	3.495114	4.219172	5.851452
9	3.627389	4.352083	5.984557
10	3.744201	4.469407	6.102911
11	3.848780	4.574353	6.207800
12	3.943443	4.669358	6.303577
13	4.029920	4.756137	6.391250
14	4.109510	4.835936	6.470261
15	4.183227	4.909784	6.544211
16	4.251905	4.978700	6.611172
17	4.316142	5.043042	6.677052
18	4.376516	5.103562	6.736930
19	4.433424	5.160591	6.793301
20	4.487303	5.214593	6.849373
21	4.538390	5.265777	6.898064
22	4.587029	5.314449	6.949203
23	4.633398	5.361064	6.994397
24	4.677700	5.405472	7.040273
25	4.720122	5.447787	7.081763
26	4.760845	5.488545	7.121935
27	4.799939	5.527653	7.160922
28	4.837596	5.565538	7.196422
29	4.873846	5.601715	7.235355
30	4.908871	5.636982	7.271269

a: equivalently given by (4.8.1)

Table 4.8.3. $\nu(k)$ and $\beta''(k)$ Values in Theorem 4.5.2

k	$\beta''(k)$	$\nu(k)$
1	−0.25500	1.49000
2	0.17159	2.68634
3	0.29915	3.79489
4	0.35769	4.86155
5	0.39047	5.90474
6	0.41113	6.93362
7	0.42524	7.95334
8	0.43544	8.96699
9	0.44314	9.97653
10	0.44916	10.98326
11	0.45400	11.98803
12	0.45798	12.99141
13	0.46130	13.99383
14	0.46413	14.99556
15	0.46656	15.99680
16	0.46868	16.99770
17	0.47054	17.99834
18	0.47219	18.99880
19	0.47366	19.99913
20	0.47498	20.99937
21	0.47618	21.99955
22	0.47727	22.99967
23	0.47826	23.99976
24	0.47916	24.99983
25	0.48000	25.99987
26	0.48077	26.99991
27	0.48148	27.99993
28	0.48214	28.99995
29	0.48276	29.99997
30	0.48333	30.99998

$\nu(k)$: defined in (4.5.8)

$\beta''(k) = k^{-1}\{\frac{1}{2}\,\nu(k)-1\}$

Table 4.8.4. Values of b_1, b_2 and A′

k	P* = .90 b_1	P* = .90 b_2	P* = .90 A′	P* = .95 b_1	P* = .95 b_2	P* = .95 A′	P* = .99 b_1	P* = .99 b_2	P* = .99 A′
2	–.100005	–.100005	–.003010	–.050005	–.050005	–.010169	–.010005	–.010005	–.006059
3	–.096420	–.089268	–.008328	–.049137	–.047413	–.010589	–.009967	–.009900	–.004966
4	–.095290	–.086039	–.009020	–.048855	–.046589	–.009625	–.009956	–.009867	–.004126
5	–.094732	–.084480	–.008820	–.048717	–.046186	–.008668	–.009949	–.009849	–.003528
6	–.094402	–.083564	–.008409	–.048635	–.045947	–.007859	–.009945	–.009838	–.003088
7	–.094181	–.082960	–.007966	–.048578	–.045786	–.007189	–.009944	–.009833	–.002752
8	–.094023	–.082530	–.007543	–.048541	–.045674	–.006631	–.009944	–.009830	–.002486
9	–.093909	–.082213	–.007155	–.048512	–.045590	–.006160	–.009943	–.009826	–.002271
10	–.093818	–.081965	–.006803	–.048487	–.045523	–.005758	–.009938	–.009820	–.002092
11	–.093745	–.081768	–.006485	–.048467	–.045468	–.005411	–.009939	–.009819	–.001942
12	–.093687	–.081609	–.006198	–.048454	–.045428	–.005109	–.009938	–.009816	–.001814
13	–.093636	–.081474	–.005937	–.048438	–.045389	–.004842	–.009939	–.009817	–.001703
14	–.093594	–.081361	–.005700	–.048431	–.045362	–.004605	–.009938	–.009815	–.001606
15	–.093562	–.081267	–.005483	–.048420	–.045334	–.004393	–.009935	–.009811	–.001519
16	–.093531	–.081183	–.005284	–.048415	–.045314	–.004202	–.009964	–.009839	–.001446
17	–.093505	–.081110	–.005101	–.048411	–.045297	–.004029	–.009950	–.009825	–.001377
18	–.093480	–.081046	–.004932	–.048404	–.045279	–.003871	–.009957	–.009831	–.001316
19	–.093461	–.080990	–.004776	–.048399	–.045264	–.003727	–.009964	–.009837	–.001260
20	–.093441	–.080938	–.004630	–.048393	–.045249	–.003594	–.009944	–.009817	–.001208

Table 4.8.4 Continued

k	P* = .90 b_1	P* = .90 b_2	P* = .90 A′	P* = .95 b_1	P* = .95 b_2	P* = .95 A′	P* = .99 b_1	P* = .99 b_2	P* = .99 A′
21	−.093425	−.080893	−.004494	−.048389	−.045237	−.003472	−.009969	−.009842	−.001163
22	−.093409	−.080851	−.004367	−.048387	−.045228	−.003359	−.009946	−.009819	−.001118
23	−.093394	−.080813	−.004248	−.048376	−.045211	−.003253	−.009958	−.009831	−.001080
24	−.093383	−.080779	−.004137	−.048372	−.045201	−.003155	−.009944	−.009817	−.001042
25	−.093372	−.080748	−.004032	−.048377	−.045199	−.003064	−.009954	−.009826	−.001009
26	−.093360	−.080718	−.003932	−.048374	−.045192	−.002978	−.009961	−.009832	−.000978
27	−.093351	−.080692	−.003839	−.048375	−.045187	−.002898	−.009963	−.009834	−.000948
28	−.093341	−.080667	−.003750	−.048364	−.045173	−.002822	−.009985	−.009855	−.000922
29	−.093334	−.080645	−.003666	−.048368	−.045172	−.002750	−.009959	−.009830	−.000894
30	−.093325	−.080623	−.003586	−.048356	−.045158	−.002683	−.009951	−.009822	−.000869

$b_i = \Sigma_{x=0}^{k-1} x^i b(k,x) q_1(xa)$, i = 1,2 where $q_1(x) = \exp(-x)I(x > 0)$, from (4.5.10)

$b(k,x) = \binom{k-1}{x}(-1)^x(x+1)^{-1}$, from (4.5.9)

$A' = \frac{1}{2} a b_2 k^{-1} - b_1\beta''$, $\beta'' = k^{-1}\{\frac{1}{2}\nu(k) - 1\}$, $\nu(k)$ from Table 4.8.3.

Table 4.8.5. Values of q from Theorem 4.6.2(iii).

	P*=.90			P*=.95			P*=.99		
k	ρ=.25	ρ=.50	ρ=.75	ρ=.25	ρ=.50	ρ=.75	ρ=.25	ρ=.50	ρ=.75
2	3.609393	1.804696	1.203131	4.302495	2.151248	1.434165	5.911523	2.955761	1.970508
3	2.732881	1.366440	.910960	3.249396	1.624698	1.083132	4.380524	2.190262	1.460175
4	2.201432	1.100716	.733810	2.610856	1.305428	.870285	3.480609	1.740305	1.160203
5	1.849826	.924913	.616609	2.189165	1.094582	.729721	2.895999	1.448000	.965333
6	1.599860	.799930	.533287	1.889969	.944984	.629990	2.485659	1.242829	.828553
7	1.412677	.706339	.470892	1.666332	.833166	.555444	2.181274	1.090637	.727091
8	1.266979	.633490	.422326	1.492501	.746250	.497500	1.946182	.973091	.648727
9	1.150139	.575069	.383380	1.353325	.676662	.451108	1.758934	.879467	.586311
10	1.054240	.527120	.351413	1.239233	.619616	.413078	1.606149	.803075	.535383
11	.974017	.487009	.324672	1.143901	.571950	.381300	1.478850	.739425	.492950
12	.905841	.452921	.301947	1.062954	.531477	.354318	1.371189	.685594	.457063
13	.847155	.423577	.282385	.993354	.496677	.331118	1.278797	.639398	.426266
14	.796058	.398029	.265353	.932782	.466391	.310927	1.198670	.599335	.399557
15	.751134	.375567	.250378	.879600	.439800	.293200	1.128496	.564248	.376165
16	.711321	.355660	.237107	.832477	.416239	.277492	1.066026	.533013	.355342
17	.675768	.337884	.225256	.790429	.395215	.263476	1.010942	.505471	.336981
18	.643818	.321909	.214606	.752673	.376337	.250891	.961305	.480652	.320435
19	.614931	.307466	.204977	.718558	.359279	.239519	.916530	.458265	.305510
20	.588688	.294344	.196229	.687582	.343791	.229194	.876231	.438115	.292077

Table 4.8.5 Continued

k	P*=.90 ρ=.25	P*=.90 ρ=.50	P*=.90 ρ=.75	P*=.95 ρ=.25	P*=.95 ρ=.50	P*=.95 ρ=.75	P*=.99 ρ=.25	P*=.99 ρ=.50	P*=.99 ρ=.75
21	.564724	.282362	.188241	.659311	.329656	.219770	.839039	.419520	.279680
22	.542758	.271379	.180919	.633403	.316702	.211134	.805490	.402745	.268497
23	.522540	.261270	.174180	.609592	.304796	.203197	.774320	.387160	.258107
24	.503864	.251932	.167955	.587593	.293797	.195865	.745826	.372913	.248609
25	.486558	.243279	.162186	.567196	.283598	.189065	.719249	.359624	.239750
26	.470477	.235238	.156826	.548265	.274132	.182755	.694619	.347310	.231540
27	.455485	.227743	.151828	.530622	.265311	.176874	.671731	.335865	.223910
28	.441481	.220741	.147160	.514167	.257084	.171389	.650219	.325109	.216740
29	.428361	.214181	.142787	.498731	.249365	.166244	.630455	.315228	.210152
30	.416050	.208025	.138683	.484277	.242139	.161426	.611801	.305900	.203934

$q = \max\{0, (k\rho)^{-1}[1 + \frac{1}{2}ab_2b_1^{-1}]\}$

b_1, b_2: Table 4.8.4

Table 4.8.6. Values of m_1^* from (4.7.7)

k	P*=.90 ρ=.25	P*=.90 ρ=.50	P*=.90 ρ=.75	P*=.95 ρ=.25	P*=.95 ρ=.50	P*=.95 ρ=.75	P*=.99 ρ=.25	P*=.99 ρ=.50	P*=.99 ρ=.75
2	3.109393	1.304696	.703131	3.802495	1.651248	.934165	5.411523	2.455761	1.470508
3	2.232881	.866440	.410960	2.749396	1.124698	.583132	3.880524	1.690262	.960175
4	1.701432	.600716	.233810	2.110856	.805428	.370285	2.980609	1.240305	.660203
5	1.349826	.424913	.116609	1.689165	.594582	.229722	2.395999	.948000	.465333
6	1.099860	.299930	.033287	1.389969	.444984	.129989	1.985659	.742829	.328553
7	.912677	.206339	–.029108	1.166332	.333166	.055444	1.681274	.590637	.227091
8	.766979	.133490	–.077674	.992500	.246250	–.002500	1.446182	.473091	.148727
9	.650139	.075069	–.116620	.853325	.176662	–.048892	1.258934	.379467	.086311
10	.554240	.027120	–.148587	.739232	.119616	–.086923	1.106149	.303075	.035383
11	.474017	–.012991	–.175328	.643901	.071950	–.118700	.978850	.239425	–.007050
12	.405841	–.047079	–.198053	.562954	.031477	–.145682	.871189	.185594	–.042937
13	.347155	–.076423	–.217615	.493354	–.003323	–.168882	.778797	.139398	–.073734
14	.296058	–.101971	–.234647	.432782	–.033609	–.189073	.698670	.099335	–.100443
15	.251134	–.124433	–.249622	.379600	–.060200	–.206800	.628496	.064248	–.123835
16	.211321	–.144340	–.262893	.332477	–.083761	–.222508	.566026	.033013	–.144658
17	.175768	–.162116	–.274744	.290429	–.104785	–.236524	.510942	.005471	–.163019
18	.143818	–.178091	–.285394	.252673	–.123663	–.249109	.461305	–.019348	–.179565
19	.114931	–.192534	–.295023	.218558	–.140721	–.260481	.416530	–.041735	–.194490
20	.088688	–.205656	–.303771	.187582	–.156209	–.270806	.376231	–.061885	–.207923

Table 4.8.6 Continued

k	P*=.90 ρ=.25	P*=.90 ρ=.50	P*=.90 ρ=.75	P*=.95 ρ=.25	P*=.95 ρ=.50	P*=.95 ρ=.75	P*=.99 ρ=.25	P*=.99 ρ=.50	P*=.99 ρ=.75
21	.064724	−.217638	−.311759	.159311	−.170344	−.280230	.339039	−.080480	−.220320
22	.042758	−.228621	−.319081	.133403	−.183298	−.288866	.305490	−.097255	−.231503
23	.022540	−.238730	−.325820	.109592	−.195204	−.296803	.274320	−.112840	−.241893
24	.003864	−.248068	−.332045	.087594	−.206203	−.304136	.245826	−.127087	−.251391
25	−.013442	−.256721	−.337814	.067196	−.216402	−.310935	.219249	−.140376	−.260250
26	−.029523	−.264762	−.343174	.048264	−.225868	−.317245	.194619	−.152690	−.268460
27	−.044515	−.272257	−.348172	.030622	−.234689	−.323126	.171731	−.164135	−.276090
28	−.058519	−.279259	−.352840	.014167	−.242916	−.328611	.150219	−.174891	−.283260
29	−.071639	−.285819	−.357213	−.001269	−.250635	−.333756	.130455	−.184772	−.289848
30	−.083950	−.291975	−.361317	−.015723	−.257861	−.338574	.111801	−.194100	−.296066

$m_1^* = \{ab_2 - b_1\rho k + 2b_1\}/(2\rho k b_1)$

b_1, b_2 : Table 4.8.4

Table 4.8.7a. Two–Stage Procedure (4.3.1) Under LFC: $m_1 = 10$, $P^* = .95$

		k=5			k=10	
C	δ^*	$\bar{n}$–C (s($\bar{n}$)) < >[1]	$\bar{p}$ (s($\bar{p}$))	δ^*	$\bar{n}$–C (s($\bar{n}$)) < >[1]	$\bar{p}$ (s($\bar{p}$))
25	.147	4.40 (.06)	.966 (.003)	.179	4.00 (.05)	.970 (.002)
50	.073	8.47 (.13)	.964 (.003)	.089	7.49 (.09)	.969 (.002)
100	.037	16.02 (.25)	.966 (.003)	.045	14.32 (.19)	.966 (.003)
200	.018	31.38 (.48)	.966 (.003)	.022	27.87 (.39)	.965 (.003)
400	.009	60.82 (.81)	.966 (.003)	.011	55.22 (.72)	.966 (.003)
800	.005	98.72 (1.39)	.966 (.003)	.006	89.71 (1.19)	.966 (.003)
		<1.04>			<1.02>	

1: The values of b^*/a; Theorem 4.3.1(iii), $b^* = 2\tau_0 k(m_1 - 1)$, τ_o from Table 4.8.1b

Note: $\bar{n}/C$ is to be compared with the asymptotic values in < >

Table 4.8.7b. Two–Stage Procedure (4.3.1) Under LFC: $m_1 = 20$, $P^* = .95$

	k=5			k=10		
C	δ^*	$\bar{n}$–C ($s(\bar{n})$) $< >^1$	$\bar{p}$ ($s(\bar{p})$)	δ^*	$\bar{n}$–C ($s(\bar{n})$) $< >^1$	$\bar{p}$ ($s(\bar{p})$)
25	.147	5.19 (.07)	.971 (.002)	.179	5.07 (.06)	.970 (.002)
50	.073	8.69 (.13)	.966 (.003)	.089	8.13 (.09)	.966 (.003)
100	.037	15.15 (.25)	.965 (.003)	.045	13.97 (.18)	.965 (.003)
200	.018	29.21 (.47)	.960 (.003)	.022	26.55 (.39)	.966 (.003)
400	.009	56.72 (.80)	.963 (.003)	.011	52.03 (.71)	.966 (.003)
800	.005	91.15 (1.35)	.966 (.003)	.006	81.24 (1.16)	.966 (.003)
		$<1.02>$			$<1.01>$	

1: The values of b^*/a; Theorem 4.3.1(iii), $b^* = 2\tau_0 k(m_1 - 1)$, τ_0 from Table 4.8.1b

Note: $\bar{n}/C$ is to be compared with the asymptotic values in $< >$

Table 4.8.7c. Two–Stage Procedure (4.3.1) Under LFC: $m_1 = 30$, $P^* = .95$

C	k=5 δ^*	k=5 $\bar{n}/C$ $(s(\bar{n}))$ < >[1]	k=5 $\bar{p}$ $(s(\bar{p}))$	k=10 δ^*	k=10 $\bar{n}/C$ $(s(\bar{n}))$ < >[1]	k=10 $\bar{p}$ $(s(\bar{p}))$
25	.147	7.14 (.07)	.973 (.002)	.179	7.01 (.06)	.971 (.002)
50	.073	9.15 (.13)	.965 (.003)	.089	9.12 (.09)	.964 (.003)
100	.037	13.14 (.25)	.966 (.003)	.045	12.19 (.17)	.966 (.003)
200	.018	27.83 (.46)	.963 (.003)	.022	24.41 (.37)	.966 (.003)
400	.009	54.58 (.80)	.964 (.003)	.011	50.17 (.70)	.964 (.003)
800	.005	89.25 (1.33)	.966 (.003)	.006	78.42 (1.14)	.965 (.003)
		<1.01>			<1.01>	

1: The values of b^*/a; Theorem 4.3.1(iii), $b^* = 2\tau_0 k(m_1 - 1)$, τ_0 from Table 4.8.1b

Note: $\bar{n}/C$ is to be compared with the asymptotic values in < >

Table 4.8.8a. Modified Two–Stage Procedure (4.4.1) – (4.4.2) Under LFC
$\sigma = .5$, $m_0^* = 10$, $P^* = .95$

k	δ^*	C	$\gamma = .2$ $\bar{n}$–C ($s(\bar{n})$) < >[1]	$\gamma = .2$ $\bar{p}$ ($s(\bar{p})$)	$\gamma = .3$ $\bar{n}$–C ($s(\bar{n})$) < >[1]	$\gamma = .3$ $\bar{p}$ ($s(\bar{p})$)	$\gamma = .4$ $\bar{n}$–C ($s(\bar{n})$) < >[1]	$\gamma = .4$ $\bar{p}$ ($s(\bar{p})$)
5	.147	12.5	2.37 (.02)	.969 (.002)	2.43 (.02)	.966 (.003)	2.75 (.03)	.961 (.003)
	.073	25	2.49 (.03)	.958 (.003)	2.78 (.04)	.961 (.003)	3.10 (.05)	.966 (.003)
	.037	50	3.63 (.05)	.960 (.003)	4.05 (.06)	.961 (.003)	4.55 (.07)	.960 (.003)
	.018	100	5.78 (.07)	.960 (.003)	6.34 (.09)	.958 (.003)	6.92 (.10)	.964 (.003)
	.009	200	9.23 (.12)	.961 (.003)	12.37 (.16)	.961 (.003)	16.64 (.20)	.963 (.003)
	.005	400	16.05 (.18)	.960 (.003)	22.88 (.21)	.960 (.003)	29.13 (.26)	.961 (.003)
10	.179	12.5	2.01 (.01)	.965 (.003)	2.19 (.01)	.966 (.003)	2.43 (.02)	.963 (.003)
	.089	25	2.06 (.02)	.967 (.003)	2.39 (.04)	.963 (.003)	2.83 (.04)	.966 (.003)
	.045	50	2.75 (.04)	.958 (.003)	3.13 (.05)	.957 (.003)	4.13 (.06)	.960 (.003)
	.022	100	4.15 (.05)	.959 (.003)	5.91 (.07)	.961 (.003)	6.25 (.08)	.961 (.003)
	.011	200	6.69 (.09)	.960 (.003)	8.33 (.13)	.960 (.003)	11.02 (.16)	.961 (.003)
	.006	400	11.29 (.13) <1>	.960 (.003)	15.18 (.16) <1>	.961 (.003)	19.39 (.19) <1>	.960 (.003)

1: $\bar{n}/C$ is to be compared with the asymptotic values in < >

Table 4.8.8b. Modified Two–Stage Procedure (4.4.1) – (4.4.2) Under LFC
$\sigma = .75$, $m_0^* = 10$, $P^* = .95$

k	δ^*	C	$\gamma = .2$		$\gamma = .3$		$\gamma = .4$	
			$\bar{n}$–C (s($\bar{n}$)) < >[1]	$\bar{p}$ (s($\bar{p}$))	$\bar{n}$–C (s($\bar{n}$)) < >[1]	$\bar{p}$ (s($\bar{p}$))	$\bar{n}$–C (s($\bar{n}$)) < >[1]	$\bar{p}$ (s($\bar{p}$))
5	.147	18.75	2.93	.969	3.23	.956	4.01	.966
			(.03)	(.002)	(.03)	(.003)	(.04)	(.003)
	.073	37.5	3.48	.960	3.39	.966	4.96	.961
			(.05)	(.003)	(.06)	(.003)	(.07)	(.003)
	.037	75	5.19	.963	5.64	.958	7.13	.958
			(.07)	(.003)	(.09)	(.003)	(.12)	(.003)
	.018	150	7.58	.961	9.18	.966	9.94	.961
			(.10)	(.003)	(.14)	(.003)	(.17)	(.003)
	.009	300	14.13	.966	15.09	.958	19.37	.961
			(.15)	(.003)	(.19)	(.003)	(.26)	(.003)
	.005	600	17.93	.960	27.68	.955	35.11	.960
			(.26)	(.003)	(.28)	(.003)	(.36)	(.003)
10	.179	18.75	2.35	.969	2.83	.966	3.21	.969
			(.01)	(.002)	(.02)	(.003)	(.03)	(.002)
	.089	37.5	2.91	.963	3.11	.961	4.03	.966
			(.02)	(.003)	(.04)	(.003)	(.06)	(.003)
	.045	75	4.07	.964	4.83	.960	5.26	.965
			(.05)	(.003)	(.06)	(.003)	(.07)	(.003)
	.022	150	5.82	.960	7.15	.961	8.25	.964
			(.07)	(.003)	(.10)	(.003)	(.13)	(.003)
	.011	300	9.24	.961	12.16	.966	13.39	.961
			(.13)	(.003)	(.16)	(.003)	(.22)	(.003)
	.006	600	18.28	.963	22.11	.962	27.34	.962
			(.20)	(.003)	(.24)	(.003)	(.28)	(.003)
			<1>		<1>		<1>	

1: $\bar{n}/C$ is to be compared with the asymptotic values in < >

Table 4.8.8c. Modified Two–Stage Procedure (4.4.1) – (4.4.2) Under LFC
$\sigma = 1$, $m_0^* = 10$, $P^* = .95$

			$\gamma = .2$		$\gamma = .3$		$\gamma = .4$	
k	δ^*	C	$\bar{n}-C$ (s($\bar{n}$)) < >[1]	$\bar{p}$ (s($\bar{p}$))	$\bar{n}-C$ (s($\bar{n}$)) < >[1]	$\bar{p}$ (s($\bar{p}$))	$\bar{n}-C$ (s($\bar{n}$)) < >[1]	$\bar{p}$ (s($\bar{p}$))
5	.147	25	3.37	.968	3.92	.964	4.53	.966
			(.05)	(.002)	(.05)	(.003)	(.06)	(.003)
	.073	50	4.52	.958	5.05	.965	5.51	.961
			(.07)	(.003)	(.08)	(.003)	(.09)	(.003)
	.037	100	6.82	.962	7.41	.964	8.11	.960
			(.10)	(.003)	(.12)	(.003)	(.14)	(.003)
	.018	200	10.81	.958	12.07	.958	14.23	.960
			(.15)	(.003)	(.18)	(.003)	(.20)	(.003)
	.009	400	19.06	.950	23.15	.957	26.42	.961
			(.22)	(.003)	(.26)	(.003)	(.29)	(.003)
	.005	800	35.42	.960	39.14	.964	44.16	.966
			(.32)	(.003)	(.37)	(.003)	(.45)	(.003)
10	.179	25	3.11	.970	3.55	.970	4.01	.963
			(.03)	(.002)	(.04)	(.002)	(.04)	(.003)
	.089	50	4.03	.968	4.63	.966	5.02	.966
			(.04)	(.002)	(.05)	(.003)	(.06)	(.003)
	.045	100	5.15	.963	5.83	.966	6.58	.964
			(.06)	(.003)	(.08)	(.003)	(.10)	(.003)
	.022	200	7.74	.954	8.60	.961	9.85	.956
			(.10)	(.003)	(.13)	(.003)	(.15)	(.003)
	.011	400	12.81	.960	14.01	.964	15.62	.962
			(.15)	(.003)	(.18)	(.003)	(.21)	(.003)
	.006	800	26.07	.955	27.61	.960	31.62	.960
			(.25)	(.003)	(.29)	(.003)	(.34)	(.003)
			<1>		<1>		<1>	

1: $\bar{n}/C$ is to be compared with the asymptotic values in < >

standard error values. Hence, via these simulation studies we provide some useful recommendations about γ for a wide range of values of C, not merely for values which are either too large or too small. In Tables 4.8.8a,b,c, we include summary findings when $\sigma = .5$, .75 and 1.0 with $\gamma = .2$, .3 and .4.

From Tables 4.8.8a,b,c, we find that, *across the board*, $(\overline{n} - C)$ values are fairly small when $\gamma = .3$ or .4. Considering both $(\overline{n} - C)$ and $\overline{p}$ values, $\gamma = .3$ or .4 appears to be a good choice in evaluating m_1 given by (4.4.1). Such a choice for γ also works well when $\sigma \geq 1$. To see this, one only needs to compare the $(\overline{n} - C)$ values in Table 4.8.8c with those in Table 4.8.8a in conjunction with a close scrutiny of the $\overline{p}$ values in relation with the target P*. If one holds k, σ, P* fixed, from Tables 4.8.8a,b,c, it becomes clear that $E(N - C)$ steadily increases as C increases. This is an undesirable feature of the modified two–stage methodology that empirically verifies the point made in Remark 4.4.2. Also, $\overline{n}/C$ is very close to the asymptotic value, that is unity, which empirically justifies Theorem 4.4.1(ii).

In Table 4.8.9, columns 2 and 3 show values of $\overline{n} - C$ and $\overline{p}$, along with their respective standard errors for the purely sequential procedure (4.5.1), while columns 4–5 and 6–7 show these figures respectively for the accelerated sequential procedure (4.6.4) – (4.6.6) and the three–stage procedure (4.7.7). For the accelerated sequential procedure, we take $\rho = .6$, the choice recommended by Mukhopadhyay and Solanky (1992b), and for the three–stage procedure we take $\rho = .75$ for $k = 5$ and $\rho = .80$ for $k = 10$. These choices seem to be the best choices. As explained in section 3.8, the *number of sampling operations* (= $\mathcal{S}$), defined as the number of steps taken by a particular sampling process to terminate, approximately equals C, ρC and 3, respectively for the purely sequential, accelerated sequential and three–stage selection methodologies. Hence, in terms of the number of sampling operations, the three–stage procedure is operationally most convenient, followed by the accelerated sequential and then the purely sequential procedure. In Table 4.8.9, we attempt to compare the second–order properties of these multistage selection methodologies empirically.

First, we address columns 2–3 in Table 4.8.9. From Theorem 4.5.2(i) and Table 4.8.3, we should expect $(\overline{n} - C)$ to be close to $\beta''(k)$, that is .39 and .45 respectively when $k = 5$ and $k = 10$, while $s(\overline{n})$ may be expected to be close to $(C/Jk)^{\frac{1}{2}}$ for large values of C, in the purely sequential case. The second assertion follows from a combination of Theorem 4.5.1(iii) and Theorem 2.4.8(iv). For $C \geq 100$, $s(\overline{n})$ values are seen to be very close to these asymptotic answers; for example, in the case $k = 5$, the expression $(C/Jk)^{\frac{1}{2}}$ gives the values .063, .089, .127, .179 respectively when $C = 100$, 200, 400 and 800. Also, when $C \geq 400$, one finds that $\beta''(k)$ is practically inside the 4–standard deviation interval, $(\overline{n} - C) \pm 4s(\overline{n})$. This is more evident for $C = 800$. Also, one will note that $(\overline{n} - C)$ stays fairly constant for fixed k, when $C \geq 100$. This is an empirical justification of second–order efficiency property for moderate sample sizes, in the presence of slow convergence.

In column 3, one finds the values of $\overline{p}$ for the purely sequential procedure. For both $k = 5$ and 10, $\overline{p}$ values appear to be near the set target P^*. In almost every case, P^* is not very far from $\overline{p} \pm 2s(\overline{p})$. The asymptotic second–order analysis in Theorem 4.5.2(ii) reveals that the values of $(\overline{p} - P^*)$ may be expected to be close to aA'/C for large C, where $A' = A'(k, P^*)$ is a constant that is evaluated in Table 4.8.4. Also, one will note that the values of aA'/C are not very far from the corresponding 2–standard deviation interval $(\overline{p} - P^*) \pm 2s(\overline{p})$ for almost all values of $C \geq 100$.

Overall, by looking at columns 2–3 in Table 4.8.9, we can safely say that the "asymptotics" kick in quite early and also *second–order correction terms* lead to useful practical guidelines in terms of what one should expect as the magnitudes of $(\overline{n} - C)$ and $(\overline{p} - P^*)$, for C–values as small as 100 or so, in the case of the purely sequential procedure (4.5.1). A similar feature was observed earlier in Chapter 3 for the purely sequential procedure (3.5.1).

Next, we consider columns 4–5 in Table 4.8.9 which give similar entries for $(\overline{n} - C)$, $\overline{p}$ and their associated estimated standard errors for accelerated sequential procedure (4.6.4) – (4.6.6) with $k = 5, 10$, $\rho = .6$ and $q = \max\{0, (k\rho)^{-1}[1 + \frac{1}{2} ab_2 b_1^{-1}]\}$, given by Theorem 4.6.2(iii). We obtain appropriate q for $\rho = .5$ by directly utilizing b_1, b_2, obtained from Table 4.8.4. As before, the number in parentheses below each $\overline{n} - C$ and $\overline{p}$ value, is the standard error of the number above it. The last row in $k = 5$ block and also in $k = 10$ block provides the asymptotic second order guidelines wherever

Table 4.8.9. Comparison of Procedures Under LFC
$m_1 = 10$, $P^* = .95$, $\sigma = 1$

	Purely Sequential Procedure (4.5.1)		Accelerated Sequential Procedure (4.6.4)–(4.6.6) ρ=.6		Three–Stage Procedure (4.7.7); k=5 ρ=.75; k=10, ρ=.80	
C^*	$\bar{n}$–C	$\bar{p}$	$\bar{n}$–C	$\bar{p}$	$\bar{n}$–C	$\bar{p}$
δ^*	($s(\bar{n})$)	($s(\bar{p})$)	($s(\bar{n})$)	($s(\bar{p})$)	($s(\bar{n})$)	($s(\bar{p})$)
	< >[1]	< >[2]	< >[3]		< >[4]	
			k = 5			
25	1.39	.959	2.74	.965	2.15	.967
.147	(.03)	(.003)	(.04)	(.003)	(.04)	(.003)
50	1.44	.952	2.72	.957	3.19	.961
.073	(.05)	(.003)	(.06)	(.003)	(.06)	(.003)
100	1.34	.954	2.83	.951	4.41	.953
.037	(.06)	(.003)	(.08)	(.003)	(.09)	(.003)
200	1.38	.951	2.71	.950	4.68	.951
.018	(.09)	(.003)	(.12)	(.003)	(.13)	(.003)
400	1.33	.952	2.64	.951	4.91	.955
.009	(.13)	(.003)	(.17)	(.003)	(.18)	(.003)
800	1.21	.951	2.57	.953	5.61	.951
.005	(.18)	(.003)	(.23)	(.003)	(.22)	(.003)
	<.39>	<–.009>	<.58, 1.58>		<.62>	
			k = 10			
25	1.43	.959	2.47	.967	2.11	.967
.179	(.02)	(.003)	(.03)	(.003)	(.04)	(.003)
50	1.44	.956	2.49	.959	3.07	.958
.089	(.03)	(.003)	(.04)	(.003)	(.07)	(.003)
100	1.44	.948	2.65	.957	4.23	.959
.045	(.05)	(.003)	(.06)	(.003)	(.10)	(.003)
200	1.37	.956	2.64	.950	4.63	.950
.022	(.06)	(.003)	(.08)	(.003)	(.13)	(.003)
400	1.30	.950	2.51	.951	4.81	.953
.011	(.09)	(.003)	(.11)	(.003)	(.12)	(.003)
800	1.07	.953	2.44	.952	5.35	.953
.006	(.13)	(.003)	(.16)	(.003)	(.14)	(.003)
	<.45>	<–.006>	<.35, 1.35>		<.34>	

1: values of $\beta''(k)$ from Table 4.8.3; $(\bar{n} - C)$ is to be compared with $\beta''(k)$

2: values of A' from Table 4.8.4; $(\bar{p} - P^*)$ is to be compared with aA'/C

3: the interval is $(q-(k\rho)^{-1}, q + 1 - (k\rho)^{-1})$, where q comes from Table 4.8.5

4: values of $\beta_0' + m_1^*$; look at (4.7.8)

available. These numbers are written as earlier in triangular brackets. For the accelerated sequential procedure (4.6.4) – (4.6.6), at the end of each block, we provide the interval $< q - (k\rho)^{-1}, q + 1 - (k\rho)^{-1} >$. From Theorem 4.6.2(i), one may observe that $E(N - C)$ theoretically falls within this interval. After computing $(\bar{n} - C) \pm 2$ (standard error), once again one notes that the simulated values for the moderate sample sizes show remarkable agreement with the interval given inside < >. The values of $\bar{p}$ are also close to the target value of .95.

In columns 6–7 in Table 4.8.9, we provide the values of $\bar{n} - C$ and $\bar{p}$ and their associated standard errors for the three–stage procedure (4.7.7). At the end of each block, we provide the values of $\beta_0' + m_1^*$, which is the asymptotic value of $E(N - C)$ by Theorem 4.7.2. Note that the values of $\bar{n} - C$ are much larger than the second order asymptotic guidelines. For example, for $k = 5$, $C = 800$, $\bar{n} - C$ is 5.61 with standard error .22, whereas its asymptotic value is .62. The explanation for such a behavior is that m_1 in our simulation of procedure (4.7.7) is a fixed number $(= 10)$, although by virtue of (4.7.1) it should grow as a function of δ^*. Such a feature was observed in chapter 3 as well for the three–stage procedure (3.7.5). The values of $(\bar{n} - C)$ do approach its asymptotic value as m_1 increases. The $\bar{p}$ values for the procedure (4.7.7) are once again above the target value of .95.

Let us now compare the procedures on Table 4.8.9 with one another. One can easily see that theoretically, the purely sequential procedure (4.5.1) is better than the other two procedures. This observation is once again along the lines of our observation made in section 3.8. Also note that as k increases the difference between the accelerated sequential procedure (4.6.4) – (4.6.6) and the three–stage procedure (4.7.7) becomes wider. The accelerated sequential procedure (4.6.4) – (4.6.6) however appears to be better than the three–stage procedure (4.7.7) in terms of less oversampling.

4.9 A Few generalizations

In this section, we consider a general situation where the aim is to select the negative exponential populations associated with the s largest location parameters, $1 \le s < k$. Among other characteristics, we also address the behavior of P(CS) under a few parameter configurations which are different from the associated LFC. Following Bechhofer's (1954) indifference zone formulation, given $\delta^* > 0$, we consider the *preference zone* $\Omega^{(s)}(\delta^*)$ for the problem of selecting the s–best populations with

$$\Omega^{(s)}(\delta^*) = \{\underset{\sim}{\theta} = (\theta_1,\ldots,\theta_k) : \theta_{[k-s+1]} - \theta_{[k-s]} \geq \delta^*\}. \tag{4.9.1}$$

Note that for $s = 1$, $\Omega^{(s)}(\delta^*)$ was earlier written as $\Omega(\delta^*)$ in (4.1.2). For a given $P^* \epsilon (\binom{k}{s}^{-1}, 1)$, we are interested in obtaining selection procedures such that P(CS) is *at least* P^* or *asymptotically* (as $\delta^* \to 0$) *at least* P^* whenever the true parameters $\underset{\sim}{\theta} \in \Omega^{(s)}(\delta^*)$.

4.9.1 PRELIMINARIES

For the moment, let us pretend that $\underset{\sim}{\sigma}$ is known. Having recorded $X_{i1},\ldots,X_{in}$ from π_i, let us write $\tilde{X}_{[1n]} \leq \cdots \leq \tilde{X}_{[sn]} \leq \cdots \leq \tilde{X}_{[kn]}$ for the ordered statistics obtained from the sample minimums $\tilde{X}_{1n},\ldots,\tilde{X}_{sn},\ldots,\tilde{X}_{kn}$. The immediate generalization of the selection rule $\mathscr{P}_n^*$ given by (4.2.1) will lead us to propose the following selection rule in order to suit the present situation.

$$\mathscr{P}_n^{*(s)}: \quad \text{Select } \pi_{i_j} \text{ in the set of s–best populations if } \tilde{X}_{i_j n} \geq \tilde{X}_{[k-sn]},\ j = 1,\ldots,s. \tag{4.9.2}$$

In other words, the selection rule $\mathscr{P}_n^{*(s)}$ picks those s populations which are associated with the s largest sample minimums, that is $\tilde{X}_{[k-s+1n]},\ldots,\tilde{X}_{[kn]}$. Note that $\mathscr{P}_n^*$ from (4.2.1) coincides with $\mathscr{P}_n^{*(1)}$.

Let us continue to write $\pi_{(i)}$ for the population associated with $\theta_{[i]}$ and let us denote $\tilde{X}_{(in)}$ for the corresponding sample minimum arising from

$\pi_{(i)}$, $i = 1,\ldots,k$. Suppose $\delta_{i1} = \theta_{[k-s+1]} - \theta_{[i]}$, $i = 1,\ldots,k-s$ and $\delta_{j2} = \theta_{[k-s+1]} - \theta_{[k-j]}$, $j = 0,1,\ldots,s-2$. Now, under the preference zone $\Omega^{(s)}(\delta^*)$ we have $\delta_{i1} \geq \delta^*$, $\delta_{j2} \leq 0$ for $i = 1,\ldots,k-s$, $j = 0, 1,\ldots,s-2$. One can easily verify that

$$P_n(CS) = s\int_0^\infty \prod_{i=1}^{k-s} [1-\exp(-y-n\delta_{i1}\sigma^{-1})] \prod_{j=0}^{s-2} [\exp(-y-n\delta_{j2}\sigma^{-1})]e^{-y}dy. \quad (4.9.3)$$

From (4.9.3), it is easy to see that

$$\inf_{\underset{\sim}{\theta}\in\Omega^{(s)}(\delta^*)\cup\Omega^{(s)c}(\delta^*)} P_n(CS)$$

$$= s\int_0^\infty [1-\exp(-y)]^{k-s}[\exp(-y)]^{s-1}e^{-y}dy = \binom{k}{s}^{-1}.$$

However, for all $\underset{\sim}{\theta} \in \Omega^{(s)}(\delta^*)$, that is when $\delta_{i1} \geq \delta^*$, $\delta_{j2} \leq 0$, $i = 1,\ldots,k-s$, and $j = 0, 1,\ldots,s-2$, one has

$$P_n(CS) \geq s\int_0^\infty \prod_{i=1}^{k-s} [1-\exp(-y-n\delta^*\sigma^{-1})] \prod_{j=0}^{s-2} [\exp(-y)]e^{-y}dy$$

and the equality holds if $\delta_{i1} = \delta^*$ and $\delta_{j2} = 0$ for all $i = 1,\ldots,k-s$, and $j = 0, 1,\ldots,s-2$. Hence,

$$\inf_{\underset{\sim}{\theta}\in\Omega^{(s)}(\delta^*)} P_n(CS) = s\int_0^\infty [1-\exp(-y-n\delta^*/\sigma)]^{k-s}[\exp(-sy)]dy, \quad (4.9.4)$$

the infimum being attained when $\theta_{[1]} = \cdots = \theta_{[k-s]} = \theta_{[k-s+1]} - \delta^* = \cdots = \theta_{[k]} - \delta^*$, a parameter configuration that is referred as the LFC. We define $a^{(s)} = a^{(s)}(k, s, P^*)$ such that

$$s\int_0^\infty [1-\exp(-y-a^{(s)})]^{k-s}[\exp(-sy)]dy = P^*. \tag{4.9.5}$$

Recall that for $s = 1$, we write "a" instead of $a^{(1)}$ in section 4.2, given by (4.2.4). Now it follows from (4.9.4) that

$$P_n(CS) \geq P^* \text{ for all } \underset{\sim}{\theta} \in \Omega^{(s)}(\delta^*),$$

if n is the smallest integer $\geq a^{(s)}\sigma/\delta^* = C^{(s)}$, say. In other words, had σ been known, one would take $[C^{(s)}]^* + 1$ samples from each π and implement the corresponding selection rule (4.9.2). The values of $a^{(s)}$ had been tabulated in Raghavachari and Starr (1970). However, σ is unknown and hence we opt for various multistage selection procedures.

Suppose in general, one devises a statistical methodology giving rise to a positive integer valued random variable N estimating $C^{(s)}$ and selects s populations via (4.9.2).

Theorem 4.9.1. *Suppose that* N *satisfies the conditions of Theorem 4.2.1. Then,*

$$\inf_{\underset{\sim}{\theta}\in\Omega^{(s)}(\delta^*)} P(CS) = sE[\int_0^\infty [1-\exp(-y-N\delta^*\sigma^{-1})]^{k-s}[\exp(-sy)]dy$$

and this infimum is attained when $\theta_{[1]} = \cdots = \theta_{[k-s]} = \theta_{[k-s+1]} - \delta^* = \cdots = \theta_{[k]} - \delta^*$.

Its proof is omitted in view of the proof of the Theorem 4.2.1 and (4.9.3) – (4.9.4).

4.9.2 PURELY SEQUENTIAL PROCEDURE

Along the lines of section 4.5, we start with $m_1(\geq 2)$ samples from each π. Then, we proceed by taking one more sample at a time from each π, according to the following stopping rule:

$$N = N(\delta^*) = \inf\{n \geq m_1 : n \geq a^{(s)} V(n) \delta^{*-1}\} \tag{4.9.6}$$

where $V(n)$, defined in (4.1.1), is an estimator of σ. It is easy to see that for fixed $\underset{\sim}{\theta}$, σ, m_1 and δ^*, $P(N < \infty) = 1$. Based on $X_{i1},\ldots,X_{iN}$ from π_i, $i = 1,\ldots,k$, we now implement the selection rule $\mathscr{R}_N^{*(s)}$ defined in (4.9.2). Throughout $\underset{\sim}{\theta}$, σ are kept fixed.

Theorem 4.9.2. *For the purely sequential procedure (4.9.6), we have as* $\delta^* \to 0$:

(i) $N/C^{(s)} \to 1$ w.p. 1;

(ii) $E(N/C^{(s)}) \to 1$;

(iii) $C^{(s)^{-\frac{1}{2}}}(N - C^{(s)}) \xrightarrow{\mathscr{L}} N(0, k^{-1})$;

(iv) $\liminf P(CS) \geq P^*$ *for all* $\underset{\sim}{\theta} \in \Omega^{(s)}(\delta^*)$;

(v) $\lim P(CS) = P^*$ *under the LFC;*

where $C^{(s)} = a^{(s)}\sigma/\delta^*$ *and* $a^{(s)}$ *comes from (4.9.5).*

The proof is similar to that of Theorem 4.5.1 and hence is omitted.

4.9.2.1 Second–Order Asymptotics

In this section, we strengthen the results given in Theorem 4.9.2(ii) and (v) by providing expansions of (a) $E(N)$ for all $\underset{\sim}{\theta} \in \Omega^{(s)}(\delta^*)$ and (b) $P(CS)$ under both the LFC as well as non LFC. Let us define

$$\delta_{i3} = a_i\delta^*, \quad \delta_{j4} = b_j\delta^* \quad \text{for } a_i \geq 1, \; b_j \leq 0;$$

$$q_1(x) = \exp(-x)\, I\,(x > 0);$$

$$b^{(s)}(k, x) = \binom{k-s}{x} (-1)^x (x + s)^{-1}; \tag{4.9.7}$$

$$b_i^{(s)} = \Sigma_{x=0}^{k-s} x^i\, b^{(s)}(k, x)\, q_1(x\, a^{(s)}), \text{ for } i = 1, 2; \tag{4.9.8}$$

$$g_{11}^{(s)}(x) = s\int_0^\infty \prod_{i=1}^{k-s} [1 - \exp(-y-a_i x)] \prod_{j=0}^{s-2} [\exp(-y-b_j x)] e^{-y} dy; \tag{4.9.9}$$

$$g_{12}^{(s)}(x) = s\int_0^\infty \{1 - \exp(-y-x)\}^{k-s} [\exp(-sy)] dy; \tag{4.9.10}$$

$i = 1,\ldots,k-s$, and $j = 0, 1,\ldots,s-2$.

Note that under the LFC, one has $\delta_{i1} = \delta_{i3}$ with $a_1 = \cdots = a_{k-s} = 1$ and $\delta_{j2} = \delta_{j4}$ with $b_0 = b_1 = \cdots = b_{s-2} = 0$. Also, under the LFC for $s = 1$, $g_{11}^{(1)}(x)$ was denoted as $g_1(y)$ in (4.5.11), and under the LFC $g_{11}^{(s)}(x)$ reduces to $g_{12}^{(s)}(x)$. Now, we have the following results. See also Problem 4.18.

Theorem 4.9.3. *For the purely sequential procedure (4.9.6), we have as* $\delta^* \to 0$:

(i) $E(N) = C^{(s)} + \beta'' + o(1)$ *for all* $\underset{\sim}{\theta} \in \Omega(\delta^*)$ *if* (a) $m_1 \geq 3$ *when* $k = 2$, (b) $m_1 \geq 2$ *when* $k \geq 3$;

(ii) $P(CS) = P^* + \delta^*\sigma^{-1}\{(1 - P^*)\beta'' - \frac{1}{4} a^{(s)} k^{-1}\} + o(\delta^*)$ *under the LFC for* $m_1 \geq 3$ *when* $k = 2$;

(iii) $P(CS) = P^* + \delta^*\sigma^{-1}\{\frac{1}{2} a^{(s)} b_2^{(s)} k^{-1} - b_1^{(s)} \beta''\} + o(\delta^*)$ *under the LFC for* $m_1 \geq 2$ *when* $k \geq 3$;

where $\beta'' = \beta''(k) = k^{-1}\{\frac{1}{2}\nu(k) - 1\}$ *with* $\nu(k)$ *defined in (4.5.8),* $a^{(s)}$ *comes from (4.9.5) and* $b_1^{(s)}, b_2^{(s)}$ *come from (4.9.8).*

Proof. The proof of part (i) and (ii) are similar to that of Theorem 4.5.2(i), (ii). Let us turn briefly now to prove part (iii). In order to prove part (iii), one writes

$$P(CS) = E[g_{12}^{(s)}(N\ \delta^*/\sigma)]$$

$$= s\ \Sigma_{j=0}^{k-s}\ b^{(s)}(k,\ j)\ E[q_1(j\ N\delta^*/\sigma)].$$

Then the proof follows along the lines of the proof of Theorem 4.5.2(iii). □

4.9.3 ACCELERATED SEQUENTIAL PROCEDURE

Along the lines of section 4.6, for technical reasons,we consider estimating σ by means of $V^*(n)$ defined in (4.1.1). We first choose and fix $0 < \rho < 1$ and $q^{(s)} \geq 0$. Along the lines of (2.5.1) – (2.5.3), we define

$$t = t(\delta^*) = \inf\{n \geq m_1 : n \geq \rho a^{(s)} V^*(n)/\delta^*\}, \tag{4.9.11}$$

$$N_1 = N_1(\delta^*) = [a^{(s)} V^*(t)\delta^{*-1} + q^{(s)}]^* + 1, \tag{4.9.12}$$

$$N = N(\delta^*) = \max\{t(\delta^*),\ N_1(\delta^*)\}, \tag{4.9.13}$$

where $m_1(\geq 2)$ is the starting sample size from each π. The accelerated sequential procedure (4.9.11) – (4.9.13) is implemented in the same way as (4.6.4) – (4.6.6). Based on $X_{i1},\ldots,X_{iN}$ from π_i, $i = 1,\ldots,k$, we then select the s–best populations via $\mathscr{R}_N^{*(s)}$ defined in (4.9.2). The asymptotic second–order properties are now summarized in the following theorem while we omit its proof for brevity. See also Problem 4.18.

Theorem 4.9.4. *For the accelerated sequential procedure (4.9.11) – (4.9.13), we have as* $\delta^* \to 0$:

(i) $q^{(s)} - (k\rho)^{-1} \leq \liminf E(N - C) \leq \limsup E(N - C) \leq q^{(s)} + 1 -$

$(k\rho)^{-1}$ *for all* $\underset{\sim}{\theta} \in \Omega^{(s)}(\delta^*)$ *if* $m_1 \geq 4$;

(ii) $\text{Var}(N) = (k\rho)^{-1}C^{(s)} + o(\delta^{*-1})$ *for all* $\underset{\sim}{\theta} \in \Omega^{(s)}(\delta^*)$ *if* $m_1 \geq 2$;

(iii) $P(CS) \geq P^* + o(\delta^*)$ *under the LFC if* $m_1 \geq 4$, *with* $q^{(s)} = \max\{0, (k\rho)^{-1}[1 + \frac{1}{2} a^{(s)} b_2^{(s)} b_1^{(s)-1}]\}$;

where $C^{(s)} = a^{(s)}\sigma/\delta^*$ *and* $a^{(s)}$ *comes from (4.9.5).*

4.9.4 THREE–STAGE PROCEDURE

In this section, a three–stage selection methodology is discussed along the lines of section 4.7. Now, we go back to estimating σ by $V(n)$. One starts with $m_1(\geq 2)$ samples from each π where it is assumed that

$$m_1 = m_1(\delta^*) = o(C^{(s)1/r}),$$

for some $r > 1$, as in (4.7.1). We also choose and fix $0 < \rho < 1$. Then, one defines

$$T = T(\delta^*) = \max\{m_1, [\rho a^{(s)} V(m_1)\delta^{*-1}]^* + 1\}, \tag{4.9.14}$$

$$m_1^{*(s)} = \{a^{(s)} b_2^{(s)} - b_1^{(s)}\rho k + 2b_1^{(s)}\}/(2\,\rho k\, b_1^{(s)}),$$

$$N = N(\delta^*) = \max\{T,\ [a^{(s)} V(T)\delta^{*-1} + m_1^{*(s)}]^* + 1\}, \tag{4.9.15}$$

The three–stage procedure (4.9.15) is implemented in the same way as in section 4.7. Based on $X_{i1},\ldots,X_{iN}$ from π_i, $i = 1,\ldots,k$, one then selects the s–best populations via $\mathscr{R}_N^{*(s)}$ defined in (4.9.2). The asymptotic second–order properties are now summarized in the following theorem while we omit its proof.

Theorem 4.9.5. *For the three–stage procedure (4.9.15), we have as* $\delta^* \to 0$:

(i) $E(N) = C^{(s)} + \beta_0' + m_1^{*(s)} + o(1)$ *for all* $\underset{\sim}{\theta} \in \Omega^{(s)}(\delta^*)$;

(ii) $\mathrm{Var}(N) = (k\rho)^{-1}\, C^{(s)} + o(\delta^{*-1})$ *for all* $\underset{\sim}{\theta} \in \Omega^{(s)}(\delta^*)$;

(iii) $P(CS) = P^* + o(\delta^*)$ *under the LFC;*

where $C^{(s)} = a^{(s)}\sigma/\delta^*$, $\beta_0' = \frac{1}{2} - (k\rho)^{-1}$ *and* $a^{(s)}$ *comes from (4.9.5).*

4.9.5 MODERATE SAMPLE PERFORMANCES AND COMMENTS

In this section we will study the moderate sample performances of the purely sequential procedure (4.9.6), accelerated sequential procedure (4.9.11) – (4.9.13) and three–stage procedure (4.9.15) for $1 \leq s < k$. We consider both the LFC as well as non LFC situations for the simulation studies in the case of purely sequential procedure (4.9.6). As in section 4.8, for the LFC case, we used IMSL (1989, Edition 10) subroutine RNEXP to generate $(k - s)$ pseudo negative exponential populations each with $\theta = 5$ and $\sigma = 1$, and the remaining s pseudo negative exponential populations with $\theta = 5 + \delta^*$ and $\sigma = 1$, where $\delta^* = a^{(s)}C^{-1}$. We consider $C = 25, 50, 100, 200, 400$ and 800, while we fix $k = 5$, $s = 2$ and $k = 10$, $s = 3$.

In Table 4.9.1, we provide our findings obtained under the LFC for the purely sequential procedure (4.9.6), accelerated sequential procedure (4.9.11) – (4.9.13) and the three–stage procedure (4.9.15). These are based on J (= 5000) replications for each entry. Under each column of $\bar{n} - C$ and $\bar{p}$ values, we provide the standard errors $s(\bar{n})$ and $s(\bar{p})$. For $k = 5$, $s = 2$ the values of $b_1^{(2)}$ and $b_2^{(2)}$ are respectively found to be –.02452 and –.02357 whereas for $k = 10$, $s = 3$ the corresponding numbers are respectively –.01629 and –.01551. One should note that the corresponding values of $\bar{n} - C$ in Tables 4.8.9 and 4.9.1 are quite similar and once again the purely sequential procedure turns out most satisfactory in terms of smaller $\bar{n} - C$ values, followed by the accelerated sequential procedure and then the three–stage procedure.

Table 4.9.1. Comparison of Procedures Under LFC
$\sigma = 1.0$, $m_1 = 10$, $P^* = .95$, $s = 2$ for $k = 5$ and $s = 3$ for $k = 10$

	Purely Sequential Procedure (4.9.6)		Accelerated Sequential Procedure (4.9.11)–(4.9.13) ρ=.6		Three–Stage Procedure (4.9.15); k=5 ρ=.75; k=10, ρ=.80	
$C^{(s)}$	$\bar{n}$–C	$\bar{p}$	$\bar{n}$–C	$\bar{p}$	$\bar{n}$–C	$\bar{p}$
δ^*	$(s(\bar{n}))$	$(s(\bar{p}))$	$(s(\bar{n}))$	$(s(\bar{p}))$	$(s(\bar{n}))$	$(s(\bar{p}))$
			k = 5			
25	1.36	.953	2.46	.955	2.37	.957
.147	(.03)	(.003)	(.04)	(.003)	(.04)	(.003)
50	1.55	.952	2.81	.952	3.11	.955
.073	(.05)	(.003)	(.06)	(.003)	(.06)	(.003)
100	1.28	.950	2.91	.952	4.52	.951
.037	(.06)	(.003)	(.08)	(.003)	(.09)	(.003)
200	1.33	.955	2.80	.951	4.69	.952
.018	(.09)	(.003)	(.12)	(.003)	(.13)	(.003)
400	1.29	.953	2.71	.950	4.88	.953
.009	(.13)	(.003)	(.17)	(.003)	(.18)	(.003)
800	1.18	.948	2.55	.953	5.71	.953
.005	(.18)	(.003)	(.24)	(.003)	(.23)	(.003)
			k = 10			
25	1.37	.955	2.39	.957	2.28	.959
.185	(.02)	(.003)	(.03)	(.003)	(.03)	(.003)
50	1.44	.951	2.63	.951	3.25	.953
.093	(.04)	(.003)	(.04)	(.003)	(.07)	(.003)
100	1.32	.953	2.66	.951	4.11	.956
.046	(.05)	(.003)	(.06)	(.003)	(.10)	(.003)
200	1.26	.953	2.60	.956	4.50	.955
.023	(.06)	(.003)	(.09)	(.003)	(.13)	(.003)
400	1.27	.953	2.50	.952	4.64	.952
.012	(.09)	(.003)	(.11)	(.003)	(.12)	(.003)
800	1.14	.951	2.41	.953	5.37	.950
.006	(.12)	(.003)	(.16)	(.003)	(.14)	(.003)

Note: δ^* values were computed correctly up to four decimal places, and then these were rounded off.

We also studied a few interesting non–LFC situations for this problem along the lines of what was included in Table 3.9.2. The general features and characteristics seen from such simulations were quite compatible with what one found from Table 3.9.2, and hence we omit the corresponding particular table for brevity.

4.10 Notes and Overview

Only the case of unknown and equal scale parameter has been addressed in this chapter. Two–stage procedures of Mukhopadhyay (1984), and Mukhopadhyay and Hamdy (1984a) were designed to handle the challenging situation when the scale parameters were unknown and unequal, and the sample sizes from the populations were also taken to be unequal. See section 8.4.2 in this context. Problems analogous to those discussed in the context of (3.10.1) also need attention in the present situation.

Further investigations of ways to properly design purely sequential, accelerated sequential and three–stage procedures would be very worthwhile in the unequal and unknown scale parameters case. In this, as well as the common and unknown scale parameter case, it will be interesting and challenging to incorporate Paulson (1964)–type elimination of "inferior" populations in the selection process in the light of second–order analysis. Section 8.2 is relevant here. Likelihood based selection rules along the lines of Bechhofer et al. (1968) have been recently addressed in Miller and Mukhopadhyay (1990) in the equal and unknown scale parameter situation. Such procedures need attention in the present perspectives. See section 8.3.3 in this context. Other types of problems such as selection of "good" populations, comparisons of treatments with a control or standard, etc. also hold much promise. See Bristol and Desu (1985) and Bristol et al. (1987), for the variety of results which are already available in these contexts. It will be truly challenging to obtain asymptotic second–order properties of the associated existing and new multistage procedures in each of these problems and other closely related ones.

Complements and Problems

4.1 Show that the sufficient conditions given in Theorem 4.2.1 are satisfied by the stopping variables N arising from the two–stage, modified two–stage, purely sequential, accelerated sequential and the three–stage selection methodologies.

[*Hints*: Use Problems 2.3 – 2.4.]

4.2 For N given by (4.5.1), show that $E(N) \le C + 0(1)$. Hence, prove Theorem 4.5.1 (ii).

[*Hints*: From (4.5.2), observe that $k(N-m_1)(N-2) \le a\delta^{*-1}\Sigma_{i=1}^{k} \Sigma_{j=1}^{N-1} (X_{ij} - \tilde{X}_{iN-1}) \le a\delta^{*-1}\Sigma_{i=1}^{k} \Sigma_{j=1}^{N-1} (X_{ij} - \theta_i) \le a\delta^{*-1}\Sigma_{i=1}^{k} \Sigma_{j=1}^{N} (X_{ij} - \theta_i)$. First, assume $E(N) < \infty$, and use Wald's first equation (Problem 2.5 (a)) to write $k(E(N))^2 - k(m_1 + 2)E(N) \le a\delta^{*-1}k\sigma E(N)$, that is $E(N) \le C + (m_1 + 2)$. Now, proceed along the lines of hints in Problem 2.6 (c).]

4.3 Prove Theorem 4.5.1 when N is replaced by that given in (4.5.6) under appropriate γ. Do the same for L given in (4.6.1).

4.4 For the selection procedure (4.5.6), show that $N \ge (a/\delta^*)^{1/(\gamma+1)}$. Can one now rationalize the choice of m_1 given in (4.4.1)?

4.5 Prove results along the lines of Theorem 4.5.2 for the selection procedure (4.6.1). Give particular attention to choices of m_1

4.6 For the purely sequential procedure (4.5.1), give a direct proof of the result that $P\{N \le \frac{1}{2}C\} = 0(C^{-k(m_1-1)})$, without appealing to Theorem 2.4.8 (i).

[*Hints*: Look at Problem 2.11.]

4.7 In the sequential selection procedure (4.5.1), suppose one uses $V^*(n)$ instead of $V(n)$ in order to estimate σ. Prove Theorem 4.5.1 for this new selection methodology. Give particular attention to sufficient conditions involving m_1.

4.8 In the accelerated sequential procedure (4.6.4) – (4.6.6), suppose one uses $V(n)$ instead of $V^*(n)$ in order to estimate σ. Prove Theorem 4.6.1 for this new selection methodology.

4.9 In the three–stage procedure (4.7.2) – (4.7.3), suppose one uses $V^*(n)$ instead of $V(n)$ in order to estimate σ. Prove Theorem 4.7.1 for this new selection methodology.

4.10 Prove Theorem 4.7.3 via Theorem 2.3.1.

[*Hints*: $P(CS) = E[g_1(N\delta^*/\sigma)]$ and note that the situation agrees with what one had in (2.3.27) where $\zeta = 1$, $L = \delta^*$, $\lambda = a/\delta^*$, $\theta^* = \sigma$ and $G^*(x) = g_1(x)$. Hence, $m_1^* = \epsilon^* = \frac{1}{2}(k\rho)^{-1}\{2 - \frac{ag_1''(a)}{g_1'(a)}\} - \frac{1}{2} = \{2g_1'(a) - ag_1''(a) - k\rho g_1'(a)\}\{2k\rho g_1'(a)\}^{-1}$, since $Var(W_1) = k^{-1}$. Now, recall b_1, b_2 defined in (4.5.10). The function $g_1(\cdot)$ had been defined in (4.5.11). Show that $g_1'(a) = -b_1$ and $g_1''(a) = b_2$.]

4.11 In the two–stage procedure (4.3.1), suppose that one uses $V^*(n)$, instead of $V(n)$, as the estimator of σ. For such a two–stage procedure, verify the properties given in Theorem 4.3.1 (i) and (iv). What happens to part (ii)?

4.12 In the modified two–stage procedure (4.4.1) – (4.4.2), suppose that one uses $V^*(n)$, instead of $V(n)$, as the estimator of σ. For such a modified two–stage procedure, verify the properties given in Theorem 4.4.1.

4.13 Prove Theorems 4.9.1 and 4.9.2.

4.14 Prove Theorem 4.9.4.

4.15 Obtain the corresponding version of Theorem 4.9.2 in the context of the three–stage procedure (4.9.14).

4.16 Prove Theorem 4.9.5.

4.17 In the case $k = 2$, show that $b_1 = b_2 = -\frac{1}{2}e^{-a} = P^* - 1$. Hence observe that the expansion given in Theorem 4.5.2(ii) follows from that given in Theorem 4.5.2(iii).

4.18 Use the definitions of $g_{11}^{(s)}(x)$, $g_{12}^{(s)}(x)$ from (4.9.9) – (4.9.10) and obtain expansions like those in Theorems 3.9.3 – 3.9.4(iii) under non–LFC situations in $\Omega^{(s)}(\delta^*)$ instead of what one has in Theorems 4.9.3 – 4.9.4(iii) for the LFC scenario.

REFERENCES

Anscombe (1952), Bain (1978), Bechhofer (1954), Bechhofer, Kiefer and Sobel (1968), Bristol, Chen and Mithongtae (1987), Bristol and Desu (1985), Chow and Robbins (1965), Desu, Narula and Villarreal (1977), Dudewicz (1971), Grubbs (1971), Hall (1981), Johnson and Kotz (1970), Lawless and Singhal (1980), Lombard and Swanepoel (1978), Miller and Mukhopadhyay (1990), Mukhopadhyay (1984, 1986, 1987), Mukhopadhyay and Hamdy (1984a), Mukhopadhyay and Solanky (1992b), Paulson (1964), Raghavachari and Starr (1970), Stein (1945), Weiner (1939), Zelen (1966).

5

ESTIMATION OF ORDERED PARAMETERS

5.1 Introduction

Suppose that one is dealing with the basic scenarios considered in Chapters 3 or 4, that is one has $k(\geq 2)$ independent populations all of which are either (a) normal having unknown means and unknown common variance, or (b) negative exponential having unknown locations and unknown common scale parameter. The problem is to estimate the largest location parameter in both situations in such a way that the final estimator can be claimed to have a *predetermined level of accuracy* in some sense, under various specific types of loss functions under consideration.

The area of estimating ordered parameters is certainly vast in its own right. See, for example, Chapter 21 of Gupta and Panchapakesan (1979) and the books of Barlow et al. (1972) and Dykstra et al. (1986). In this chapter, however, we explore and summarize a few interesting ordered parameter estimation problems where the unified theories discussed in Chapter 2 have fruitfully led to interesting *asymptotic second-order results* for the associated purely sequential and other multistage estimators. We may add that our present specific endeavors may seem somewhat narrow at first, and yet the future scope of research in this otherwise broad area seems virtually endless.

5.2 Fixed-width confidence intervals

The construction of *fixed-width confidence intervals* for parameters of interest is of considerable practical value. Traditionally, the usual confidence intervals have random widths and in certain situations, such random widths can exceed "specifications" with "unacceptable" levels of probabilities. Hence, the experimenter may like to specify a predetermined width and associated confidence coefficient for the confidence interval that is to be constructed. Along the lines of Dantzig (1940), it is easy to see the non–existence of any fixed–sample size procedure for our two specific types of problems on hand. See also Dudewicz (1971) for some of the details. Thus, one has to pursue various multistage and purely sequential procedures for such estimation problems. Again, the emphasis lies in the area of *second–order asymptotics.*

5.2.1 LARGEST MEAN OF NORMAL POPULATIONS

Inference on the largest mean of $k(\geq 2)$ populations is often of interest in statistics, especially in the areas of quality control, reliability analysis and multiple comparisons. Let $\pi_1,\ldots,\pi_k$ be independent populations with the distribution of the ith population being $N(\mu_i, \sigma^2)$, $i = 1,\ldots,k$. We assume that the means $\mu_1,\ldots,\mu_k$ and the common variance σ^2 are *all unknown parameters.* We write $\mu_{[k]} = \max\{\mu_1,\ldots,\mu_k\}$ and $\underset{\sim}{\mu} = (\mu_1,\ldots,\mu_k)$ as in Chapter 3. Having recorded i.i.d. random variables $X_{i1},\ldots,X_{in}$ from π_i, we write $\overline{X}_{in}$, S^2_{in}, $U(n)$ and $U^*(n)$ as in (3.1.1), $i = 1,\ldots,k$. We also let

$$\overline{X}^*_n = \max\{\overline{X}_{1n},\ldots,\overline{X}_{kn}\} \tag{5.2.1}$$

and $\overline{X}^*_n$ is taken to be the estimator of $\mu_{[k]}$. For fixed $d(>0)$ and $0 < a < 1$, the goal is to develop procedures so that the width of the confidence interval for $\mu_{[k]}$ is $2d$ and the coverage probability is at least $(1-a)$ or approximately $(1-a)$, up to various orders of approximations. Kuo and Mukhopadhyay (1990a) had strengthened Tong's (1970) results by obtaining second–order expansions of various characteristics for his purely sequential fixed–width symmetric confidence interval procedure. Kuo and Mukhopadhyay (1990a) had also discussed similar properties for the corresponding three–stage procedure. Recently, Mukhopadhyay et al. (1993)

have proposed an accelerated sequential procedure for the same problem. We should add that such procedures have not yet been developed for non–symmetric intervals similar to those of Tong (1973).

Given $d(>0)$, we consider $I_n = [\overline{X}_n^* - d, \overline{X}_n^* + d]$ as the fixed–width confidence interval for $\mu_{[k]}$ when we have recorded $n(\geq 2)$ samples from each π. Saxena and Tong (1969) proved that

$$\inf_{\underset{\sim}{\mu}} P\{\mu_{[k]} \in I_n\} = \{\Phi(dn^{\frac{1}{2}}/\sigma)\}^k - \{\Phi(-dn^{\frac{1}{2}}/\sigma)\}^k \tag{5.2.2}$$

for every fixed d, σ and n, the infimum in (5.2.2) is attained when $\mu_1 = \cdots = \mu_k$. Therefore, to achieve the desired coverage probability, that is $P\{\mu_{[k]} \in I_n\} \geq 1 - \alpha$, we first determine "a" such that $\Phi^k(a) - \Phi^k(-a) = 1 - \alpha$, and then n needs to be the smallest integer $\geq a^2\sigma^2/d^2 = C$, say. Since C is unknown, we utilize purely sequential and multistage sampling techniques as we did in Chapters 3 and 4. Various second order approximations are then carried out asymptotically as $d \to 0$. In passing, let us mention that C is called the *optimal fixed sample size* needed from each population, had σ been known.

Suppose, in general, one devises a statistical methodology that leads to a positive integer valued random variable N estimating C, thereby estimating $\mu_{[k]}$ by means of the confidence interval

$$I_N = [\overline{X}_N^* \pm d]. \tag{5.2.3}$$

Theorem 5.2.1. *Suppose that* N *satisfies the following conditions. The distribution of* N *does not depend on* $\underset{\sim}{\mu}$; $I(N = n)$ *is independent of* $(\overline{X}_{1n},\ldots,\overline{X}_{kn})$ *for every fixed* $n \geq 2$; $P(N < \infty) = 1$. *Then*

$$\inf_{\underset{\sim}{\mu}} P\{\mu_{[k]} \in I_N\} = E[\{\Phi(N^{\frac{1}{2}}d\sigma^{-1})\}^k - \{\Phi(-N^{\frac{1}{2}}d\sigma^{-1})\}^k]$$

and this infimum is attained when $\mu_1 = \cdots = \mu_k$.

Proof. Suppose that $m_1(\geq 2)$ is the starting sample size from each π giving rise to N. The condition $P(N < \infty) = 1$ guarantees that $\overline{X}_N^*$ is well–defined. Now,

$$P\{\mu_{[k]} \in I_N\} = \Sigma_{n=m_1}^{\infty} P\{\mu_{[k]} \in I_N \mid N=n\} P(N=n)$$

$$= \Sigma_{n=m_1}^{\infty} P\{\mu_{[k]} \in I_n\} P(N=n), \tag{5.2.4}$$

since $I(N = n)$ and $(\overline{X}_{1n},\ldots,\overline{X}_{kn})$ are independent for all $n \geq m_1$. Next, utilize (5.2.2) and note that $P\{\mu_{[k]} \in I_n\}$ is minimized for all fixed $n(\geq m_1)$ at the *same parameter configuration*, namely when $\mu_1 = \cdots = \mu_k$. Also, $P(N = n)$ does not depend upon $\underset{\sim}{\mu}$. Hence, (5.2.4) leads to

$$\inf_{\underset{\sim}{\mu}} P\{\mu_{[k]} \in I_N\} = \Sigma_{n=m_1}^{\infty} \inf_{\underset{\sim}{\mu}} P\{\mu_{[k]} \in I_n\} P(N=n) = E[\psi_1(d^2N/\sigma^2)], \tag{5.2.5}$$

where we let

$$\psi_1(x) = \{\Phi(x^{\frac{1}{2}})\}^k - \{\Phi(-x^{\frac{1}{2}})\}^k \tag{5.2.6}$$

for $x > 0$. Obviously, the infimum in (5.2.5) is attained when one has $\mu_1 = \cdots = \mu_k$. □

In what follows, one by one we provide different multistage estimation techniques.

5.2.1.1 Two–Stage Procedure

The procedures of Chen and Dudewicz (1973) and Tong (1970) are summarized here. Let $G_{k,\eta^0}(x_1,\ldots,x_k)$ be the d.f. of k–dimensional multivariate t having η^0 degrees of freedom and zero equicorrelation. Let $m_1(\geq 2)$ be the starting sample size from each π and let us define $\tau_1 = \tau_1(m_1, d, a)$ in such a way that

$$G_{k,\eta^0}(\tau_1,\ldots,\tau_1) - G_{k,\eta^0}(-\tau_1,\ldots,-\tau_1) = 1-a, \tag{5.2.7}$$

with $\eta^0 = k(m_1 - 1)$. Define

$$N = N(d) = \max\{m_1, [\tau_1^2 U(m_1)d^{-2}]^* + 1\}. \tag{5.2.8}$$

If $N = m_1$, we do not take any more samples from any population. However, if $N > m_1$, then we sample the difference in the second stage, that is, we take $(N - m_1)$ more samples from each π in the second stage. Based on the totality of all samples, that is having $X_{i1},\ldots,X_{iN}$ from π_i, $i = 1,\ldots,k$ we now estimate $\mu_{[k]}$ by the fixed–width confidence interval I_N given by (5.2.3). Throughout, we keep $\underset{\sim}{\mu}$ and σ fixed.

Theorem 5.2.2. *For the two–stage procedure (5.2.8), with* τ_1 *defined in (5.2.7), we have:*

(i) $P\{\mu_{[k]} \in I_N\} \geq 1 - a$ *for all* $\underset{\sim}{\mu}$;

(ii) $\tau_1^2\sigma^2 d^{-2} \leq E(N) \leq m_1 + \tau_1^2\sigma^2 d^{-2}$;

(iii) $E(N/C) \to \tau_1^2/a^2$ *as* $d \to 0$;

(iv) $\liminf P\{\mu_{[k]} \in I_N\} \geq 1 - \alpha$ *for all* $\underset{\sim}{\mu}$, *as* $d \to 0$;

where $C = a^2\sigma^2/d^2$ *and* $\Phi^k(a) - \Phi^k(-a) = 1 - \alpha$.

Parts (i) – (iii) were proved in Tong (1970). For completeness, however, we give a fairly simple proof.

Proof. The basic inequality (2.2.3) turns out to be

$$\tau_1^2 U(m_1) d^{-2} \leq N \leq m_1 + \tau_1^2 U(m_1) d^{-2}, \tag{5.2.9}$$

and now taking expectations throughout leads to part (ii), since $E\{U(m_1)\} = \sigma^2$. Now, dividing throughout the expressions in part (ii) by C and taking limits as $d \to 0$ leads to part (iii). In order to prove part (i), note from (5.2.9) that $Nd^2\sigma^{-2} \geq \tau_1^2 U(m_1)\sigma^{-2}$ w.p. 1 and hence we get from (5.2.5),

$$\inf_{\underset{\sim}{\mu}} P\{\mu_{[k]} \in I_N\} \geq E[\psi_1(\tau_1^2 U(m_1)\sigma^{-2})]$$

$$= E[P\{Y_i \leq \tau_1 U^{\frac{1}{2}}(m_1)\sigma^{-1}, i = 1,\ldots,k\} - P\{Y_i \leq -\tau_1 U^{\frac{1}{2}}(m_1)\sigma^{-1},$$

$$i = 1,\ldots,k\} \Big| U(m_1)]$$

where $Y_1,\ldots,Y_k$ are i.i.d. $N(0, 1)$ and these are chosen to be independent of $U(m_1)$. Let $Z_i = Y_i U^{-\frac{1}{2}}(m_1)\sigma$, $i = 1,\ldots,k$ and note that $(Z_1,\ldots,Z_k)$ is distributed as the k–dimensional multivariate t having degrees of freedom $\eta^0 = k(m_1 - 1)$ and zero equicorrelation. Hence, part (i) follows by the choice of τ_1.

Next, observe that $Nd^2\sigma^{-2} \to \tau_1^2 U(m_1)\sigma^{-2}$ w.p. 1 as $d \to 0$, which follows from (5.2.9). Hence, for all $\underset{\sim}{\mu}$, from (5.2.5) one obtains by way of dominated convergence theorem,

$$\liminf_{d\to 0} P\{\mu_{[k]} \in I_N\} \geq \lim_{d\to 0} E[\psi_1(d^2N/\sigma^2)]$$

$$= E[\psi_1(\tau_1^2 U(m_1)/\sigma^2)] = 1 - \alpha,$$

as before. □

Remark 5.2.1. Tong (1970) gave an approximation of τ_1 based on the corresponding percentiles of a Student's t–distribution. The numerical value of τ_1 seems to exceed the corresponding "a", in other words, the limiting ratio given in Theorem 5.2.2 (iii) does appear to be larger than unity, that is the two–stage procedure (5.2.8) seems to oversample, even asymptotically. Recall that we noticed such undesirable features in sections 3.3 and 4.3 of Chapters 3 and 4, respectively as well.

5.2.1.2 Modified Two–Stage Procedure

Here, a modified two–stage procedure is proposed along the lines of sections 2.2.2, 3.4 and 4.4 so that one can achieve *asymptotic first–order efficiency property* (3.4.7). The actual construction is motivated along the lines of (3.4.1) – (3.4.2).

We choose $m_0^*(\geq 2)$, $\gamma \in R^+$ and let

$$m_1 = m_1(d) = \max\{m_0^*, [(a^2d^{-2})^{1/(\gamma+1)}]^* + 1\}, \tag{5.2.10}$$

where recall that $\Phi^k(a) - \Phi^k(-a) = 1 - \alpha$. This m_1 is then the experimenter's choice of the starting sample size from each population. With τ_1 defined in (5.2.7), let

$$N = N(d) = \max\{m_1, [\tau_1^2 U(m_1) d^{-2}]^* + 1\}, \tag{5.2.11}$$

and from a practical point of view, we implement the modified version (5.2.10) – (5.2.11) in quite the same way as (5.2.8) and finally propose the fixed–width confidence interval I_N given in (5.2.3) as the estimator of $\mu_{[k]}$. One holds $\underset{\sim}{\mu}$ and σ fixed throughout.

Theorem 5.2.3. *For the modified two–stage procedure (5.2.10) – (5.2.11), with* τ_1 *defined in (5.2.7), we have:*

(i) $P\{\mu_{[k]} \in I_N\} \geq 1 - \alpha$ *for all* $\underset{\sim}{\mu}$;

(ii) $N/C \to 1$ *w.p.* 1 *and* $E(N/C) \to 1$ *as* $d \to 0$;

(iii) $\liminf P\{\mu_{[k]} \in I_N\} \geq 1 - \alpha$ *for all* $\underset{\sim}{\mu}$ *as* $d \to 0$;

where $C = a^2\sigma^2/d^2$ *and* $\Phi^k(a) - \Phi^k(-a) = 1 - \alpha$.

Proof. Part (i) is immediate from Theorem 5.2.2 (i). Now, we turn to part (ii). Note that $m_1(d) \to \infty$ and $m_1(d)/C \to 0$ as $d \to 0$, whatever $\gamma \in R^+$. Also, note that $\tau_1 = \tau_1(m_1, k, \alpha)$ is actually a function of d and $\tau_1 \to a$ as $d \to 0$. Hence, one now can write down a basic inequality along the line of (3.4.3) and prove part (ii) in the same way as Theorem 3.4.1 (ii). Proof of part (iii) then follows from a combination of ideas found in the proof of Theorem 5.2.2 (iv) and Theorem 3.4.1 (iii). □

Remark 5.2.2. The two–stage and modified two–stage procedures are both *consistent* and *asymptotically consistent.* The modified two–stage procedure is *asymptotically first–order efficient* also. See (3.4.5) – (3.4.8) in this context.

5.2.1.3 Purely Sequential Procedure

In this and following sections, we consider procedures where one can decisively claim that $E(N - C)$ remains bounded as $d \to 0$. We also discuss other types associated *second-order characteristics* of the proposed methodologies. First, let us look at the purely sequential estimation procedure of Tong (1970).

We start with $m_1(\geq 2)$ samples from each π. Then, we keep taking one more sample at a time from each population according to the following stopping rule:

$$N = N(d) = \inf\{n \geq m_1 \ : \ n \geq a^2 U(n) d^{-2}\}. \tag{5.2.12}$$

For fixed $\underset{\sim}{\mu}$, σ^2, m_1 and d, we have $P(N < \infty) = 1$. Based on the totality of all samples, that is having $X_{i1},\ldots,X_{iN}$ from π_i, $i = 1,\ldots,k$, we now estimate $\mu_{[k]}$ by the fixed-width confidence interval I_N given by (5.2.3). Throughout, we keep $\underset{\sim}{\mu}$ and σ fixed.

Theorem 5.2.4. *For the purely sequential procedure (5.2.12), we have as* $d \to 0$:

(i) $N/C \to 1$ *w.p.* 1;

(ii) $E(N/C) \to 1$;

(iii) $C^{-\frac{1}{2}}(N - C) \xrightarrow{\mathscr{L}} N(0, 2/k)$;

(iv) $\liminf P\{\mu_{[k]} \in I_N\} \geq 1 - \alpha$ *for all* $\underset{\sim}{\mu}$;

where $C = a^2\sigma^2/d^2$ *and* $\Phi^k(a) - \Phi^k(-a) = 1 - \alpha$.

This theorem can be verified along the lines of Theorem 3.5.1 with obvious modifications. Details are omitted for brevity. The rate of convergence for the result given in Theorem 5.2.4 (iii) in the non–normal case can also be derived if we follow Remark 3.5.2. To be specific, with $0 < \gamma < \infty$, let

$$N = N(d) = \inf\{n \geq m_1 \ : \ n \geq a^2 d^{-2}[U(n) + n^{-\gamma}]\}.$$

Theorem 2.4.4 implies that for $\gamma > \frac{1}{2}$, we have

$$C^{-\frac{1}{2}}(N - C) \xrightarrow{\mathscr{L}} N(0, 2/k),$$

while Theorem 2.4.7 leads to

$$\sup_x \mid P[(2C/k)^{-\frac{1}{2}}(N - C) \leq x] - \Phi(x) \mid = 0(d^{\frac{1}{2}-\lambda}),$$

for $\gamma > 3/4$ and $0 < \lambda < \frac{1}{2}$, as $d \to 0$.

We are now in a position to address the results of Kuo and Mukhopadhyay (1990a) and discuss various asymptotic second–order properties of the purely sequential estimation procedure (5.2.12). Note, however, that this procedure is analogous to the purely sequential procedure given in (3.5.1) and thus one can write $N = Q + 1$ where Q comes from (3.5.6). Let $U^* = (N - C)^2/C$. Then, similar to Theorem 3.5.2 (i), we can claim here that

> $E(N) = C + \beta + o(1)$ for all $\underset{\sim}{\mu}$ if (a) $m_1 \geq 3$ when $k = 2$, (b) $m_1 \geq 2$ when $k \geq 3$ where $\beta = \beta(k) = k^{-1}\{\nu(k)-2\}$, $\nu(k)$ being defined in equation (3.5.7). (5.2.13)

Also, from (3.5.8), one obtains

$$U^* \text{ is uniformly integrable when } m_1 > 1 + 2k^{-1}. \qquad (5.2.14)$$

A combination of Theorem 5.2.4 (iii) and (5.2.14) leads to

$$E(U^*) = 2k^{-1} + o(1) \text{ if } m_1 > 1 + 2k^{-1}. \tag{5.2.15}$$

The asymptotic expansion of the coverage probability is given in the following theorem. Recall that $\inf_{\underset{\sim}{\mu}} P\{\mu_{[k]} \in I_N\}$ is attained when $\mu_1 = \cdots = \mu_k$, a parameter configuration that is, as before, referred as the LFC.

Theorem 5.2.5. *For the purely sequential procedure (5.2.12), we have as* $d \to 0$:

$P\{\mu_{[k]} \in I_N\} = (1-\alpha) + a^2C^{-1}\{\beta\psi_1'(a^2) + a^2k^{-1}\psi_1''(a^2)\} + o(d^2)$ *under the LFC if* (a) $m_1 \geq 4$ *when* $k = 2$, (b) $m_1 \geq 3$ *when* $k = 3, 4, 5$, (c) $m_1 \geq 2$ *when* $k \geq 6$;

where $C = a^2\sigma^2/d^2$, $\Phi^k(a) - \Phi^k(-a) = 1 - \alpha$, *and* $\psi_1(x)$ *and* β *respectively come from (5.2.6) and (5.2.13).*

Proof. From Theorem 5.2.1, under the LFC, one has

$$P\{\mu_{[k]} \in I_N\} = (1-\alpha) + a^2\psi_1'(a^2)C^{-1}E(N-C) + \tfrac{1}{2}a^4C^{-1}E\{U^*\psi_1''(T^*)\} \tag{5.2.16}$$

where T^* is a suitable random variable between Nd^2/σ^2 and a^2. Now, observe that $\psi_1''(x) \leq a_1x^{-\frac{1}{2}} + a_1x^{-1} + a_2x^{-3/2} = g^*(x)$, say, for $x > 0$ where $a_1 = \frac{1}{2}k(2\pi)^{-\frac{1}{2}}$ and $a_2 = \frac{1}{4}k(k-1)\pi^{-1}$ are two positive constants. It is easy to see that $T^* > \frac{1}{2}a^2$ on the set $N > \frac{1}{2}C$, and thus $|U^*\psi_1''(T^*)I(N > \frac{1}{2}C)| \leq g^*(\frac{1}{2}a^2)U^*$, so that the left hand side is uniformly integrable if $m_1 > 1 + 2k^{-1}$. See equation (5.2.15). Thus, for $m_1 > 1 + 2k^{-1}$, we have

$$E\{U^*\psi_1''(T^*)I(N > \tfrac{1}{2}C)\} = 2k^{-1} + o(1). \tag{5.2.17}$$

On the other hand, $T^* > Nd^2/\sigma^2$ on the set $N \leq \frac{1}{2}C$. Thus,

$$E\{|U^*\psi_1''(T^*)I(N \le \tfrac{1}{2}C)|\} \le E\{C(1-NC^{-1})^2 g^*(T^*)I(N \le \tfrac{1}{2}C)\}$$

$$\le CE\{g^*(Nd^2/\sigma^2)I(N \le \tfrac{1}{2}C)\}. \tag{5.2.18}$$

Now, the slowest term in (5.2.18) is of the form $CE\{(C/N)^{3/2}I(N \le \frac{1}{2}C)\} \le C^{\frac{5}{2}}P(N \le \frac{1}{2}C) = 0(C^{\frac{5}{2} - \frac{1}{2}(m_1-1)k}) = o(1)$ if $m_1 > 1 + 5k^{-1}$. Hence combining this with (5.2.16) and (5.2.17), we get the theorem. □

5.2.1.4 Three–Stage Procedure

In order to reduce sampling operations needed by the purely sequential procedure (5.2.12) and yet maintain asymptotic second–order characteristics, Kuo and Mukhopadhyay (1990a) had also proposed a suitable three–stage procedure for the problem. Recall that $C = a^2\sigma^2/d^2$ where $\Phi^k(a) - \Phi^k(-a) = 1 - \alpha$. Similar to the methodology discussed in section 3.7, we start with an initial sample of size $m_1 = m_1(\delta^*)$ (≥ 2) from each population π where

$$m_1 = m_1(\delta^*) = 0(C^{1/r})$$

for some $r > 1$. We choose and fix $0 < \rho < 1$ and let

$$T = T(d) = \max\{m_1, [\rho a^2 U(m_1)d^{-2}]^* + 1\}, \tag{5.2.19}$$

$$N = N(d) = \max\{T, [a^2 U(T)d^{-2}]^* + 1\}. \tag{5.2.20}$$

The procedure (5.2.19) – (5.2.20) is implemented as we did in the case of (3.7.2) – (3.7.3). That is, we sample the difference $(T - m_1)$ and $(N - T)$ respectively in the second and third stage from each π. Then, based on the totality of all samples $X_{i1},\ldots,X_{iN}$ from π_i, $i = 1,\ldots,k$, $\mu_{[k]}$ is now estimated by the fixed–width confidence interval I_N given by (5.2.3).

Throughout, we keep $\underset{\sim}{\mu}$ and σ fixed.

Theorem 5.2.6. *For the three-stage procedure (5.2.19) – (5.2.20), we have as* $d \to 0$:

(i) $N/C \to 1$ *w.p.* 1;

(ii) $E(N/C) \to 1$;

(iii) $C^{-\frac{1}{2}}(N - C) \xrightarrow{\mathscr{L}} N(0, 2(k\rho)^{-1})$;

(iv) $\liminf P\{\mu_{[k]} \in I_N\} \geq 1 - \alpha$ *for all* $\underset{\sim}{\mu}$;

where $C = a^2\sigma^2/d^2$ *and* $\Phi^k(a) - \Phi^k(-a) = 1 - \alpha$.

The first–order asymptotic properties listed in this theorem can be proved quite analogously as we did in the case of Theorem 3.7.1. Details are omitted. Also, along the lines of Theorem 3.7.2, one can prove that

$$E(N) = C + \beta' + o(1) \text{ and } Var(N) = 2(k\rho)^{-1}C + o(d^{-2})$$
$$\text{for all } \underset{\sim}{\mu} \text{ where } \beta' = \tfrac{1}{2} - 2(k\rho)^{-1}. \qquad (5.2.21)$$

We now modify the three–stage procedure (5.2.19) – (5.2.20) so as to be able to conclude that $P\{\mu_{[k]} \in I_N\} = (1 - \alpha) + o(d^2)$ under the LFC as $d \to 0$. Such modified estimation procedure is defined as follows. Let

$$T = T(d) = \max\{m_1, [\rho a^2 \mathrm{U}(m_1)d^{-2}]^* + 1\},$$

$$m_1^* = \{2\psi_1'(a^2) - a^2\psi_1''(a^2)\}\{\rho k\psi_1'(a^2)\}^{-1} - \tfrac{1}{2},$$

$$N = N(d) = \max\{T, [a^2\mathrm{U}(T)d^{-2} + m_1^*]^* + 1\}, \qquad (5.2.22)$$

and propose the confidence interval I_N given by (5.2.3) for $\mu_{[k]}$. The moment expansions given in (5.2.21) immediately lead to

$$E(N) = C + \beta' + m_1^* + o(1) \text{ and } Var(N) = 2(k\rho)^{-1}C + o(d^{-2})$$

$$\text{for all } \underset{\sim}{\mu}, \text{ when N comes from (5.2.22).} \tag{5.2.23}$$

Theorem 5.2.7. *For the three–stage procedure (5.2.22), we have as* $d \to 0$:

$$P\{\mu_{[k]} \in I_N\} = (1 - \alpha) + o(d^2)$$

under the LFC.

Proof. One can give a direct proof of this result along the lines of Theorem 3.7.3. We however derive the result from Theorem 2.3.4. From Theorem 5.2.1, under the LFC, one gets

$$P\{\mu_{[k]} \in I_N\} = E[\psi_1(d^2N/\sigma^2)], \tag{5.2.24}$$

the ψ_1–function being defined in (5.2.6).

The original three–stage procedure (5.2.19) – (5.2.20) has the same representations given in (2.3.1) – (2.3.2) where Y's are i.i.d. $\sigma^2 k^{-1}\chi_k^2$. Now, (5.2.24) has the same form as in (2.3.27) with $\zeta = 1$, $L = d^2$ and $\theta^* = \sigma^2$. Also, $G^*(\cdot) = \psi_1(\cdot)$ and $Var(W_1) = 2k^{-1}$. Hence, the choice of m_1^* coincides with the choice of ϵ^* in Theorem 2.3.4. One only has to replace "a" by "a^2" in the expression of ϵ^*. □

5.2.1.5 Accelerated Sequential Procedure

In this section, we refer to Mukhopadhyay et al. (1993) and propose an accelerated version of the purely sequential procedure (5.2.12). See sections 2.5 and 3.6 for some basic ideas. Note that $C = a^2\sigma^2/d^2$ and we now use

$U^*(n)$ defined in (3.1.1) as the estimator of σ^2.

We first choose and fix $0 < \rho < 1$ and $q \geq 0$ and let

$$t = t(d) = \inf\{n \geq m_1 : n \geq \rho a^2 U^*(n)/d^2\}, \tag{5.2.25}$$

$$N_1 = N_1(d) = [a^2 U^*(t) d^{-2} + q]^* + 1, \tag{5.2.26}$$

$$N = N(d) = \max\{t(d), N_1(d)\}. \tag{5.2.27}$$

We start the experiment with $m_1(\geq 2)$ samples from each π. Then, we sample one at a time from each π sequentially and end up with $X_{i1},\ldots,X_{it}$ from π_i, $i = 1,\ldots,k$. Note that $t(d)$ estimates ρC, a fraction of C. Now, $N_1(d)$ estimates C based on t samples observed from each π. If $N = t$, we do not take any more samples. If $N > t$, we sample the difference $(N - t)$ from each π in one single batch at this stage. Then, based on the totality of all samples $X_{i1},\ldots,X_{iN}$ from π_i, $i = 1,\ldots,k$, $\mu_{[k]}$ is estimated by the fixed–width confidence interval I_N given by (5.2.3). Throughout, we keep $\underset{\sim}{\mu}$ and σ fixed. Observe that Theorem 5.2.1 again holds.

Theorem 5.2.8. *For the accelerated sequential procedure (5.2.25) – (5.2.27), we have as* $d \to 0$:

(i) $N/C \to 1$ *w.p.* 1;

(ii) $E(N/C) \to 1$;

(iii) $C^{-\frac{1}{2}}(N - C) \xrightarrow{\mathscr{L}} N(0, 2(k\rho)^{-1})$;

(iv) $\liminf P\{\mu_{[k]} \in I_N\} \geq 1 - \alpha$ *for all* $\underset{\sim}{\mu}$;

where $C = a^2\sigma^2/d^2$ *and* $\Phi^k(a) - \Phi^k(-a) = 1 - \alpha$.

This theorem can be proved along the lines of Theorem 3.6.1. We omit further details. Now, similar to Theorem 3.6.2 (i) – (ii), one can claim that

$q - 2(k\rho)^{-1} \leq \liminf E(N-C) \leq \limsup E(N-C) \leq q + 1 - 2(k\rho)^{-1}$ if $m_1 \geq 4$, and $Var(N) = 2(k\rho)^{-1}C + o(d^{-2})$ if (a) $m_1 \geq 3$ when $k = 2$, (b) $m_1 \geq 2$ when $k \geq 3$. (5.2.28)

Theorem 5.2.9. *For the accelerated sequential procedure (5.2.25) – (5.2.27), we have as* $d \to 0$:

$$P\{\mu_{[k]} \in I_N\} = (1 - \alpha) + o(d^2)$$

under the LFC when $m_1 \geq 4$ *if* $q = \max\{0, (k\rho)^{-1}[2 - \frac{1}{2}(a^2 + 1) + \frac{1}{2}(k-1)a\phi(a)\frac{\Phi^{k-2}(a)-\Phi^{k-2}(-a)}{\Phi^{k-1}(a)+\Phi^{k-1}(-a)}]\}$, *where* $\Phi^k(a) - \Phi^k(-a) = 1 - \alpha$.

Proof. Recall $\psi_1(x)$ from (5.2.6). Let $l = d^2/\sigma^2$. From Theorem 5.2.1, under the LFC, one has

$$P\{\mu_{[k]} \in I_N\} = E[\psi_1(lN)]$$

$$= \psi_1(l\,E(N)) + \tfrac{1}{2}l^2 E\{(N-E(N))^2\}\psi_1''(l\,E(N)) + r_1(l), \text{ say,} \quad (5.2.29)$$

where $|r_1(l)| = 0(l^3 E(|N-E(N)|^3))$. Now, note that the accelerated sequential procedure (5.2.25) – (5.2.27) is covered by the unified theory discussed in section 2.5.1 where m is *replaced* by $m_1 - 1$ and its $\delta = 1$, $h = k/C$, $\theta = k$, $\tau^2 = 2k$, $a = \frac{1}{2}k$, $p = 2/k$, $n_0^* = C$, and $m_o(\delta) = 2.02$ or 2 according as $k = 2$ or $k \geq 3$ respectively. Also, Theorem 2.5.2 (iii) immediately gives $|r_1(l)| = 0(l^3 l^{-3/2}) = 0(l^{3/2}) = o(l)$ if $m_1 > 1 + 3k^{-1}$. Again, since $\psi_1(a^2) = 1 - \alpha$, we write for a suitable $z(>0)$,

$$\psi_1(l\,E(N)) = (1-\alpha) + \{l\,E(N)-a^2\}\psi_1'(a^2) + r_2(l) \qquad (5.2.30)$$

where $|r_2(l)| = \frac{1}{2}|(l\,E(N)-a^2)^2\psi_1''(z)| = O(l^2) = o(l)$, if one has $m_1 > 1 + \max\{m_0(\delta), 3k^{-1}\}$, in view of Theorem 2.5.2 (ii) and (5.2.28). Now, combining (5.2.28) – (5.2.30), we get

$$P\{\mu_{[k]} \in I_N\} \geq (1-\alpha) + l\psi_1'(a^2)\{q-2(k\rho)^{-1} + o(1)\}$$

$$+ l\{a^2(k\rho)^{-1}\psi_1''(a^2) + o(1)\} \qquad (5.2.31)$$

if $m_1 > 1 + \max\{m_0(\delta), 3k^{-1}\}$. Thus, from (5.2.31), we obtain,

$$P\{\mu_{[k]} \in I_N\} \geq (1-\alpha) + o(d^2), \qquad (5.2.32)$$

if $q = \max[0, (k\rho)^{-1}\{2 - a^2 \dfrac{\psi_1''(a^2)}{\psi_1'(a^2)}\}]$. Hence, the theorem, once one simplifies $\psi_1''(a^2)/\psi_1'(a^2)$ in terms of $\Phi(a)$ and $\phi(a)$. □

Remark 5.2.3. Note that the purely sequential, accelerated sequential and three–stage estimation procedures are all *asymptotically second–order efficient.* See (3.4.5) – (3.4.8) in this context.

5.2.2 Largest Location Among Negative Exponential Populations

Consider $k(\geq 2)$ independent negative exponential populations $\pi_1,\ldots,\pi_k$ where the ith population's p.d.f. is given by $f(x;\, \theta_i,\, \sigma)$ where $f(x;\, \mu,\, \sigma)$ was defined in (2.2.5). It is assumed that $\theta_1,\ldots,\theta_k$ are *unknown locations* $(\in \mathbb{R})$ and $\sigma(\in \mathbb{R}^+)$ is the *unknown scale* parameter, as in Chapter 4. We write $\theta_{[k]} = \max\{\theta_1,\ldots,\theta_k\}$ and $\underset{\sim}{\theta} = (\theta_1,\ldots,\theta_k)$ as in Chapter 4. Having recorded i.i.d. random variables $X_{i1},\ldots,X_{in}$ from π_i, we write $\tilde{X}_{in}$, $V(n)$

and $V^*(n)$ as in (4.1.1), i = 1,...,k. We also let

$$T_n^* = \max\{\tilde{X}_{1n},\ldots,\tilde{X}_{kn}\} \tag{5.2.33}$$

which estimates $\theta_{[k]}$. For fixed $d(>0)$ and $0 < \alpha < 1$, the goal is to develop procedures so that the width of the confidence interval for $\theta_{[k]}$ is d and the coverage probability is at least $(1 - \alpha)$ or approximately $(1 - \alpha)$. Here, we discuss various multistage and sequential fixed–width confidence interval procedures as developed in Mukhopadhyay and Kuo (1988). For completeness, new results for the corresponding accelerated sequential methodology have also been included. For $k = 1$, such estimation problems were carried out in Ghurye (1958), Mukhopadhyay (1974, 1982), Swanepoel and van Wyk (1982), and Mukhopadhyay and Mauromoustakos (1987). See the recent articles of Mukhopadhyay (1988, 1993b) for a review.

Note that $T_n^* \geq \theta_{[k]}$ w.p. 1, for all n. Hence, given $d(>0)$, we consider the fixed–width confidence interval $I_n = [T_n^* - d, T_n^*]$ for $\theta_{[k]}$ when we have recorded $n(\geq 2)$ samples from each π. Then, one has

$$\begin{aligned} P\{\theta_{[k]} \in I_n\} &= P\{\theta_{[k]} \leq T_n^* \leq \theta_{[k]} + d\} \\ &= P\{\tilde{X}_{in} \leq \theta_{[k]} + d,\ i = 1,\ldots,k\} \\ &= \prod_{i=1}^{k} [1 - \exp\{-n(d + \{\theta_{[k]} - \theta_i\})/\sigma\}], \end{aligned} \tag{5.2.34}$$

and hence one can see that

$$P\{\theta_{[k]} \in I_n\} \geq \{1 - \exp(-nd/\sigma)\}^k \tag{5.2.35}$$

for all $\underset{\sim}{\theta}$, while the equality in (5.2.35) holds when $\theta_1 = \cdots = \theta_k$, a parameter configuration referred in the sequel as LFC. That is,

$$\operatorname*{Inf}_{\underset{\sim}{\theta}} P\{\theta_{[k]} \in I_n\} = [1 - \exp(-nd/\sigma)]^k \tag{5.2.36}$$

for every fixed d, σ and n. The infimum in (5.2.36) is of course attained at the LFC. Now, in order to achieve the desired coverage probability, that is $P\{\theta_{[k]} \in I_n\} \geq 1 - \alpha$, we observe that n has to be the smallest integer $\geq a\sigma/d = C$, say, where one defines $a = -\ln[1 - (1-\alpha)^{1/k}]$. The optimal fixed sample size C is indeed unknown and thus the experimenter will need to implement suitable multistage and sequential procedures as before in order to tackle this problem as well.

Suppose, in general, one devises a statistical methodology giving rise to a positive integer valued random variable N estimating C, thereby estimating $\theta_{[k]}$ by means of the confidence interval

$$I_N = [T_N^* - d,\ T_N^*]. \tag{5.2.37}$$

Theorem 5.2.10. *Suppose that* N *satisfies the following conditions. The distribution of* N *does not depend on* $\underset{\sim}{\theta}$; $I(N = n)$ *is independent of* $(\underset{\sim}{X}_{1n},\ldots,\underset{\sim}{X}_{kn})$ *for every fixed* $n \geq 2$; $P(N < \infty) = 1$. *Then*

$$\operatorname*{Inf}_{\underset{\sim}{\theta}} P\{\theta_{[k]} \in I_N\} = E[\{1 - \exp(-Nd/\sigma)\}^k]$$

and this infimum is attained when $\theta_1 = \cdots = \theta_k$.

Proof. Suppose that $m_1(\geq 2)$ is the starting sample size from each π giving rise to N. The condition $P(N < \infty) = 1$ guarantees that T_N^* is well–defined. Now,

$$P\{\theta_{[k]} \in I_N\} = \Sigma_{n=m_1}^{\infty} P\{\theta_{[k]} \in I_n \big| N=n\}P(N=n)$$

$$= \Sigma_{n=m_1}^{\infty} P\{\theta_{[k]} \in I_n\} P(N=n), \tag{5.2.38}$$

since $I(N = n)$ and $(\tilde{X}_{1n},\ldots,\tilde{X}_{kn})$ are independent for all $n \geq m_1$. Now, utilize (5.2.36) and observe that $P\{\theta_{[k]} \in I_n\}$ is minimized for all $n \geq m_1$ at the same parameter configuration, namely the LFC. Also, $P(N = n)$ does not depend on $\tilde{\theta}$. Hence, (5.2.38) leads to

$$\operatorname*{Inf}_{\tilde{\theta}} P\{\theta_{[k]} \in I_N\} = E[\psi_2(Nd/\sigma)] \tag{5.2.39}$$

where we let

$$\psi_2(x) = \{1 - \exp(-x)\}^k, \tag{5.2.40}$$

for $x > 0$. Obviously, the infimum in (5.2.39) is attained when one has $\theta_1 = \cdots = \theta_k$. □

5.2.2.1 Two–Stage Procedure

Let $m_1(\geq 2)$ be the starting sample size from each π and define $\tau^* = \tau^*(m_1, k, \alpha)$ such that

$$P\{F_i \leq \tau^*, \ i = 1,\ldots,k\} = 1 - \alpha, \tag{5.2.41}$$

where $(F_1,\ldots,F_k)$ has Kimbal's (1951) multivariate F–distribution with the degrees of freedom 2 and $2k(m_1 - 1)$. Define

$$N = N(d) = \max\{m_1, [\tau^* V(m_1) d^{-1}]^* + 1\}. \tag{5.2.42}$$

The implementation aspects are similar to those discussed in the context of (5.2.8). Based on the totality of all samples, that is having $X_{i1},\ldots,X_{iN}$ from

π_i, $i = 1,...,k$, we now estimate $\theta_{[k]}$ by the fixed–width confidence interval I_N given by (5.2.37). Throughout, we keep $\underset{\sim}{\theta}$ and σ fixed.

Theorem 5.2.11. *For the two–stage procedure (5.2.42), with* τ^* *defined in (5.2.41), we have*

(i) $P\{\theta_{[k]} \in I_N\} \geq 1 - \alpha$ *for all* $\underset{\sim}{\theta}$;

(ii) $\tau^*\sigma d^{-1} \leq E(N) \leq \tau^*\sigma d^{-1} + m_1$;

(iii) $E(N/C) \to \tau^*/a$ *as* $d \to 0$;

(iv) $\liminf P\{\theta_{[k]} \in I_N\} \geq 1 - \alpha$ *for all* $\underset{\sim}{\theta}$ *as* $d \to 0$;

where $C = a\sigma/d$ *and* $a = -\ln\{1 - (1-\alpha)^{1/k}\}$.

Proof. Proofs of (i) – (iii) are similar to those of Theorem 5.2.2 (i) – (iii). In view of Theorem 5.2.10, since $N \geq \tau^*V(m_1)/d$ w.p. 1, one gets

$$\inf_{\underset{\sim}{\theta}} P\{\theta_{[k]} \in I_N\} = E[\psi_2(Nd/\sigma)] \geq E[\psi_2(\tau^*V(m_1)/\sigma)]$$

$$= E[P\{Y_i \leq \tau^*V(m_1)\sigma^{-1}, i = 1,...,k \mid V(m_1)\}] \tag{5.2.43}$$

where $Y_1,...,Y_k$ are i.i.d. standard exponentials (i.e. the p.d.f. is $f(y; 0, 1)$ given by (2.2.5)) chosen independently of $V(m_1)$. Let $F_i = \sigma Y_i/V(m_1)$, $i = 1,...,k$. Then, $(F_1,...,F_k)$ has the k–dimensional multivariate F–distribution. Hence, by the choice of τ^* given in (5.2.41), part (iv) follows from (5.2.43). □

Remark 5.2.4. For several values of α, k and m_1, the upper percentage point

τ^* of the multivariate F–distribution can be obtained from Finney (1941), Gupta and Sobel (1962) and Krishnaiah and Armitage (1964, 1970). We also noticed that τ^* always exceeded "a", and this indicates that the two–stage procedure (5.2.42) overestimates C, even asymptotically.

5.2.2.2 Modified Two–Stage Procedure

Along the lines of sections 2.2.2 and 5.2.1.2, a modified two–stage procedure is now proposed so that one can achieve *asymptotic first–order efficiency property* (3.4.7). One chooses $m_0^*(\geq 2)$, $\gamma \in R^+$, and defines

$$m_1 = m_1(d) = \max\{m_0^*, [(a/d)^{1/(\gamma+1)}]^* + 1\}. \qquad (5.2.44)$$

This m_1 is then the experimenter's starting sample size from each π. Now, let

$$N = N(d) = \max\{m_1, [\tau^* V(m_1) d^{-1}]^* + 1\}. \qquad (5.2.45)$$

This methodology is implemented in quite the same way as (5.2.10) – (5.2.11). One proposes the fixed–width confidence interval I_N given in (5.2.37) as the estimator of $\theta_{[k]}$.

Theorem 5.2.12. *For the modified two–stage procedure (5.2.44) – (5.2.45), with* τ^* *defined in (5.2.41), we have:*

(i) $P\{\theta_{[k]} \in I_N\} \geq 1 - \alpha$ *for all* $\underset{\sim}{\theta}$;

(ii) $N/C \to 1$ *w.p.* 1 *and* $E(N/C) \to 1$ *as* $d \to 0$;

(iii) $\liminf P\{\theta_{[k]} \in I_N\} \geq 1 - \alpha$ *for all* $\underset{\sim}{\theta}$ *as* $d \to 0$;

where $C = a\sigma/d$ *and* $a = -\ln\{1 - (1-\alpha)^{1/k}\}$.

Proof. Parts (i) – (ii) are similar to Theorem 5.2.3 (i) – (ii). For part (iii), observe that $m_1(d) \to \infty$ as $d \to 0$, and in (5.2.41), $(F_1,\ldots,F_k) \xrightarrow{\mathscr{L}} (Y_1,\ldots,Y_k)$ as $m_1 \to \infty$ where Y's are i.i.d. standard exponentials, that is $\tau^* \to a$ as $d \to 0$. Hence, part (iii) clearly follows. □

Obviously, Remark 5.2.2 holds in this context also. Now, we discuss methodologies which will allow us to claim that $E(N - C)$ remains bounded as $d \to 0$.

5.2.2.3 Purely Sequential Procedure

We start with $m_1(\geq 2)$ samples from each π. Then, we keep taking one more sample at a time from each population according to the following stopping rule:

$$N = N(d) = \inf\{n \geq m_1 \ : \ n \geq aV(n)d^{-1}\}. \tag{5.2.46}$$

For fixed $\underset{\sim}{\theta}$, σ^2, m_1 and d, we have $P(N < \infty) = 1$. Based on the totality of all samples, that is having $X_{i1},\ldots,X_{iN}$ from π_i, $i = 1,\ldots,k$, we now estimate $\theta_{[k]}$ by means of the fixed–width confidence interval I_N given by (5.2.37). We keep $\underset{\sim}{\theta}$, σ fixed throughout.

Recall that there is a basic difference between the sequences of $U(n)$ and $V(n)$ statistics defined respectively in (3.1.1) and (4.1.1). By employing Helmert's orthogonal transformations (see Problem 2.2), $U(n)$ could be represented as the sample mean of i.i.d. random variables such as $(n-1)^{-1}\Sigma_{i=1}^{n-1} Z_i$. However, in this exponential case, weighted spacings between successive order statistics play the crucial role of Z_i's, but then we can write $V(n) = (n-1)^{-1}\Sigma_{i=1}^{n-1} Z_{in}$ where $Z_{1n},\ldots,Z_{n-1n}$ are i.i.d. for every fixed $n \geq 2$. See Problem 2.4. Such structural differences were pointed out in Chapter 4 as well. However, the embedding techniques of Lombard and Swanepoel (1978) come to rescue again in this setup since the two sequences $\{V(n) \ : \ n \geq m_1\}$ and $\{(n-1)^{-1}\Sigma_{i=1}^{n-1} Y_i \ : \ n \geq m_1\}$ have identical probability distributions where the p.d.f. of Y is given by

$$g_2(y) = (k/\sigma)^k \{\Gamma(k)\}^{-1} y^{k-1} \exp(-ky/\sigma) I(y > 0). \qquad (5.2.47)$$

Theorem 5.2.13. *For the purely sequential procedure (5.2.46), we have as* $d \to 0$:

(i) $N/C \to 1$ *w.p.* 1;

(ii) $E(N/C) \to 1$;

(iii) $C^{-\frac{1}{2}}(N-C) \xrightarrow{\mathscr{L}} N(0, k^{-1})$;

(iv) $\liminf P\{\theta_{[k]} \in I_N\} \geq 1 - \alpha$ *for all* $\underset{\sim}{\theta}$;

where $C = a\sigma/d$ *and* $a = -\ln\{1 - (1-\alpha)^{1/k}\}$.

Use the representation of N as Q + 1 with

$$Q = \inf\{n \geq m_1 - 1 : \Sigma_{i=1}^n W_i \leq 2kn(n+1)C^{-1}\},$$

where W's are i.i.d. χ^2_{2k} in view of (5.2.47). Then the theorem follows along the lines of Theorem 3.5.1 with obvious modifications. Further details are omitted. The "ν" given by (2.4.9), however, simplifies to

$$\nu = \nu(k) = (1+k) - \Sigma_{n=1}^{\infty} n^{-1} E[\max(0, \chi^2_{2nk} - 4nk)]. \qquad (5.2.48)$$

This is exactly the same as ν given in (4.5.8). Then, similar to Theorem 3.5.2 (i), we can claim here that

$E(N) = C + \beta'' + o(1)$ for all $\underset{\sim}{\theta}$ if $m_1 \geq 2$

where $\beta'' = \beta''(k) = k^{-1}\{\frac{1}{2}\nu(k) - 1\}$. (5.2.49)

Also, from (3.5.8), with $U^* = C^{-\frac{1}{2}}(N-C)$, one obtains

$$U^{*2} \text{ is uniformly integrable if } m_1 \geq 2, \tag{5.2.50}$$

and now Theorem 5.2.13 (iii) leads to

$$E(U^{*2}) = k^{-1} + o(1) \text{ if } m_1 \geq 2. \tag{5.2.51}$$

Theorem 5.2.14. *For the purely sequential procedure (5.2.46), we have as* $d \to 0$:

$$P\{\theta_{[k]} \in I_N\} = (1-\alpha) + aC^{-1}\{\beta''p(a, k) + q(a, k)\} + o(d)$$

under the LFC if $m_1 \geq 2$, *where* $p(a,k) = \Sigma_{j=0}^{k}(-1)^{j+1}\binom{k}{j}je^{-aj}$, $q(a, k) = \Sigma_{j=0}^{k} \frac{1}{2}(-1)^j\binom{k}{j}j^2e^{-aj}ak^{-1}$, β'' *comes from (5.2.49) and* $a = \ln\{1-(1-\alpha)^{1/k}\}$.

Proof. From (5.2.39) note that, under the LFC,

$$P\{\theta_{[k]} \in I_N\} = E[\psi_2(Nd/\sigma)] = \Sigma_{j=0}^{k}(-1)^j\binom{k}{j}E[\exp(-Ndj/\sigma)]. \tag{5.2.52}$$

Now, we use the machineries from section 2.4.2 along the lines of the proof of Theorem 5.2.5 to obtain

$$E[\exp(-Ndj/\sigma)] = e^{-aj} - e^{-aj}jd\sigma^{-1}\{\beta'' + o(1)\}$$
$$+ \tfrac{1}{2}j^2da(k\sigma)^{-1}\{e^{-aj} + o(1)\}, \tag{5.2.53}$$

for $m_1 \geq 2$ and all fixed $j = 0, 1, \ldots, k$. The theorem follows by combining (5.2.52) – (5.2.53). □

Remark 5.2.5. From (5.2.49), one can see that $E(N-C) \to k^{-1}(\frac{1}{2}\nu-1)$ as $d \to 0$

if $m_1 \geq 2$, and this is a considerably stronger conclusion than Theorem 5.2.13 (ii). Note that p(a, k) and q(a, k) can be respectively simplified to $ke^{-a}(1-e^{-a})^{k-1}$ and $\frac{1}{2}ae^{-a}(1-e^{-a})^{k-2}(ke^{-a}-1)$.

5.2.2.4 Three–Stage Procedure

Now, a suitable three–stage procedure is developed here along the lines of our discussions in section 5.2.1.4. We start with an initial sample of size $m_1 = m_1(\delta^*)$ (≥ 2) from each population π with

$$m_1 = m_1(\delta^*) = 0(C^{1/r})$$

for some $r > 1$ where $C = a\sigma/d$ and $a = -\ln\{1 - (1-\alpha)^{1/k}\}$. Choose and fix $0 < \rho < 1$ and let

$$m_1^* = \{2p(a,k) - p(a,k)\rho k - 2kq(a,k)\}/\{2p(a,k)\rho k\},$$

$$T = T(d) = \max\{m_1, [\rho aV(m_1)d^{-1}]^* + 1\},$$

$$N = N(d) = \max\{T, [aV(T)d^{-1} + m_1^*]^* + 1\}, \tag{5.2.54}$$

where p(a,k) and q(a,k) come from Theorem 5.2.14. The procedure (5.2.54) is then implemented as we did in the case of (3.7.2) – (3.7.3). Then, based on the totality of all samples $X_{i1},\ldots,X_{iN}$ from π_i, $i = 1,\ldots,k$, $\theta_{[k]}$ is estimated by the fixed–width confidence interval I_N given in (5.2.37). Note that the LFC still remains the same.

The asymptotic first–order properties as $d \to 0$ can be summarized as follows:

$$N/C \to \text{w.p. } 1,\ E(N/C) \to 1; \tag{5.2.55}$$

$$C^{-\frac{1}{2}}(N-C) \xrightarrow{\mathscr{L}} N(0, (k\rho)^{-1}); \text{ and} \tag{5.2.56}$$

$$\liminf P\{\theta_{[k]} \in I_N\} \geq 1-\alpha \text{ for all } \underset{\sim}{\theta}. \tag{5.2.57}$$

We now state, without proof, the associated second-order asymptotic properties. See, Mukhopadhyay and Kuo (1988) for details. Also, recall that the embedding technique that led to (5.2.47) will again truly be essential.

Theorem 5.2.15. *For the three-stage procedure (5.2.54), we have as* $d \to 0$:

(i) $E(N) = C - q(a,k)\{\rho p(a,k)\}^{-1} + o(1)$ *for all* $\underset{\sim}{\theta}$ *if* $m_1 \geq 2$;

(ii) $Var(N) = (\rho k)^{-1}C + o(d^{-1})$ *for all* $\underset{\sim}{\theta}$ *if* $m_1 \geq 2$;

(iii) $P\{\theta_{[k]} \in I_N\} = (1-\alpha) + o(d)$ *under the LFC, if* $m_1 \geq 2$;

where $C = a\sigma/d$, $a = -\ln\{1-(1-\alpha)^{1/k}\}$ *and* $p(a,k)$, $q(a,k)$ *come from Theorem 5.2.14.*

5.2.2.5 Accelerated Sequential Procedure

We first choose and fix $0 < \rho < 1$ and $q \geq 0$ and let

$$t = t(d) = \inf\{n \geq m_1 : n \geq \rho a V^*(n)/d\}, \tag{5.2.58}$$

$$N_1 = N_1(d) = [aV^*(t)d^{-1} + q]^* + 1, \tag{5.2.59}$$

$$N = N(d) = \max\{t(d), N_1(d)\}, \tag{5.2.60}$$

where $V^*(n)$, the estimator of σ, was defined in (4.1.1). Such a procedure is implemented very similarly as it was done in the case of (5.2.25) – (5.2.27). Then, based on the totality of all samples $X_{i1},\ldots,X_{iN}$ from π_i, $i = 1,\ldots,k$, $\theta_{[k]}$ is estimated by the fixed-width confidence interval I_N given in (5.2.37).

Note again that the LFC still remains the same.

The asymptotic first–order properties are then the same as in (5.2.55) – (5.2.57) and these can be verified along the lines of Theorem 3.6.1. Using the embedding technique that led to (5.2.47), one can see that the accelerated sequential procedure (5.2.58) – (5.2.60) agrees with that given in (2.5.1) – (2.5.3) with $\delta = 1$, $h^* = 2k/C$, $m = m_1 - 1$, W's are i.i.d. χ^2_{2k}, $\theta = 2k$, $\tau^2 = 4k$, $p = 1/k$, $b = k$ and $n_0^* = C$. Also, $m_0(\delta)$ turns out to be 2 whatever k. Now, similar to Theorem 3.6.2 (i) – (ii), one can claim that

$$q - (k\rho)^{-1} \le \liminf E(N-C) \le \limsup E(N-C) \le q + 1 - (k\rho)^{-1} \text{ if } m_1 \ge 4, \text{ and } Var(N) = (k\rho)^{-1}C + o(d^{-1}) \text{ if } m_1 \ge 2. \tag{5.2.61}$$

Theorem 5.2.16. *For the accelerated sequential procedure (5.2.58) – (5.2.60), we have as* $d \to 0$:

$$P\{\theta_{[k]} \in I_N\} = (1-\alpha) + o(d)$$

under the LFC if $q = \max\{0, (k\rho)^{-1}[1 - a \dfrac{\psi_2''(a)}{2\psi_2'(a)}]\}$ *when* $m_1 \ge 4$. *The function* ψ_2 *had come from (5.2.40) and* $a = -\ln\{1-(1-\alpha)^{1/k}\}$.

This theorem is analogous to Theorem 5.2.9 and hence we omit its proof. Note also that the comments made earlier in Remark 5.2.3 hold in this problem as well.

5.3 Bounded Maximal Risk and Minimax Risk Point Estimators

Here we discuss the point estimation problems for the largest (a) mean of $k(\ge 2)$ normal populations having unknown and common variance σ^2 and (b) location of $k(\ge 2)$ negative exponential populations having unknown and common scale parameter σ. It is also assumed that the populations are independent. Under specific types of loss functions, to be defined in each problem, point estimators are constructed with the goal that either the *maximal risk* is bounded from above in a certain setup or the *minimum maximal* (minimax) *risk* is attained, at least asymptotically, in certain other scenarios.

5.3.1 LARGEST MEAN OF NORMAL POPULATIONS

As in section 5.2.1, let $\pi_1,\ldots,\pi_k$ be independent populations with the distribution of the ith population being $N(\mu_i, \sigma^2)$, $i = 1,\ldots,k$. All the parameters are assumed *unknown*. As before, we write $\mu_{[k]} = \max\{\mu_1,\ldots,\mu_k\}$ and $\underset{\sim}{\mu} = (\mu_1,\ldots,\mu_k)$. Having recorded i.i.d. random variables $X_{i1},\ldots,X_{in}$ from π_i, we again use the notations $\overline{X}_{in}$, S^2_{in}, $U(n)$ and $U^*(n)$ as defined in (3.1.1), $i = 1,\ldots,k$ and $\overline{X}^*_n$ is the same as in (5.2.1). Suppose that the loss incurred in estimating $\mu_{[k]}$ by $\overline{X}^*_n$ is given by

$$L_n = A|\overline{X}^*_n - \mu_{[k]}|^s \tag{5.3.1}$$

where A and s are known positive numbers.

Under the loss function (5.3.1) we first determine the *maximal risk* when the sample size is fixed in advance. Next, under some mild assumptions, we derive a similar result when the sample size is a positive integer valued random variable. Then, we devise the methodologies of sequential and other multistage sampling. Let us write $R(\underset{\sim}{\mu}, n)$ for $E(L_n)$. The theorem given below follows from Theorem 3.1 of Cohen and Sackrowitz (1982). However, for completeness, we give a simple and direct proof.

Theorem 5.3.1. *Given fixed positive numbers* σ, A, s *and positive integer* n, *under the loss function (5.3.1),* $R(\underset{\sim}{\mu}, n)$ *is maximized when* $\mu_1 = \cdots = \mu_k$, *and the associated maximal risk is then given by* $B\sigma^2/n^{\frac{1}{2}s}$ *where one has*

$$B = kA\int_{-\infty}^{\infty} |y|^s \Phi^{k-1}(y)\phi(y)dy.$$

Proof. Let $\delta_i = (\mu_{[k]} - \mu_i)n^{\frac{1}{2}}/\sigma$, $i = 1,\ldots,k$ and $Y_n = n^{\frac{1}{2}}(\overline{X}_n^* - \mu_{[k]})/\sigma$. Now, for $y \epsilon R$, we write $F_{Y_n}(y) = P(Y_n < y) = \prod_{i=1}^{k} \Phi(\delta_i + y)$, and

$$E\{|\overline{X}_n^* - \mu_{[k]}|^s\} = \sigma^s n^{-\frac{1}{2}s} E\{|Y_n|^s\}. \tag{5.3.2}$$

Let

$$f^*(\delta_1,\ldots,\delta_k) = E\{|Y_n|^s\} = \int_{-\infty}^{\infty} |y|^s \frac{\partial}{\partial y}\{\prod_{i=1}^{k} \Phi(\delta_i+y)\}dy \tag{5.3.3}$$

and without any loss of generality, we assume that $\delta_k = 0$. Observe that the proof will be complete by showing that

$$f^*(\delta_1,\ldots,\delta_k) = C^* - \int_{-\infty}^{\infty} sy^{s-1}\{\prod_{i=1}^{k} \Phi(\delta_i+y) - \prod_{i=1}^{k} \Phi(\delta_i-y)\}dy \tag{5.3.4}$$

where C^* is a constant independent of $\delta_1,\ldots,\delta_{k-1}$, since Saxena and Tong (1969) did prove in their Theorem 1 that $\prod_{i=1}^{k} \Phi(\delta_i+y) - \prod_{i=1}^{k} \Phi(\delta_i-y)$ is minimized when $\delta_1 = \cdots = \delta_{k-1} = 0$ for every fixed $y \epsilon R$.

Now, to verify (5.3.4), we first consider $\frac{\partial}{\partial \delta_1} f^*(\delta_1,\ldots,\delta_k)$ in (5.3.3) and then integrate it over δ_1. Since $\frac{\partial}{\partial \delta_1}\{\frac{\partial}{\partial y}\prod_{i=1}^{k} \Phi(\delta_i+y)\} = \frac{\partial}{\partial y}\{\frac{\partial}{\partial \delta_1}\prod_{i=1}^{k} \Phi(\delta_i+y)\}$, which is also continuous, the derivative $\frac{\partial}{\partial \delta_1} f^*(\delta_1,\ldots,\delta_k)$ can be passed inside the integral sign in (5.3.3). See page 283 of Apostol (1974). Hence,

$$\frac{\partial}{\partial \delta_1} f^*(\delta_1,\ldots,\delta_k) = \int_{-\infty}^{\infty} |y|^s \frac{\partial}{\partial y}\{\frac{\partial}{\partial \delta_1}\prod_{i=1}^{k} \Phi(\delta_i+y)\}dy$$

$$= -\int_{-\infty}^{\infty} s|y|^{s-1}\text{sgn}(y)\frac{\partial}{\partial\delta_1}\{\prod_{i=1}^{k}\Phi(\delta_i+y)\}dy, \qquad (5.3.5)$$

using integration by parts. Here $\text{sgn}(y) = 1$ or -1 according as $y \geq 0$ or $y < 0$. Therefore, (5.3.5) leads to

$$f^*(\delta_1,\ldots,\delta_k) = C^* - \int_{-\infty}^{\infty} s|y|^{s-1}\text{sgn}(y)\prod_{i=1}^{k}\Phi(\delta_i+y)dy$$

$$= C^* - \int_{0}^{\infty} sy^{s-1}\{\prod_{i=1}^{k}\Phi(\delta_i+y) - \prod_{i=1}^{k}\Phi(\delta_i-y)\}dy.$$

Thus, (5.3.4) follows. Now, combine (5.3.2) and (5.3.3) to obtain

$$\sup_{\underset{\sim}{\mu}} R(\underset{\sim}{\mu}, n) = A\sigma^s n^{-\frac{1}{2}s}\int_{-\infty}^{\infty}|y|^s k\Phi^{k-1}(y)\phi(y)dy = B\sigma^s n^{-\frac{1}{2}s}, \qquad (5.3.6)$$

with the given choice of "B". □

Our goal is to achieve a *maximal risk* which is at most $W^*(>0)$, a preassigned number. In other words, by looking at (5.3.6), one can see that what we want is $B\sigma^s n^{-\frac{1}{2}s} \leq W^*$, that is n should be the smallest integer $\geq (B/W^*)^{2/s}\sigma^2 = n^*$, say. Since n^* is unknown, we propose to utilize purely sequential and multistage sampling techniques in order to make the associated maximal risk "close" to the preassigned upper bound W^* up to appropriate orders of approximation as $A \to \infty$.

Suppose that in general a statistical methodology gives rise to a positive integer valued random variable N estimating n^* subsequently followed by the estimator $\overline{X}_N^*$ of $\mu_{[k]}$.

Theorem 5.3.2. *Suppose that* N *satisfies the following conditions. The distribution of* N *does not depend on* $\underset{\sim}{\mu}$; $I(N=n)$ *is independent of* $(\overline{X}_{1n},\ldots,\overline{X}_{kn})$ *for every fixed* $n \geq 2$; $P(N < \infty) = 1$. *Then, under the loss function (5.3.1), the maximal risk is given by*

$$\sup_{\underset{\sim}{\mu}} E(L_N) = B\sigma^s E(N^{-\frac{1}{2}s})$$

where B *is the same as in Theorem 5.3.1, and this supremum is attained when* $\mu_1 = \cdots = \mu_k$.

Proof. Suppose that $m_1(\geq 2)$ is the starting sample size from each π, leading eventually to N. The condition $P(N < \infty) = 1$ guarantees that $\overline{X}_N^*$ is well–defined. Now,

$$\begin{aligned} E(L_N) &= \Sigma_{n=m_1}^{\infty} E(L_n \big| N=n)\, P(N=n) \\ &= A\Sigma_{n=m_1}^{\infty} E[|\overline{X}_n^* - \mu_{[k]}|^s \big| N=n]\, P(N=n) \\ &= A\Sigma_{n=m_1}^{\infty} E[|\overline{X}_n^* - \mu_{[k]}|^s]\, P(N=n), \end{aligned}$$

since $I(N=n)$ and $(\overline{X}_{1n},\ldots,\overline{X}_{kn})$ are independent. That is

$$E(L_N) = \Sigma_{n=m_1}^{\infty} E(L_n)\, P(N=n)$$

and now use the facts that $P(N=n)$ does not depend on $\underset{\sim}{\mu}$ and $E(L_n)$ attains its supremum for all n when $\mu_1 = \cdots = \mu_k$. Hence,

$$\sup_{\underset{\sim}{\mu}} E(L_N) = \Sigma_{n=m_1}^{\infty} B\sigma^s n^{-\frac{1}{2}s} P(N=n) = B\sigma^s E(N^{-\frac{1}{2}s}), \tag{5.3.7}$$

and this supremum is attained when $\mu_1 = \cdots = \mu_k$, which is referred to as the *least favorable configuration* (LFC). □

Naturally, n^* is unknown since σ is unknown. First, we discuss the two–stage and modified two–stage procedures of Mukhopadhyay et al. (1993). Then, the purely sequential and three–stage procedures of Kuo and Mukhopadhyay (1990a) are summarized, followed by the accelerated sequential methodology of Mukhopadhyay et al. (1993). In each case, Theorem 5.3.2 holds.

5.3.1.1 Two–Stage Procedure

We start with $m_1(\geq 2 \text{ and } > 1 + sk^{-1})$ observations from each population to begin the experiment. Let

$$h_0 = h_0(m_1) = \{\tfrac{1}{2}k(m_1-1)\}\{\Gamma(\frac{k(m_1-1)-s}{2})/\Gamma(\frac{k(m_1-1)}{2})\}^{2/s}, \quad (5.3.8)$$

and define

$$N = N(A) = \max\{m_1, [(B/W^*)^{2/s}h_0U(m_1)]^* + 1\}. \quad (5.3.9)$$

If $N = m_1$, then we do not take any more samples from π_i's in the second stage. On the other hand, if $N > m_1$, then we sample the difference $(N - m_1)$ from each π_i. Finally, based on $X_{i1},\ldots,X_{iN}$ from π_i, we estimate $\mu_{[k]}$ by $\overline{X}^*_N$. This Stein (1945)–type procedure has the following properties.

Theorem 5.3.3. *For the two–stage procedure (5.3.9), under the loss function (5.3.1), we have:*

(i) $\sup_{\underset{\sim}{\mu}} E(L_N) \leq W^*$ *for every fixed* $W^*(>0)$;

(ii) $E(N/n^*) \to h_0(>1)$ *as* $A \to \infty$;

(iii) $\sup_{\underset{\sim}{\mu}} E(L_N) \to W^*$ *as* $A \to \infty$;

where $n^* = (B/W^*)^{2/s}\sigma^2$, B *comes from Theorem 5.3.1, and* h_0 *comes from (5.3.8).*

Proof. In view of Theorem 5.3.2, we get

$$\sup_{\underset{\sim}{\mu}} E(L_N) = B\sigma^s E(N^{-\frac{1}{2}s}). \tag{5.3.10}$$

From (5.3.9), it is immediate to note that $N \geq (B/W^*)^{2/s} h_0 U(m_1)$ w.p. 1 and hence (5.3.10) leads to (since $m_1 \geq 2$ and $m_1 > 1 + sk^{-1}$)

$$\sup_{\underset{\sim}{\mu}} E(L_N) \leq W^*\{h_0^{-1}k(m_1-1)\}^{\frac{1}{2}s} E[\{\chi^2_{k(m_1-1)}\}^{-\frac{1}{2}s}]$$

$$= W^*\{\tfrac{1}{2}h_0^{-1}k(m_1-1)\}^{\frac{1}{2}s}\{\Gamma(\frac{k(m_1-1)-s}{2})/\Gamma(\frac{k(m_1-1)}{2})\}$$

$$= W^*,$$

by the choice of "h_0". Hence, we have part (i). For part (ii), observe the basic inequality

$$h_0(B/W^*)^{2/s}U(m_1) \leq N \leq h_0(B/W^*)^{2/s}U(m_1) + m_1 \tag{5.3.11}$$

and hence

$$h_0 \leq E(N)/n^* \leq h_0 + mn^{*-1},$$

which implies that $\lim_{A\to\infty} E(N)/n^* = h_0$. Obviously, h_0 is strictly larger than unity which follows from Lemma 2(b) (see Problem 5.25 (b)) of

Mukhopadhyay and Hilton (1986).

In order to prove part (iii), first note from (5.3.10) that

$$\sup_{\underset{\sim}{\mu}} E(L_N) = W^* E\{(n^*/N)^{\frac{1}{2}s}\}. \qquad (5.3.12)$$

From (5.3.10), we immediately obtain $\lim_{A\to\infty}(N/n^*)^{-\frac{1}{2}s} = (h_0^{-1}\sigma^2/U(m_1))^{\frac{1}{2}s}$ w.p. 1. Hence, one obtains $\liminf_{A\to\infty} E\{(n^*/N)^{\frac{1}{2}s}\} \geq \{\frac{k(m_1-1)}{2}\}^{\frac{1}{2}s} E[\{\chi^2_{k(m_1-1)}\}^{-\frac{1}{2}s}]$ $= 1$ by the obvious choice of "h_0" and Fatou's Lemma. On the other hand, $E\{(n^*/N)^{\frac{1}{2}s}\} \leq E\{(h_0^{-1}\sigma^2/U(m_1))^{\frac{1}{2}s}\} = 1$, by our choice of "$h_0$" as before. Thus part (iii) follows from (5.3.12). □

Remark 5.3.1. The goal of *bounding the maximal risk* has been met *exactly* (Theorem 5.3.3 (i)), however, the characteristic depicted by Theorem 5.3.3 (ii) is disturbing since it implies that the two–stage procedure (5.3.9) *oversamples* (in comparison with n^*), even *asymptotically.* In order to remedy this drawback, we now move on to implement a modified version of the two–stage procedure (5.3.9).

5.3.1.2 Modified Two–Stage Procedure

We let m_1, the starting sample size, depend on A and thereby let it "grow" as A gets larger, however at a rate slower than that of n^*. Along the lines of section 2.2.2, we define

$$m_1 = m_1(A) = \max\{m_0^*, [1+sk^{-1}]^*+1, [A^{2/(s+\gamma)}]^* + 1\} \quad (5.3.13)$$

for arbitrary, but fixed and chosen, $\gamma(>0)$ and $m_0^*(\geq 2)$. With this $m_1 = m_1(A)$ and h_0 from (5.3.8), we now define the following two–stage procedure: Let

$$N = N(A) = \max\{m_1, [(B/W^*)^{2/s} h_0 U(m_1)]^* + 1\} \tag{5.3.14}$$

and we estimate $\mu_{[k]}$ by $\overline{X}^*_N$. The following properties hold.

Theorem 5.3.4. *For the modified two-stage procedure (5.3.13) – (5.3.14), with* h_0 *defined in (5.3.8), under the loss function (5.3.1), we have:*

(i) $\sup_{\underset{\sim}{\mu}} E(L_N) \le W^*$ *for every fixed* $W^*(>0)$;

(ii) $N/n^* \to 1$ *w.p. 1 and* $E(N/n^*) \to 1$ *as* $A \to \infty$;

(iii) $\sup_{\underset{\sim}{\mu}} E(L_N) \to W^*$ *as* $A \to \infty$;

where $n^* = (B/W^*)^{2/s}\sigma^2$, B *comes from Theorem 5.3.1, and* $0 < \gamma < \infty$ *is arbitrary, but fixed.*

Proof. Part (i) follows from Theorem 5.3.3 (i). To prove part (ii), observe that $m_1 = m_1(A) \to \infty$, but $m_1(A)/n^* \to 0$ as $A \to \infty$. Hence, $h_0 = h_0(A) \to 1$ as $A \to \infty$. See Lemma 1 (see Problem 5.25 (a)) in Mukhopadhyay and Hilton (1986). Now, part (ii) follows along the lines of Theorem 3.4.1 (ii). Part (iii) is similar to Theorem 5.3.3 (iii). □

It is clear that the modified two–stage procedure (5.3.13) – (5.3.14) maintains the same properties given in Theorem 5.3.3 (i) and (iii), and additionally we note that this procedure is also *asymptotically first–order efficient* in the sense of (3.4.7). But, in view of Lemma 1 (see Problem 5.25 (a)) of Mukhopadhyay and Hilton (1986), one notes that

$$E(N) - n^* \ge (B/W^*)^{2/s}\sigma^2\{h_0(A) - 1\} = 0(A^{2/s})0(m_1^{-1})$$

$$= 0(A^{-2/(s+\gamma)+2/s}) = 0(A^{2\gamma/s(s+\gamma)}) \to \infty,$$

as $A \to \infty$, since $\gamma \in R^+$. That is, the modified two–stage methodology is *not asymptotically second–order efficient* in the sense of (3.4.8).

An arbitrary multistage or sequential estimation procedure will be referred as

(i) *risk efficient* in the Robbins (1959) sense if

$$E(L_N) \le W^* \text{ for all } \underset{\sim}{\mu}; \tag{5.3.15}$$

(ii) *asymptotically risk efficient* or *asymptotically first–order risk efficient* in the Robbins (1959) or Ghosh and Mukhopadhyay (1981) sense respectively if

$$W^{*-1} \sup_{\underset{\sim}{\mu}} E(L_N) \to 1 \text{ as } A \to \infty; \tag{5.3.16}$$

(iii) *asymptotically second–order risk efficient* in the Ghosh and Mukhopadhyay (1981) sense if

$$B_1 n^{*-1} + o(n^{*-1}) \le W^{*-1} \sup_{\underset{\sim}{\mu}} E(L_N) - 1$$

$$\le B_2 n^{*-1} + o(n^{*-1}) \text{ as } A \to \infty, \tag{5.3.17}$$

for some fixed constants (independent of n^*) B_1 and B_2.

Now, we discuss a few other estimation procedures which are *asymptotically* both *second–order efficient* and *second–order risk efficient* in the sense of (3.4.8) and (5.3.17) respectively.

5.3.1.3 Purely Sequential Procedure

We start with $m_1(\ge 2)$ samples from each π. Then, we take one sample at a time from each π according to the following stopping rule. Let

$$N = N(A) = \inf\{n \ge m_1 : n \ge (B/W^*)^{2/s} U(n)\}. \tag{5.3.18}$$

It is easy to check that $P(N < \infty) = 1$ for all fixed $\underset{\sim}{\mu}$, σ^2, m_1 and A. We estimate $\mu_{[k]}$ by $\overline{X}^*_N$. Now, similar to Theorem 5.2.4 (i) – (iii), one can prove that as $A \to \infty$:

$$N/n^* \to 1 \text{ w.p. } 1, \;\; E(N/n^*) \to 1;$$

$$n^{*-\frac{1}{2}}(N-n^*) \xrightarrow{\mathscr{L}} N(0, 2k^{-1}). \tag{5.3.19}$$

Theorem 5.3.5. *For the purely sequential procedure (5.3.18), under the loss function (5.3.1), we have as* $A \to \infty$:

(i) $\sup_{\underset{\sim}{\mu}} E(L_N) \to W^*$ *if* $m_1 > 1 + sk^{-1}$;

(ii) $E(N) = n^* + \beta + o(1)$ *if* (a) $m_1 \geq 3$ *when* $k = 2$, (b) $m_1 \geq 2$ *when* $k \geq 3$, *for all* $\underset{\sim}{\mu}$;

(iii) $\mathrm{Var}(N) = 2k^{-1}n^* + o(A^{2/s})$ *if* (a) $m_1 \geq 3$ *when* $k = 2$, (b) $m_1 \geq 2$ *when* $k \geq 3$, *for all* $\underset{\sim}{\mu}$;

(iv) $\sup_{\underset{\sim}{\mu}} E(L_N) = W^* + \frac{1}{2}sW^*(3 + \frac{1}{2}s - \nu)(kn^*)^{-1} + o(A^{-2/s})$ *if* (a) $m_1 \geq 4$ *when* $k = 2, s = 2$, (b) $m_1 \geq 3$ *when* $k = 3, 4, s = 2$, (c) $m_1 \geq 2$ *when* $k \geq 5, s = 2$, (d) $m_1 > 1 + (s+6)k^{-1}$ *when* $s \in R^+ - \{2\}$;

where $n^* = (B/W^*)^{2/s}\sigma^2$, $\beta = \beta(k) = k^{-1}\{\nu(k)-2\}$ *with* $\nu = \nu(k)$ *defined in (3.5.7), and* B *comes from Theorem 5.3.1.*

Proof. Note that N can be represented as $Q + 1$ where

$$Q = \inf\{n \geq m_1 - 1 \; : \; kn^2(1+n^{-1})n^{*-1} \geq \Sigma_{i=1}^n W_i\},$$

W_i's being i.i.d. χ_k^2. Hence, parts (ii) and (iii) immediately follow from Theorem 3.5.2 (i) and (3.5.8).

Let us turn to prove parts (i) and (iv). From Theorem 5.3.2, note that

$$\sup_{\underset{\sim}{\mu}} E(L_N) = W^* E\{(n^*/N)^{\frac{1}{2}s}\}. \tag{5.3.20}$$

Since $N/n^* \to 1$ w.p. 1, $\liminf E\{(n^*/N)^{\frac{1}{2}s}\} \geq 1$ by Fatou's Lemma. On the other hand,

$$E[(n^*/N)^{\frac{1}{2}s}] = E[(n^*/N)^{s/2} I(N \leq \tfrac{1}{2}n^*)] + E[(n^*/N)^{s/2} I(N > \tfrac{1}{2}n^*)]$$

and $(n^*/N)^{s/2} I(N > \frac{1}{2}n^*) \leq 2^{s/2}$, that is

$$E[(n^*/N)^{s/2} I(N > \tfrac{1}{2}n^*)] = 1 + o(1) \text{ if } m_1 \geq 2. \tag{5.3.21}$$

But, one observes that $E[(n^*/N)^{s/2} I(N \leq \frac{1}{2}n^*)] \leq (n^*/m_1)^{s/2} P(N \leq \frac{1}{2}n^*) = 0(n^{*s/2})0(n^{*\,-\frac{1}{2}k(m_1-1)})$, by Theorem 2.4.8 (i). Thus, $E[(n^*/N)^{s/2} I(N \leq \frac{1}{2}n^*)] = o(1)$ if $m_1 > 1 + sk^{-1}$. Combine this with (5.3.20) – (5.3.21) and then part (i) follows. Next, one writes

$$\begin{aligned}
\sup_{\underset{\sim}{\mu}} E(L_N) &= W^* E\{(n^*/N)^{\frac{1}{2}s}\} = W^* E[\{n^*/(Q+1)\}^{\frac{1}{2}s}] \\
&= W^* E\{(n^*/Q)^{\frac{1}{2}s}\} - \tfrac{1}{2}s W^* n^{*-1} E\{(n^*/T^*)^{\frac{1}{2}s+1}\} \\
&= W^* J_{11} - \tfrac{1}{2}s W^* n^{*-1} J_{12}, \text{ say,}
\end{aligned} \tag{5.3.22}$$

where T^* is a suitable random variable $\in (Q, Q+1)$. From Theorem 2.4.8 (v), we obtain

$$J_{11} = 1 + \tfrac{1}{2}s n^{*-1}\{\tfrac{1}{2}(s+2)k^{-1} - (\nu-2)k^{-1} + 1\} + o(n^{*-1}) \tag{5.3.23}$$

if (i) $m_1 > 1 + 4k^{-1}$ when $s = 2$ and (ii) $m_1 > 1 + (s+6)k^{-1}$ when $s \in (0, \infty) - \{2\}$. On the other hand, along the lines of part (i), one can see that $E\{(n^*/T^*)^{\frac{1}{2}s+1}\} = 1 + o(1)$ if $m_1 > 1 + (s+2)k^{-1}$. Hence, part (iv) follows by combining (5.3.22) and (5.3.23). □

Remark 5.3.2: Technically speaking, when $k = 1$, parts (ii) and (iv) of Theorem 5.3.5 agree with Mukhopadhyay's (1985a) Theorem 2 in the context of bounded risk point estimation of the mean of a normal distribution having unknown variance.

5.3.1.4 Three–Stage Procedure

Here, we propose a three–stage sampling procedure for operational convenience of batch sampling, and derive the second–order approximations of the associated risk function and other characteristics. We start with an initial sample of size $m_1 = m_1(A)(\geq 2)$ from each population π where

$$m_1 = m_1(A) = 0(n^{*1/r}),$$

for some $r > 1$ and $n^* = (B/W^*)^{2/s}\sigma^2$. Choose and fix $0 < \rho < 1$ and let

$$T = T(A) = \max\{m_1, [\rho(B/W^*)^{2/s}U(m_1)]^* + 1\},$$

$$N = N(A) = \max\{T, [(B/W^*)^{2/s}U(T)]^* + 1\}. \tag{5.3.24}$$

The procedure is implemented as we did in the case of (3.7.2) – (3.7.3). Then, based on the totality of all samples $X_{i1},\ldots,X_{iN}$ from π_i, $i = 1,\ldots,k$, $\mu_{[k]}$ is estimated by $\overline{X}^*_N$. One can verify that as $A \to \infty$:

$$N/n^* \to 1 \text{ w.p. } 1, \quad E(N/n^*) \to 1;$$

$$n^{*-\frac{1}{2}}(N-n^*) \xrightarrow{\mathscr{L}} N(0, 2(k\rho)^{-1}). \quad (5.3.25)$$

Theorem 5.3.6. *For the three–stage procedure (5.3.24), under the loss function (5.3.1), we have as* $A \to \infty$:

(i) $E(N) = n^* + \frac{1}{2} - 2(k\rho)^{-1} + o(1)$ *for all* $\underset{\sim}{\mu}$;

(ii) $Var(N) = 2(k\rho)^{-1}n^* + o(A^{2/s})$ *for all* $\underset{\sim}{\mu}$;

(iii) $\sup_{\underset{\sim}{\mu}} E(L_N) = W^* - \frac{1}{4}sW^*n^{*-1}\{1 - (s+6)(k\rho)^{-1}\} + o(A^{-2/s})$;

where $n^* = (B/W^*)^{2/s}\sigma^2$ *and* B *comes from Theorem 5.3.1.*

Proof. The three–stage procedure (5.3.24) agrees with the representations given in (2.3.1) – (2.3.2) where $\lambda = (B/W^*)^{2/s}$ and Y's are i.i.d. $\sigma^2k^{-1}\chi^2_k$. Part (i) now follows from Theorem 2.3.2. Part (ii) follows from Theorem 2.3.3 (ii). For part (iii), observe that Theorem 5.3.2 holds and thus

$$\sup_{\underset{\sim}{\mu}} E(L_N) = W^*E\{(n^*/N)^{\frac{1}{2}s}\},$$

hence, part (iii) follows from Theorem 2.3.3 (iii). □

5.3.1.5 Accelerated Sequential Procedure

We first choose and fix $0 < \rho < 1$ and let

$$t = t(A) = \inf\{n \geq m_1 : n \geq \rho(B/W^*)^{2/s}U^*(n)\}, \quad (5.3.26)$$

$$N_1 = N_1(A) = [(B/W^*)^{2/s}U^*(t)]^* + 1, \quad (5.3.27)$$

$$N = N(A) = \max\{t(A), N_1(A)\}, \quad (5.3.28)$$

where $m_1(\geq 2)$ and $U^*(n)$, the estimator of σ^2, was defined in (3.1.1). Such a procedure is implemented very similarly as it was done in (5.2.25) – (5.2.27). Then, based on the totality of all samples $X_{i1},\ldots,X_{iN}$ from π_i, $i = 1,\ldots,k$, we estimate $\mu_{[k]}$ by $\overline{X}_N^*$. The results given in (5.3.25) hold here as well if $m_1 \geq 2$. Some second–order properties are given in the following theorem.

Theorem 5.3.7. *For the accelerated sequential procedure (5.3.26) – (5.3.28), under the loss function (5.3.1), we have as* $A \to \infty$:

(i) $-2(k\rho)^{-1} \leq \liminf E(N-n^*) \leq \limsup E(N-n^*) \leq 1 - 2(k\rho)^{-1}$ *for all* $\underset{\sim}{\mu}$, *if* $m_1 \geq 4$;

(ii) $\mathrm{Var}(N) = 2(k\rho)^{-1}n^* + o(A^{2/s})$ *for all* $\underset{\sim}{\mu}$, *if* (a) $m_1 \geq 3$ *when* $k = 2$, (b) $m_1 \geq 2$ *when* $k \geq 3$;

(iii) $W^* + s(s+6-2k\rho)\, W^*(4k\rho n^*)^{-1} + o(A^{-2/s}) \leq \sup_{\underset{\sim}{\mu}} E(L_N) \leq W^*+s(s+6)\, W^*(4k\rho n^*)^{-1} + o(A^{-2/s})$ *if* (a) $m_1 \geq 4$ *when* $s = 2$, (b) $m_1 > 1 + \max\{2.02, (6+s)k^{-1}\}$ *when* $s \in R^+ - \{2\}$;

where $n^* = (B/W^*)^{2/s}\sigma^2$ *and* B *comes from Theorem 5.3.1.*

Proof. The accelerated sequential procedure (5.3.26) – (5.3.28) falls under the general framework of section 2.5 with m *replaced by* $m_1 - 1$, and $\delta = 1$, $\theta = k$, $\tau^2 = 2k$, $a = \frac{1}{2}k$, $p = 2/k$, $m_0(\delta) = 2.02$ if $k = 2$ and 2 if $k \geq 3$, $h^* = k/n^*$, and $q = 0$. Hence, parts (i) and (ii) follow immediately from Corollary 2.5.1 and Remark 2.5.2 respectively.

The maximal risk is again of the form given in (5.3.20) where N is given by (5.3.28). Here, we will utilize Corollary 2.5.2 with $l = -\frac{1}{2}s$. Note that $\eta'' = 2/k\rho$. When $s = 2$, we need $m_1 - 1 > \max\{m_0(\delta), \frac{2}{\delta a}\} =$

$\max\{m_0(\delta), 4/k\}$, that is $m_1 \geq 4$. When $s \in R^+ - \{2\}$, the sufficient condition on m_1 becomes $m_1 > 1 + \max\{2.02, (6+s)k^{-1}\}$. □

Remark 5.3.3. Observe that the purely sequential, accelerated sequential and three–stage estimation procedures are all *asymptotically second–order efficient* and *asymptotically second–order risk efficient.* See (3.4.5) – (3.4.8) and (5.3.15) – (5.3.17) in this context.

5.3.2 LARGEST LOCATION AMONG NEGATIVE EXPONENTIAL POPULATIONS

Consider $k(\geq 2)$ independent negative exponential populations $\pi_1, \ldots, \pi_k$ as in section 5.2.2. Having recorded i.i.d. random variables $X_{i1}, \ldots, X_{in}$, we use the same notations $\tilde{X}_{in}$, T_n^*, $V(n)$ (or $V^*(n)$) for estimators of θ_i, $\theta_{[k]}$, and σ respectively. The methodologies and associated results discussed here have been adapted from Kuo and Mukhopadhyay (1990b). Suppose that the loss incurred in estimating $\theta_{[k]}$ by T_n^* is given by

$$L_n = A(T_n^* - \theta_{[k]})^s + cn^{\omega} \tag{5.3.29}$$

where A, c, s and ω are all known positive real numbers. Here "c" is thought of as the cost per "unit" sample. With $s = 2$ and $\omega = 1$, one gets the more conventional squared error plus linear cost scenario. We deliberately take a loss function that is quite different from the one considered in (5.3.1) for the corresponding normal problem. The idea is to show "diversity" within the types of problems where asymptotic second–order theory has been fruitfully applied.

Under the loss function (5.3.29), we first determine the fixed sample size n^* required to *minimize* the *maximal risk* over the parameters $\tilde{\theta}$. Naturally, we determine this *minimax* procedure pretending, of course, that σ

is known. In reality, however, n^* is unknown, since σ is unknown. Thus, we devise sequential and other methodologies for this problem. Let us write $R(\underset{\sim}{\theta}, n)$ for $E(L_n)$.

Theorem 5.3.8. *Given fixed positive numbers* σ, A, s, ω, c *and positive integer* n, *under the loss function (5.3.29),* $R(\underset{\sim}{\theta}, n)$ *is maximized when* $\theta_1 = \cdots = \theta_k$, *and the associated maximal risk is given by*

$$R_n(c) = B_o\sigma^s(sn^s)^{-1} + cn^\omega \tag{5.3.30}$$

where

$$B_o = Ask\int_0^\infty u^s e^{-u}(1 - e^{-u})^{k-1}du. \tag{5.3.31}$$

Proof. Let $U_n = n(T_n^* - \theta_{[k]})/\sigma$. For $u > 0$, we write

$$F_{U_n}(u) = P(U_n \le u) = \prod_{i=1}^{k} [1 - \exp\{-u - \tfrac{n}{\sigma}(\theta_{[k]} - \theta_i)\}]. \tag{5.3.32}$$

Let $\delta_i = \theta_{[k]} - \theta_i$, $i = 1,\ldots,k$, and one utilizes (5.3.32) to obtain

$$\begin{aligned} AE\{(T_n^* - \theta_{[k]})^s\} &= A(\sigma/n)^s\int_0^\infty u^s \frac{\partial}{\partial u}[\prod_{i=1}^{k}\{1 - e^{-u - \frac{n\delta_i}{\sigma}}\}]du \\ &= g^*(\delta_1,\ldots,\delta_k), \text{ say.} \end{aligned} \tag{5.3.33}$$

For any fixed i, observe that $\frac{\partial}{\partial\delta_i}[\prod_{j=1}^{k}\{1 - e^{-u - \frac{n\delta_j}{\sigma}}\}]$ is continuous in R^{+k}, and therefore differentiation of $g^*(\delta_1,\ldots,\delta_k)$ with respect to δ_i can be passed inside the integral sign in (5.3.33). See page 283 of Apostol (1974). Hence,

$$\frac{\partial}{\partial \delta_i} g^*(\delta_1,\ldots,\delta_k) = A(\sigma/n)^s \int_0^\infty u^s \frac{\partial}{\partial u}\Big[\frac{n}{\sigma} e^{-u - \frac{n\delta_i}{\sigma}} \prod_{j \neq i}^{k} \{1 - e^{-u - \frac{n\delta_j}{\sigma}}\}\Big] du$$

$$= -As(\sigma/n)^{s-1} e^{-\frac{n\delta_i}{\sigma}} \int_0^\infty u^{s-1} e^{-u} \prod_{j \neq i}^{k} \{1 - e^{-u - \frac{n\delta_j}{\sigma}}\} du,$$

using integration by parts. Note that $\delta_i \geq 0$, and $\frac{\partial}{\partial \delta_i} g^*(\delta_1,\ldots,\delta_k)$ is negative for all fixed i. Therefore, the maximum of $g^*(\delta_1,\ldots,\delta_k)$ is attained when $\delta_1 = \cdots = \delta_k = 0$, and thus

$$\sup_{\underset{\sim}{\theta}} AE\{(T_n^* - \theta_{[k]})^s\} = g^*(0,\ldots,0) = kA(\sigma/n)^s \int_0^\infty u^s e^{-u}(1 - e^{-u})^{k-1} du.$$

Hence, the maximal value of $R(\underset{\sim}{\theta}, n)$ is given by (5.3.30) with the choice of B_0 given in (5.3.31). □

Our goal is to achieve the *minimum* of the *maximal possible risk* $R_n(c)$ given by (5.3.30). Pretend that σ is known and for convenience, treat "n" as a continuous variable. $R_n(c)$ is a strict convex function in n and $R_n(c) \to \infty$ as $n \to 0$ or $n \to \infty$. Thus, $R_n(c)$ is minimized by choosing $n = \{B_0\sigma^s/(c\omega)\}^{1/(s+\omega)} = n^*$, say. However, n^* is unknown since σ is unknown. Hence, we first discuss a purely sequential methodology. Then, a three-stage procedure is presented in order to construct a *minimax estimator* of $\theta_{[k]}$ where the loss function under consideration is taken to be a weighted analog of (5.3.29). Suppose that in general a statistical methodology gives rise to a positive integer valued random variable N estimating n^* followed by the estimator of T_N^* of $\theta_{[k]}$.

Theorem 5.3.9. *Suppose that* N *satisfies the following conditions. The distribution of* N *does not depend on* $\underset{\sim}{\theta}$; $I(N=n)$ *is independent of*

$(\tilde{X}_{1n},\ldots,\tilde{X}_{kn})$ *for every fixed* $n \geq 2$; $P(N < \infty) = 1$. *Then, under the loss function (5.3.29), the maximal risk is*

$$\sup_{\tilde{\theta}} E(L_N) = \{B_0 \sigma^s / s\} E(N^{-s}) + cE(N^{\omega})$$

under the loss function (5.3.29) where B_0 *is given by (5.3.31), and this supremum is attained when* $\theta_1 = \cdots = \theta_k$.

Proof. The proof is quite similar to the proof of Theorem 5.3.2. Thus, we only give a few lines of justification. One writes.

$$E(L_N) = A\Sigma_{n=m_1}^{\infty} E[(T_n^* - \theta_{[k]})^s] P(N=n) + cE(N^{\omega})$$

and then Theorem 5.3.8 leads to

$$\sup_{\tilde{\theta}} E(L_N) = \Sigma_{n=m_1}^{\infty} \sup_{\tilde{\theta}} E[A(T_n^* - \theta_{[k]})^s] P(N=n) + cE(N^{\omega})$$

$$= (B_0 \sigma^s / s) E(N^{-s}) + cE(N^{\omega}),$$

since the maximal risk in (5.3.30) is attained when $\theta_1 = \cdots = \theta_k$, whatever n. Further details are omitted for brevity. □

5.3.2.1 Purely Sequential Procedure

The *minimax risk* associated with (5.3.30) is given by

$$R^*(A) = R_{n^*}(A) = c(1 + \omega s^{-1}) n^{*\omega}. \tag{5.3.34}$$

Now, we start with $m_1 (\geq 2)$ samples from each π and let

$$N = N(A) = \inf\{n \geq m_1 : n \geq [B_0(c\omega)^{-1}V^s(n)]^{1/(s+\omega)}\}. \quad (5.3.35)$$

It is easy to see that $P(N < \infty) = 1$ for all fixed $\underset{\sim}{\theta}$, σ, m_1 and A. We estimate $\theta_{[k]}$ by T_N^* obtained from the combined samples $X_{i1},...,X_{iN}$ from π_i, $i = 1,...,k$. Theorem 5.3.9 will apply and hence the maximal risk associated with the estimator T_N^* is given by

$$R(A) = \sup_{\underset{\sim}{\theta}} E(L_N)$$

$$= c[\omega s^{-1} n^{*s+\omega} E(N^{-s}) + E(N^{\omega})]. \quad (5.3.36)$$

The *risk efficiency* and *regret* of a particular procedure in the Robbins (1959) sense are respectively defined as

$$e_1(A) = R(A)/R^*(A) = (1+\omega s^{-1})^{-1}[\omega s^{-1}E\{(n^*/N)^s\}+E\{(N/n^*)^{\omega}\}] \quad (5.3.37)$$

and

$$e_2(A) = R(A) - R^*(A). \quad (5.3.38)$$

In the present problem a procedure is called

(i) *asymptotically minimax risk efficient* or *asymptotically first-order minimax risk efficient* in the Robbins (1959) sense or Ghosh and Mukhopadhyay (1981) sense respectively if

$$e_1(A) \to 1 \text{ as } A \to \infty \quad (5.3.39)$$

(ii) *asymptotically second-order minimax risk efficient* in the Ghosh and Mukhopadhyay (1981) sense if

$$e_2(A) = k_0 A^{-(s+1)/(s+\omega)} + o(A^{-1}) \text{ as } A \to \infty, \quad (5.3.40)$$

for some fixed constant k_0 depending on ω, s, k and σ only.

Along the lines of Mukhopadhyay (1974), one can verify that

$$N/n^* \to 1 \text{ w.p. } 1, \quad E(N/n^*) \to 1 \tag{5.3.41}$$

as $A \to \infty$ if $m_1 \geq 2$ and that

$$e_1(A) \to 1 \tag{5.3.42}$$

as $A \to \infty$ if $m_1 > 1 + s^2/(k\omega+ks)$. The results in (5.3.41) and (5.3.42) show that the sequential procedure (5.3.35) is asymptotically *first–order efficient and first–order minimax risk efficient.* See (3.4.7) and (5.3.40). We now state the following stronger assertion.

Theorem 5.3.10. *For the sequential procedure (5.3.35), under the loss function (5.3.29) we have as* $A \to \infty$:

$$e_2(A) = \tfrac{1}{2}c\omega k^{-1}s^2(s+\omega)^{-1}n^{*\omega-1} + o(A^{-1}),$$

if (a) $m_1 > 1 + 2s(ks + k\omega)^{-1}$ *when* $s = 1$ *and* $\omega \in [2, \infty) \cup \{1\}$, (b) $m_1 > 1 + s(3+s)(ks + k\omega)^{-1}$ *when* $s \in (0, 1) \cup (1, \infty)$ *and* $\omega \in R^+$, *and* (c) $m_1 > \max\{1 + 2s(ks + k\omega)^{-1}, 1 + s(3-t)(ks + k\omega)^{-1}\}$ *when* $s = 1$ *and* $\omega \in (0, 1) \cup (1, 2)$ *where* $n^* = \{B_0\sigma^s/(c\omega)\}^{1/(s+\omega)}$ *and* B_0 *comes from (5.3.31).*

This theorem can be proved along the lines of Theorem 2.4.8. The proof is omitted for brevity.

5.3.2.2 Three–Stage Procedure

Suppose that in order to estimate $\theta_{[k]}$ by T_n^*, we have the following weighted loss function instead of (5.3.29). Let

$$L_n = A\sigma^{\omega}(T_n^* - \theta_{[k]})^s + cn^{\omega} \qquad (5.3.43)$$

where A, s, c and ω are exactly the same as these were in (5.3.29). Using Theorem 5.3.8 with obvious modifications, the *maximal risk* associated with (5.3.43) for fixed n, A, s, c, ω and σ can be shown to be

$$R_n(A) = B_0\sigma^{s+\omega}(sn^s)^{-1} + cn^{\omega}, \qquad (5.3.44)$$

where B_0 comes from (5.3.31). Let $B_1 = \{B_0(c\omega)^{-1}\}^{1/(s+\omega)}$. The *maximal* risk in (5.3.44) is *minimized* by $n = n^* = B_1\sigma$ and the associated minimax risk is given by

$$R^*(A) = R_{n^*}(A) = c(1 + \omega s^{-1})n^{*\omega}.$$

We start with an initial sample of size $m_1 = m_1(A)(\geq 2)$ from each population π where

$$m_1 = m_1(A) = 0(n^{*1/r}),$$

with some $r > 1$. Choose and fix $0 < \rho < 1$ and let

$$T = T(A) = \max\{m_1, [\rho B_1 V(m_1)]^* + 1\},$$

$$N = N(A) = \max\{T, [B_1 V(T)]^* + 1\}. \qquad (5.3.45)$$

This procedure is implemented as we did in the case of (3.7.2) – (3.7.3). Then, based on the totality of all samples $X_{i1},...,X_{iN}$ from π_i, $i = 1,...,k$, we estimate $\theta_{[k]}$ by T_N^*. An appropriate modification of Theorem 5.3.9 holds which shows that the maximal risk associated with T_N^* is given by

$$R(A) = (B_0\sigma^{s+\omega}/s)E(N^{-s}) + cE(N^{\omega})$$

$$= c[\omega s^{-1}n*^{s+\omega}E(N^{-s}) + E(N^{\omega})]. \tag{5.3.46}$$

The *regret* function is given by

$$e_2(A) = R(A) - R^*(A). \tag{5.3.47}$$

Theorem 5.3.11. *For the three–stage procedure (5.3.45), under the loss function (5.3.43), we have as* $A \to \infty$:

$$e_2(A) = \tfrac{1}{2}c\omega(s+\omega)(k\rho)^{-1}n*^{\omega-1} + o(A^{-(s+1)/(s+\omega)})$$

where $n^* = B_1\sigma$, $B_1 = \{B_0(c\omega)^{-1}\}^{1/(s+\omega)}$. *Here,* B_0 *and* $e_2(A)$ *respectively come from (5.3.31) and (5.3.47).*

In our two–parameter exponential problems, one needs to exploit Lombard and Swanepoel's (1978) embedding techniques along the lines of our discussions in section 5.2.2.3. Then, one uses the theories developed in section 2.3 to prove Theorem 5.3.11. See Kuo and Mukhopadhyay (1990b) for further details.

5.4 Notes and overview

Only the case of unknown and equal variance has been considered, whether the populations are normal or negative exponential. The analogous problems with unequal and unknown variances will be of considerable interest. At the same time, one should investigate the roles of asymmetric versions of the fixed–width confidence intervals mentioned in section 5.2.1 hold much promise in the context of second–order asymptotics. See Tong (1973), Gupta and Panchapakesan (1979), and others for reviews. One may also look at the remarks made in sections 3.10 and 4.10 in this same light as well. The area of parametric sequential point estimation problem has been reviewed in

Mukhopadhyay (1991). In either problem area, various aspects of obtaining elimination type estimation techniques with associated second–order asymptotics needs attention. One should also investigate the possibilities of bringing in the ideas of asymptotically pointwise optimal (APO) rules and exploiting the literature on hierarchical and empirical Bayes sequential estimation problems in the present situation. The review articles of Ghosh (1991), and Ghosh and Hoekstra (1991) will be most valuable in this context.

The analogous problems in the case of exponential populations having only the unknown scale parameters would also be very challenging. See Saxena and Tong (1971) for some of the basic ideas.

Complements and Problems

5.1 Show that the stopping variable N arising from the two–stage, modified two–stage, purely sequential, three–stage and accelerated sequential estimation procedures of section 5.2.1 satisfy the sufficient conditions given in Theorem 5.2.1.

5.2 For the purely sequential estimation procedure (5.2.12), show that $E(N) \leq C + 0(1)$.

5.3 (a) Suppose that $V(n)$ is replaced by $V^*(n)$ in (5.2.46). Prove Theorem 5.2.13 for the associated stopping variable N.

(b) How will the results in (5.2.49) and Theorem 5.2.14 then be modified?

5.4 In the accelerated sequential procedure (5.2.58) – (5.2.60), suppose that one uses $V(n)$ instead of $V^*(n)$ for estimating σ. Obtain the first–order asymptotic properties (e.g. Theorem 5.2.13) for such a modified procedure.

5.5 Show that h_0 given in (5.3.8) is strictly larger than unity.

5.6 Derive the asymptotic first– and second–order properties of the purely sequential procedure (5.3.18) when one replaces U(n) by U*(n) as the estimator of σ^2. Do you obtain anything different in Theorem 5.3.5 (ii) – (iv) for the modified procedure?

5.7 In the three–stage procedure (5.3.24) suppose one uses U*(n) instead of U(n). Do you obtain the same asymptotic first–order properties?

5.8 Prove Theorem 5.2.4.

5.9 Prove Theorem 5.2.6. Also verify (5.2.21).

5.10 Prove Theorem 5.2.8. Also verify (5.2.28).

5.11 Prove Theorem 5.2.13. Also verify (5.2.49), (5.2.50) and (5.2.51).

5.12 Prove (5.2.55), (5.2.56) and (5.2.57).

5.13 Prove Theorem 5.2.15.

5.14 Verify (5.2.61) and prove Theorem 5.2.16.

5.15 Verify (5.3.19).

5.16 Verify (5.3.25).

5.17 Show that (5.3.25) holds for the accelerated sequential procedure (5.3.26) – (5.3.28).

5.18 Verify (5.3.41) – (5.3.42).

5.19 Prove Theorem 5.3.10.

5.20 Prove Theorem 5.3.11.

5.21 Verify that the expression of q obtained after (5.2.32) agrees with that given in the statement of Theorem 5.2.9.

5.22 In the two–stage procedure (5.2.8) and the modified two–stage procedure (5.2.10) – (5.2.11), suppose that one uses $U^*(n)$, instead of $U(n)$, as the estimator of σ^2. Verify the properties given in Theorem 5.2.2 (i), (iv) and Theorem 5.2.3 respectively, for such improvised procedures.

5.23 In the two–stage procedure (5.2.42) and the modified two–stage procedure (5.2.44) – (5.2.45), suppose that one uses $V^*(n)$, instead of $V(n)$, as the estimator of σ. Verify the properties given in Theorem 5.2.11 (i), (iv) and Theorem 5.2.12 respectively, for such improvised procedures.

5.24 In the modified two–stage procedure (5.3.13) – (5.3.14), suppose the one uses $U^*(n)$, instead of $U(n)$, in order to estimate σ^2. Prove Theorem 5.3.4 (ii) – (iii) for such a procedure.

5.25 Prove the following results involving gamma functions.

(a) For every fixed real number $b(> -a)$,

$$\Gamma(a + b)a^{-b}\{\Gamma(a)\}^{-1} = 1 + 0(a^{-1})$$

as $a \to \infty$.

(b) For every real number m, prove that

$$(m - 1)^{-s}\Gamma(m-1+s)\{\Gamma(m - 1)\}^{-1} < 1 \text{ if } 0 < s < 1 \text{ and } m > 1,$$

and

$$(m - 1)^{s}\Gamma(m-1-s)\{\Gamma(m - 1)\}^{-1} > 1 \text{ if } s > 0 \text{ and } m > 1 + s.$$

[*Hints*: See Lemmas 1 and 2 in Mukhopadhyay and Hilton (1986).]

REFERENCES

Apostol (1974), Barlow, Bartholomew, Bremner and Brunk (1972), Chen and Dudewicz (1973), Cohen and Sackrowitz (1982), Dantzig (1940), Dudewicz (1971), Dykstra, Robertson and Wright (1986), Finney (1941), Ghosh (1991), Ghosh and Hoekstra (1991), Ghosh and Mukhopadhyay (1981), Ghurye (1958), Gupta and Panchapakesan (1979), Gupta and Sobel (1962), Kimbal (1951), Krishnaiah and Armitage (1964, 1970), Kuo and Mukhopadhyay

(1990a,b), Lombard and Swanepoel (1978), Mukhopadhyay (1974, 1982, 1985a, 1988, 1991, 1993b), Mukhopadhyay, Chattopadhyay and Sahu (1993), Mukhopadhyay and Hilton (1986), Mukhopadhyay and Kuo (1988), Mukhopadhyay and Mauromoustakos (1987), Robbins (1959), Saxena and Tong (1969, 1971), Stein (1945), Swanepoel and vanWyk (1982), Tong (1970, 1973).

6

SELECTING THE BEST COMPONENT IN A MULTIVARIATE NORMAL POPULATION

6.1 INTRODUCTION

The area of multivariate selection and ranking has not quite developed as steadily as its univariate counterparts. The available results are almost exclusively for certain types of multivariate normal populations and some associated ranking problems. Refer to section 7.3 of Gupta and Panchapakesan (1979) and Dudewicz and Taneja (1981).

We consider selection procedures relating to a single k–dimensional normal population $N_k(\underset{\sim}{\mu}, \Sigma)$, $\underset{\sim}{\mu}$ being the mean vector consisting of $\mu_1,\ldots,\mu_k$ and $\Sigma = (\sigma_{ij})$ where $\sigma_{ii} = \sigma_i^2$, $\sigma_{ij} = \rho_{ij}\sigma_i\sigma_j$, $i \neq j = 1,\ldots,k$ with $k \geq 2$. It is assumed that all the parameters are unknown where $\mu_i \in R$, $\sigma_i \in R^+$, Σ is positive definite (p.d.), and *the pairwise correlations* ρ_{ij}*'s are nonnegative*, that is $\rho_{ij} \in [0, 1)$. We now formulate the problem and introduce some basic notation. Let $\underset{\sim}{X}_1, \underset{\sim}{X}_2,\ldots$ be i.i.d. $N_k(\underset{\sim}{\mu}, \Sigma)$. We write $\underset{\sim}{X}_j' = (X_{1j},\ldots,X_{kj})$, $j = 1, 2,\ldots$. The random samples $X_{i1}, X_{i2},\ldots$ are then i.i.d. $N(\mu_i, \sigma_i^2)$ arising from the ith component, referred to as π_i, $i = 1,\ldots,k$. Let $\mu_{[1]} \leq \cdots \leq \mu_{[k]}$ be the ordered μ–values. Our goal is to select the component associated with $\mu_{[k]}$. Such a component is called the *best* component in this situation and we assume that there is only one such component.

Along the lines of Chapters 3 and 4, we again adopt Bechhofer's (1954) *indifference zone* approach, and given $\delta^*(>0)$, we let

$$\Omega(\delta^*) = \{\underset{\sim}{\mu} : \mu_{[k]} - \mu_{[k-1]} \geq \delta^*\}. \tag{6.1.1}$$

For a given $P^* \epsilon (k^{-1}, 1)$, we consider selection procedures such that P(CS) is at least P^* or asymptotically (as $\delta^* \to 0$) at least P^* whenever $\underset{\sim}{\mu} \epsilon \Omega(\delta^*)$. In the case when $\rho_{ij} = \rho$, $\sigma_i^2 = \sigma^2$ for all $i \neq j = 1,\ldots,k$, σ being known, Frischtak (1973) gave a solution for this problem. See also Gupta (1963) and Gupta et al. (1973) in this context. In the problem we have on hand, there are many more nuisance parameters to tackle in comparison with what Frischtak (1973) had, because Σ is arbitrary p.d. with $\rho_{ij} \geq 0$ for all $i \neq j = 1,\ldots,k$. Various sequential and multistage selection procedures and their associated asymptotics are discussed here along the lines of Mukhopadhyay and Chou (1984) and Mukhopadhyay (1985b).

Having recorded $\underset{\sim}{X}_1,\ldots,\underset{\sim}{X}_n$ we let $Y_j = \Sigma_{i=1}^k X_{ij}$, $\overline{Y}_n = n^{-1}\Sigma_{j=1}^n Y_j$, $S^2(n) = (n-1)^{-1}\Sigma_{j=1}^n (Y_j - \overline{Y}_n)^2$, $\overline{X}_{in} = n^{-1}\Sigma_{j=1}^n X_{ij}$ for $n \geq 2$, $i = 1,\ldots,k$ and $j = 1,\ldots,n$. A natural selection rule is simply to pick the component associated with the largest sample mean $\overline{X}_{in}$, that is

$$\mathscr{P}_n^{**} : \text{Select } \pi_j \text{ as the best component if } \overline{X}_{jn} = \max_{1 \leq i \leq k} \overline{X}_{in}. \tag{6.1.2}$$

6.2 Two–stage procedure

Let $m_1(\geq 2)$ be the starting sample size from $N_k(\underset{\sim}{\mu}, \Sigma)$ and consider $a_1 = a_1(m_1, k, P^*)$, a positive real number to be specified later. Define

$$N = N(\delta^*) = \max\{m_1, [a_1^2 S^2(m_1)\delta^{*-2}]^* + 1\}. \tag{6.2.1}$$

If $N = m_1$, we do not take any more samples from $N_k(\underset{\sim}{\mu}, \Sigma)$. However, if N

$> m_1$, then we take $(N - m_1)$ more samples from $N_k(\underset{\sim}{\mu}, \Sigma)$ in the second stage. Based on the totality of all samples, that is having $\underset{\sim}{X}_1,\ldots,\underset{\sim}{X}_N$, one now implements the selection procedure $\mathscr{P}_N^{**}$ given by (6.1.2). Throughout, $\underset{\sim}{\mu}$ and Σ are kept fixed.

Theorem 6.2.1. *For the two-stage procedure (6.2.1), with "a_1" defined in (6.2.5), we have:*

(i) $P(CS) \geq P^*$ *for all* $\underset{\sim}{\mu} \in \Omega(\delta^*)$;

(ii) $a_1^2 \sigma^{*2} \delta^{*-2} \leq E(N) \leq m_1 + a_1^2 \sigma^{*2} \delta^{*-2}$;

(iii) $E(N/C) \to a_1^2 / a^{*2}$ *as* $\delta^* \to 0$;

(iv) $\liminf P(CS) \geq P^*$, *for all* $\underset{\sim}{\mu} \in \Omega(\delta^*)$, *as* $\delta^* \to 0$;

where $\sigma^{*2} = \underset{\sim}{1}' \Sigma \underset{\sim}{1}$, a^* *is given by* $\Phi(a^*) = P^{*1/(k-1)}$ *and* $C = a^{*2} \sigma^{*2} / \delta^{*2}$.

Proof. Parts (ii) – (iii) can be proved as in Theorem 3.3.1 (ii) – (iii). Now, let us turn to part (i) and write $\pi_{(i)}$ for the component having the mean $\mu_{[i]}$, variance $\sigma_{(i)}^2$, giving rise to the overall sample mean $\overline{X}_{(iN)}$, $i = 1,\ldots,k$. The correlation coefficient between $\pi_{(i)}$ and $\pi_{(j)}$ is denoted by $\rho_{(ij)}$, $i \neq j = 1,\ldots,k$. Then, we have

$$P(CS) = P\{\overline{X}_{(iN)} < \overline{X}_{(kN)},\ i = 1,\ldots,k-1\}$$

$$= P\{Z_{(i)}(N) < (\mu_{[k]} - \mu_{[i]})(N/C_{ik})^{\frac{1}{2}}, i = 1,\ldots,k-1\}, \qquad (6.2.2)$$

where $C_{ik} = \sigma^2_{(i)} + \sigma^2_{(k)} - 2\rho_{(ik)}\sigma_{(i)}\sigma_{(k)}$, and $Z_{(i)}(N) = \{\overline{X}_{(iN)} - \overline{X}_{(kN)} + \mu_{[k]} - \mu_{[i]}\}(N/C_{ik})^{\frac{1}{2}}$ for $i = 1,\ldots,k-1$. One can see that the conditional distribution of $Z_{(1)}(N),\ldots,Z_{(k-1)}(N)$ given $S^2(m_1)$ is $(k-1)$–dimensional normal with zero means, unit variances, and $\mathrm{Cov}(Z_{(i)}(N), Z_{(j)}(N)) = \sigma^2_{(k)}(C_{ik}C_{jk})^{-\frac{1}{2}}$, $i \neq j = 1,\ldots,k-1$. From (6.2.1) we obtain $N \geq a_1^2S^2(m_1)\delta^{*-2}$ w.p. 1, and this implies $\{\mu_{[k]} - \mu_{[i]}\}(N/C_{ik})^{\frac{1}{2}} \geq a_1S(m_1)\{\mu_{[k]} - \mu_{[i]}\}C_{ik}^{-\frac{1}{2}}\delta^{*-1} = Q_i$, say. Hence, from (6.2.2) we obtain

$$P(CS) \geq P\{Z_{(i)}(N) < Q_i, i = 1,\ldots,k-1\}. \tag{6.2.3}$$

Now, given $S(m_1)$, Q_i can be thought of as being constants, and thus using Slepian's (1962) inequality, we get

$$P\{Z_{(i)}(N) < Q_i, i=1,\ldots,k-1 \,\big|\, S(m_1)\} > \prod_{i=1}^{k-1} P\{Z_{(i)}(N) < Q_i \,\big|\, S(m_1)\} = \prod_{i=1}^{k-1} \Phi(Q_i).$$

For some details on Slepian's inequality, one may also look at (8.4.10) – (8.4.11). It is easy to see that $S^2(m_1)$ estimates the variance of Y_j which is $\underset{\sim}{1}'\Sigma\underset{\sim}{1} = \sigma^{*2}$, say, and also that $\sigma^{*2} \geq C_{ik}$ for all $i = 1,\ldots,k-1$, since ρ_{ij}'s are all assumed to be non–negative. Thus, we have

$$P(CS) \geq E\{\prod_{i=1}^{k-1} \Phi(Q_i)\}$$

$$\geq \int_0^{\infty} \{\prod_{i=1}^{k-1} \Phi[a_1u^{\frac{1}{2}}(m_1-1)^{-\frac{1}{2}}(\mu_{[k]}-\mu_{[i]})/\delta^*]\}f^0(u)du \tag{6.2.4}$$

where $f^0(\cdot)$ is the probability density function of a $\chi^2_{m_1-1}$ random variable.

Since we are working under the condition $\mu_{[1]} \leq \cdots \leq \mu_{[k-1]} \leq \mu_{[k]} - \delta^*$, the integral in (6.2.4) has the infimum when we have $\mu_{[1]} = \cdots = \mu_{[k-1]} = \mu_{[k]} - \delta^*$, which is referred as the *least favorable configuration* (LFC). Hence, we immediately obtain

$$P(CS) \geq P^* \text{ for all } \underset{\sim}{\mu} \in \Omega(\delta^*),$$

if we choose "a_1" such that

$$\int_0^\infty [\Phi\{a_1 u^{\frac{1}{2}}(m_1-1)^{-\frac{1}{2}}\}]^{k-1} f^0(u)du = P^*. \tag{6.2.5}$$

In fact, for such values of "a_1", the infimum of $P(CS)$ for $\underset{\sim}{\mu} \in \Omega(\delta^*)$ is also at least P^*. □

Remark 6.2.1. In case σ^* is known, one can easily show that by taking $n = [C]^* + 1$ samples from $N_k(\underset{\sim}{\mu}, \Sigma)$ and implementing the corresponding selection rule $\mathscr{P}_n^{**}$ from (6.1.2), one has $P(CS) \geq P^*$ for all $\underset{\sim}{\mu} \in \Omega(\delta^*)$. That is, C can be viewed as an analog of the optimal fixed sample size required from each component, had σ^* been known. This is the reason why we compare E(N) with C in Theorem 6.2.1 (iii).

6.2.1 THE INDEPENDENT CASE

Suppose that $\rho_{ij} = 0$ for all $i \neq j = 1,\ldots,k$, and thus the k components are independent. We can still implement the two-stage procedure (6.2.1) without any change whatsoever. Suppose the situation is such that we can obtain samples of unequal sizes, if necessary, from the k components $\pi_1,\ldots,\pi_k$. Thus, one can also implement Dudewicz and Dalal's (1975) procedure P_R. This procedure is discussed at some length later in section 8.4.1.2 of Chapter 8. For our purposes at present though, the same methodology as discussed in Rinott (1978) is summarized as follows. Let $h_1 = h_1(m_1, k, P^*)$ be an appropriate positive number.

P_R : One starts with an initial sample of size $m_1(\geq 2)$ from each component. Write $\overline{X}_i(m_1) = m_1^{-1}\Sigma_{j=1}^{m_1} X_{ij}$, $S_i^2(m_1) = (m_1-1)^{-1}\Sigma_{j=1}^{m_1} (X_{ij} - \overline{X}_i(m_1))^2$, $\eta_i^o = h_1^2 S_i^2(m_1)\delta^{*-2}$, $N_i = \max\{m_1, [\eta_i^o]^* + 1\}$, $i = 1,\ldots,k$. Then, we compute $\overline{X}_{iN_i}$ based on $X_{i1},\ldots,X_{iN_i}$, and we select as best the component associated with the largest $\overline{X}_{iN_i}$ value, $i = 1,\ldots,k$.

Rinott (1978) showed that if h_1 is determined so as to satisfy

$$\int_0^\infty \left[\int_0^\infty \Phi\{h_1[(m_1-1)(x^{-1}+y^{-1})]^{-\frac{1}{2}}\} f^0(x)dx\right]^{k-1} f^0(y)dy = P^*, \tag{6.2.6}$$

then, $P(CS) \geq P^*$ whenever $\underset{\sim}{\mu}\epsilon\Omega(\delta^*)$. Recall that $f^0(\cdot)$ is the p.d.f. of $\chi^2_{m_1-1}$. See also Mukhopadhyay (1979).

A natural question is then how does the procedure (6.2.1) compare with P_R in the contexts when both such methodologies are applicable. For the procedure P_R, the total sample size $N_R = \Sigma_{i=1}^k N_i$, while for (6.2.1), the total sample size is $N_S = kN$. It can be easily seen that

$$\lim_{\delta^*\to 0} \{E(N_S)/E(N_R)\} = ka_1^2/h_1^2 = \beta_1^*, \text{ say.} \tag{6.2.7}$$

We use the β_1^*–values for comparing the two–stage procedures P_R and (6.2.1). In being able to compute β_1^*, we naturally need h_1 for various values of k, m_1 and P^*. In order to obtain the h_1–values, one should note that Tables I – III of Mukhopadhyay and Chou (1984) have provided values of the integral given in (6.2.6) for $k = 3, 4, 5,$ $m_1 = 3(1)10(5)30$. The

h_1–values for $k = 3, 4, 5$, $m_1 = 5(1)10(5)30$, $P^* = .50, .75, .80, .85, .90, .95, .975, .99$ are provided in Tables IV – VI in that same paper of Mukhopadhyay and Chou (1984). For $k = 2$, the solutions "h_1" for (6.2.6),

Table 6.2.1. Values of β_1^* from (6.2.7) for k = 2

	P^*				
m_1	.75	.90	.95	.975	.99
5	.845	.896	.940	.993	1.063
10	.932	.956	.986	1.012	1.052
15	.958	.971	.992	1.009	1.039
20	.966	.977	.996	1.007	1.029
25	.977	.983	.999	1.006	1.026
30	.971	.982	.994	1.008	1.018

given P^* and m_1, are tabled in Dudewicz and Dalal (1975) and also in the appendix A.9 of Gibbons et al. (1977).

From Table 6.2.1, one finds that for $k = 2$, the two–stage procedure (6.2.1) can be more efficient asymptotically than the procedure P_R of Dudewicz and Dalal (1975) when $P^* \leq .95$. However, P_R becomes more efficient when $P^* = .975, .99$ for $k = 2$. In the case $k \geq 3$, the procedure P_R turns out to be more efficient than (6.2.1).

Remark 6.2.2. We add that the procedure (6.2.1) is the only one available thus far in the case when all the component variances are completely unknown and all of the ρ_{ij}'s are non–negative but otherwise completely unknown, $1 \leq i \neq j \leq k$. It is of some tangential interest only that the two–stage procedure (6.2.1) is also found to be asymptotically *superior* to P_R for $k = 2$, and $P^* \leq 0.95$ while for $k \geq 3$ the procedure P_R is the better one, assuming additionally in either case that $\rho_{ij} = 0$ for all $i \neq j$. This comparison may not seem too appropriate to some readers after all, since P_R is designed to handle the case of *all zero correlations only*. See section 8.4.1.2 for further details on the procedure P_R.

Remark 6.2.3. From (6.2.5) it is fairly easy to see that "a_1" must be the solution of

$$P\{T_1 \leq a_1, \ldots, T_{k-1} \leq a_1\} = P^*, \tag{6.2.8}$$

where $(T_1, \ldots, T_{k-1})$ has the equicorrelated multivariate t–distribution with $(m_1 - 1)$ degrees of freedom, with the common correlation coefficient $\rho = 0$. The Tables 1, 11, 21, 31 in Krishnaiah and Armitage (1965) give values of "a_1" for $P^* = .90, .95, .975, .99$, $k = 2(1)11$, $m_1 = 6(1)36$. For $k = 2$, one can use the usual Student's t–table given by Smirnov (1961) for other values of P^*. For $k = 3$, Mukhopadhyay and Chou (1984) present a_1–values (Table VII) for $P^* = .50, .60, .70, .75, .80$ and $m_1 = 5(5)30$.

6.3 MODIFIED TWO–STAGE PROCEDURE

Along the lines of section 3.4 and Mukhopadhyay (1985b), we now discuss a modified version of the two–stage procedure (6.2.1) so that one is able to claim *asymptotic first–order efficiency* in the sense of (3.4.7). Choose and fix $m_0^*(\geq 2)$, $\gamma \in R^+$ and let

$$m_1 = m_1(\delta^*) = \max\{m_0^*, [(a^{*2}\delta^{*-2})^{1/(\gamma+1)}]^* + 1\}, \tag{6.3.1}$$

where recall that $\Phi(a^*) = P^{*1/(k-1)}$. This m_1 is then the experimenter's choice for the starting sample size, and let

$$N = N(\delta^*) = \max\{m_1, [a_1^2 S^2(m_1)\delta^{*-2}]^* + 1\}, \tag{6.3.2}$$

where "a_1" comes from (6.2.5). The modified two–stage procedure (6.3.1) – (6.3.2) is now implemented in exactly the same way as (6.2.1). Along the lines of Theorem 3.4.1, one can show that

$$P(CS) \geq P^* \text{ for all } \underset{\sim}{\mu} \in \Omega(\delta^*); \tag{6.3.3}$$

$$N/C \to 1 \text{ w.p. } 1 \text{ and } E(N/C) \to 1 \text{ as } \delta^* \to 0; \tag{6.3.4}$$

$$\liminf P(CS) \geq P^*, \text{ for all } \underset{\sim}{\mu} \in \Omega(\delta^*), \text{ as } \delta^* \to 0; \tag{6.3.5}$$

with $C = a^{*2}\sigma^{*2}/\delta^{*2}$, $\sigma^{*2} = \underset{\sim}{1}'\Sigma\underset{\sim}{1}$.

On the other hand, using the Cornish–Fisher expansion of "a_1" in terms of "a^*", available in Gupta and Sobel (1957), one can show that

$$\liminf_{\delta^* \to 0} E(N - C) = \infty \tag{6.3.6}$$

for the modified two–stage procedure. See the proof of (3.4.10). Thus, we can claim that both the two–stage and modified two–stage selection procedures are *consistent* and *asymptotically first–order consistent.* The modified version is, however, *asymptotically first–order efficient,* but it fails to have asymptotic second–order efficiency property. Recall (3.4.5) – (3.4.8) in this context.

6.4 PURELY SEQUENTIAL PROCEDURE

We now discuss a purely sequential selection procedure that turns out to be *asymptotically second–order efficient.* For details, see Mukhopadhyay (1985b). One starts with $m_1(\geq 2)$ samples. Then, we take one sample at a time according to the following stopping rule:

$$N = N(\delta^*) = \inf\{n \geq m_1 : n \geq a^{*2}S^2(n)\delta^{*-2}\}. \tag{6.4.1}$$

Based on the totality of all samples, that is having $\underset{\sim}{X}_1, \ldots, \underset{\sim}{X}_N$, one now implements the selection procedure $\mathscr{P}_N^{**}$ given by (6.1.2). Throughout, $\underset{\sim}{\mu}$ and Σ are kept fixed. Along the lines of Theorem 3.5.1, one can easily verify that (6.3.3) – (6.3.5) holds for our purely sequential procedure. Additionally, we now have the *asymptotic second–order efficiency* property as well.

Theorem 6.4.1. *For the purely sequential procedure (6.4.1), we have as* $\delta^* \to 0$:

$$E(N) = C + \beta + o(1) \text{ for all } \underset{\sim}{\mu} \epsilon \Omega(\delta^*), \text{ if } m_1 \geq 4;$$

where $C = a^{*2}\sigma^{*2}/\delta^{*2}$, $\beta = \nu(1) - 2$ *with* $\nu(k)$ *defined in (3.5.7),* $\sigma^{*2} = \underset{\sim}{1}'\Sigma\underset{\sim}{1}$, *and* $\Phi(a^*) = P^{*1/(k-1)}$.

Proof. First note that N given in (6.4.1) can be equivalently written as

$$N = \inf\{n \geq m_1 : \Sigma_{i=1}^{n-1} W_i \leq n(n-1)C^{-1}\}$$

and hence $N = Q + 1$ where

$$Q = \inf\{n \geq m_1 - 1 : n^2(1 + n^{-1})C^{-1} \geq \Sigma_{i=1}^{n} W_i\},$$

W's being i.i.d. χ_1^2. This form agrees with the representation given in (3.5.6) with its "k" being unity. Let $\nu = \nu(1)$ and $\beta = \beta(1) = \nu - 2$ where $\nu(1)$ comes from (3.5.7). Now, the result follows from Theorem 2.4.8 (v) with $\omega = 1$, $m = m_1 - 1$, $b = \frac{1}{2}$. □

6.4.1 RATES OF CONVERGENCES IN THE CASE OF TWO COMPONENTS

In this section, we consider the same basic problem as discussed earlier.

We, however, wish to *relax the condition of correlation coefficients being positive*, but we do this only when $k = 2$. In the section 6.4.1.1 we still work under normality, and derive second–order expansions for P(CS) while using a purely sequential procedure along the lines of (6.4.1). In section 6.4.1.2, we modify this sequential procedure further to incorporate the situation of non–normality. Theorems 2.4.2 – 2.4.7 are used in this second topic.

6.4.1.1 Bivariate Normal with Unknown Σ

Here, we do not assume anything specific about the sign of ρ_{12}. We however, use the same formulation as in section 6.1. We start the experiment with an initial sample of size $m_1(\geq 2)$, that is to start with $\underset{\sim}{X}_1, \ldots, \underset{\sim}{X}_{m_1}$. Let $U_j = X_{1j} - X_{2j}$, $j = 1,2,\ldots,m_1$, and write $\overline{U}_1(m_1) = m_1^{-1}\Sigma_{j=1}^{m_1} U_j$, $U_1^2(m_1) = (m_1-1)^{-1}\Sigma_{j=1}^{m_1}(U_j - \overline{U}_1(m_1))^2$. Similarly, we define $U_1^2(n)$ when we have recorded n samples, for $n \geq m_1$. Let "a*" be the same as before, that is, $\Phi(a^*) = P^*$. The sequential procedure is defined by the following stopping time:

$$N = N(\delta^*) = \inf\{n \geq m_1 : n \geq a^{*2}U_1^2(n)\delta^{*-2}\}. \tag{6.4.2}$$

The only difference between (6.4.1) and (6.4.2) is that we are now using $U_1^2(n)$ instead of $S^2(n)$ as the estimator of σ^{*2} in the sequential rule. Now, $P(N < \infty)$ being unity, we choose as best the component associated with the larger $\overline{X}_{iN}$, $i = 1, 2$ value via $\mathscr{P}_N^{**}$ in (6.1.2). One advantage of taking $U_1^2(n)$ instead of $S^2(n)$ is that we do not need the positivity of ρ_{12} in this section. An important technical difference lies in the fact that under the LFC, that is $\mu_{[2]} = \mu_{[1]} + \delta^*$, $\lim_{\delta^* \to 0}$ P(CS) turns out to be P* for the procedure given in (6.4.2), while the same limit is only at least P* for (6.4.1). Without any more deliberations, let us proceed with the second–order expansions of

E(N) and P(CS) for the rule (6.4.2).

Theorem 6.4.2. *For the purely sequential procedure (6.4.2), we have as* $\delta^* \to 0$:

(i) $D^{-\frac{1}{2}}(N - D) \xrightarrow{\mathscr{L}} N(0, 2)$, *if* $m_1 \geq 2$;

(ii) $E(N) = D + \beta + o(1)$, *for all* $\underset{\sim}{\mu} \in \Omega(\delta^*)$, *if* $m_1 \geq 4$;

(iii) $P(CS) = P^* + a^{*2}D^{-1}\{\beta\psi_3'(a^{*2}) + \frac{1}{2}a^{*2}\psi_3''(a^{*2})\} + o(\delta^{*2})$ *under the LFC, if* $m_1 \geq 4$;

where β *is the same as in Theorem 6.3.1,* $\psi_3(x) = \Phi(x^{\frac{1}{2}})$ *for* $x > 0$, $D = a^{*2}\sigma^2/\delta^{*2}$ *with* $\sigma^2 = \sigma_1^2 + \sigma_2^2 - 2\rho_{12}\sigma_1\sigma_2$.

Proof. Parts (i) and (ii) follow along the lines of Theorem 3.5.1 (iii) and Theorem 6.3.1 respectively. For part (iii), under the LFC, we have

$$P(CS) = E[\Phi(N^{\frac{1}{2}}\delta^*\sigma^{-1})] = E\{\psi_3(a^{*2}N/D)\},$$

with $\psi_3(x) = \Phi(x^{\frac{1}{2}})$ for $x > 0$. If one proceeds as in the proof of Theorem 3.5.2 (ii) by utilizing Taylor's series expansion, one will obtain

$$E\{\psi_3(a^{*2}N/D)\} = P^* + a^{*2}D^{-1}\{\beta\psi_3'(a^{*2}) + \tfrac{1}{2}a^{*2}\psi_3''(a^{*2})\} + o(\delta^{*2}),$$

as $\delta^* \to 0$, for $m_1 \geq 4$. This is part (iii). □

Mukhopadhyay (1985b) also proposed the following two–stage procedure. We start the experiment with $m_1(\geq 2)$ samples, that is with $\underset{\sim}{X}_1, \ldots, \underset{\sim}{X}_{m_1}$. Let h_2 be the upper $100(1 - P^*)\%$ point of the student's t–distribution with $(m_1 - 1)$ degrees of freedom. Define

$$N = N(\delta^*) = \max\{m_1, [h_2^2 U_1^2(m_1)\delta^{*-2}]^* + 1\}. \tag{6.4.3}$$

Then, we choose as best the component having the larger $\overline{X}_{iN}$ value, $i = 1, 2$ via $\mathscr{P}_N^{**}$ in (6.1.2). Now we verify some useful properties of the two–stage procedure (6.4.3) in the following theorem.

Theorem 6.4.3. *For the two–stage procedure (6.4.3), we have:*

(i) $P(CS) \geq P^*$ *for all* $\underset{\sim}{\mu} \in \Omega(\delta^*)$;

(ii) $E(N/D) \to h_2^2/a^{*2}$ (>1) *as* $\delta^* \to 0$;

(iii) $\lim P(CS) = P^*$, *under the LFC, as* $\delta^* \to 0$;

where h_2 *is the upper* $100(1 - P^*)\%$ *point of the Student's t–distribution with* $(m_1 - 1)$ *degrees of freedom,* $D = a^{*2}\sigma^2/\delta^{*2}$ *as in Theorem 6.4.2.*

We omit its proof for brevity. See Theorem 3 in Mukhopadhyay (1985b) for further details.

6.4.1.2 Bivariate Unknown Distribution

This is the only section in this chapter where we *relax* the *assumption of multivariate normality.* We suppose that we can obtain i.i.d. random variables $\underset{\sim}{X}_1, \underset{\sim}{X}_2, \ldots$ having some unspecified distribution. The Σ matrix is assumed finite and p.d., but unknown. In this case, to accommodate discrete situations, we use the following sequential rule instead of (6.4.2): With the starting sample size $m_1 (\geq 2)$, let

$$N = N(\delta^*) = \inf\{n \geq m_1 : n \geq a^{*2}(U_1^2(n) + n^{-1})\delta^{*-2}\}, \qquad (6.4.4)$$

and we choose as best the component associated with the larger $\overline{X}_{iN}$, $i = 1, 2$ value via the selection rule $\mathscr{P}_N^{**}$ in (6.1.2). Following Robbins et al. (1968) or Mukhopadhyay (1983a,b), it easily follows from (6.4.4) that $\lim_{\delta^* \to 0} P(CS) = P^*$ under the LFC. However, we state the following stronger assertion.

Theorem 6.4.4. *For the purely sequential procedure (6.4.4), if* $E(|U_1|^q) < \infty$ *for* $q > 5(2\gamma)^{-1} - 1$, *then we have as* $\delta^* \to 0$:

$$P(CS) = P^* + 0(\delta^{*\frac{1}{2}-\gamma}),$$

under the LFC, with $\gamma \epsilon (0, \frac{1}{2})$.

Proof. Under the LFC, one can write

$$P(CS) = P\{Z_{1N} < N^{\frac{1}{2}}\delta^*\sigma^{-1}\}$$

where $Z_{1N} = N^{\frac{1}{2}}\{\overline{X}_{(1N)} - \overline{X}_{(2N)} + \delta^*\}/\sigma$. Now, note that Z_{1N} is a standardized sample mean that has been randomly stopped and $U_1^2(n)$ is a U–statistic. Then, along the lines of Theorems 1 and 2 of Mukhopadhyay (1981), one can obtain

$$\sup_x |P\{Z_{1N} \leq x\} - \Phi(x)| = 0(\delta^{*\frac{1}{2}-\gamma}),$$

and

$$P\{|ND^{-1} - 1| > c_1 D^{-\frac{1}{2}+\gamma}\} = 0(\delta^{*\frac{1}{2}-\gamma}),$$

for every constant $c_1(>0)$. Then, one will follow along the lines of Csenki

(1980) and Mukhopadhyay (1983b) to conclude the result. Also, refer to Mukhopadhyay and Vik (1985) for some of the details. □

6.5 Notes and Overview

The results described here are of very special nature. Certainly there are other types of interesting and useful multivariate selection and ranking problems. Refer to Chapter 7 of Gupta and Panchapakesan (1979), and Dudewicz and Taneja (1981) for an array of such problems. One should investigate these other problems and examine how far one can go in terms of obtaining analogous second–order asymptotics. This area appears absolutely wide open for such investigations.

Complements and Problems

6.1 Verify (6.3.3), (6.3.4) and (6.3.5).

6.2 Verify (6.3.6).

6.3 Consider the general k–dimensional problem where $\rho_{ij} \geq 0$ for all $1 \leq i \neq j \leq k$ and implement the selection procedure $\mathscr{P}_N^{**}$ from (6.1.2). Show that $P_n(CS) \geq \prod_{i=1}^{k-1} \Phi(n^{\frac{1}{2}}[\mu_{[k]} - \mu_{[i]}]/\sigma^*)$. The lower bound is minimized when $\mu_{[1]} = \cdots = \mu_{[k-1]} = \mu_{[k]} - \delta^*$. Under the LFC, one then has $P_n(CS) = \{\Phi(n^{\frac{1}{2}}\delta^*/\sigma^*)\}^{k-1}$. See Remark 6.2.1 in this context.

6.4 Suppose that a general sampling procedure eventually leads to N, an appropriate stopping variable. Assume that $I(N = n)$ is independent of $(\overline{X}_{1n},\ldots,\overline{X}_{kn})$; the distribution of N does not depend upon $\underset{\sim}{\mu}$; and $P(N < \infty) = 1$. If one implements $\mathscr{P}_N^{**}$ from (6.1.2), then show that $P(CS) \geq E[\{\Phi(N^{\frac{1}{2}}\delta^*/\sigma^*)\}^{k-1}]$ for all $\underset{\sim}{\mu}$.

6.5 Consider the purely sequential selection procedure (6.4.1). In each problem, give particular attention to sufficient conditions involving m_1.

(a) Show that $N/C \to 1$ w.p. 1.

(b) Show that $E(N/C) \to 1$ as $\delta^* \to 0$.

(c) Show that $(C/N)^{\omega}$ is uniformly integrable for all $\omega > 0$.

(d) Show that $\liminf_{\delta^* \to 0} P(CS) \geq P^*$ for all $\underset{\sim}{\mu} \in \Omega(\delta^*)$.

(e) Show that $P(CS) \geq P^* + a_1^* C^{-1} + o(\delta^{*2})$ for all $\underset{\sim}{\mu} \in \Omega(\delta^*)$. Here, determine the expression for a_1^*.

[*Hints*: In (b), show that $E(N) \leq C + 0(1)$. In (c), first show uniform integrability of $(C/N)^{\omega} I(N > \frac{1}{2}C)$. Then, write $E[(C/N)^{\omega} I(N \leq \frac{1}{2}C)] \leq (C/m_1)^{\omega} P(N \leq \frac{1}{2}C) = o(1)$, by Theorem 2.4.8 (i). In (e), use the lower bound obtained in Problem 6.4, and follow along the lines of Theorem 3.5.2 (ii), say.]

6.6 *Three-Stage Procedure*: Consider Problem 6.3. Suppose $m_1 (\geq 2)$ is the starting sample size and $T = \max\{m_1, [\rho^* a^{*2} S^2(m_1)\delta^{*-2}]^* + 1\}$, $N = \max\{T, [a^{*2} S^2(T)\delta^{*-2} + q]^* + 1\}$ where $q \in R$ and $0 < \rho^* < 1$ are chosen and fixed in advance. Here $m_1 = 0(C^{1/r})$ for some $r > 1$. One then implements $\mathcal{P}_N^{**}$ from (6.1.2).

(a) Show that $N/C \to 1$ w.p. 1 as $\delta^* \to 0$.

(b) Show that $E(N/C) \to 1$ as $\delta^* \to 0$.

(c) Find $\lim_{\delta^* \to 0} E(N - C)$.

(d) Show that $(C/N)^{\omega}$ is uniformly integrable for all $\omega > 0$.

(e) Show that $\liminf_{\delta^* \to 0} P(CS) \geq P^*$ for all $\underset{\sim}{\mu} \in \Omega(\delta^*)$.

(f) Determine q as a function of a^*, ρ^* and k such that $P(CS) \geq P^* + o(\delta^{*2})$ for all $\underset{\sim}{\mu} \in \Omega(\delta^*)$.

[*Hints*: In (c), use Theorem 2.3.2. In (d), proceed as in Problem 6.5 (c) and use Theorem 2.3.3 (i). In (f), expand the lower bound in Problem 6.3 and proceed along Theorem 2.3.4.]

6.7 *Accelerated Sequential Procedure*: Consider Problem 6.3. Suppose $m_1(\geq 2)$ is the starting sample size and define $t = \inf\{n \geq m_1 : n \geq \rho^* a^{*2} S^{*2}(n)/\delta^{*2}\}$, $N_1 = [a^{*2} S^{*2}(t)\delta^{*-2} + q]^* + 1$, $N = \max\{N_1, t\}$ where $q \geq 0$ and $0 < \rho^* < 1$ are chosen and fixed in advance, and $S^{*2}(n) = n(n-1)^{-1}S^2(n)$. One then implements $\mathcal{P}_N^{**}$ from (6.1.2) in this purely sequential followed up by batch sampling methodology. In each problem, give particular attention to sufficient conditions involving m_1.

(a) Show that $N/C \to 1$ w.p. 1 as $\delta^* \to 0$.

(b) Show that $E(N/C) \to 1$ as $\delta^* \to 0$.

(c) Show that $(C/N)^{\omega}$ is uniformly integrable for all $\omega > 0$.

(d) Show that $E(N - C)$ is bounded as $\delta^* \to 0$.

(e) Show that $\liminf_{\delta^* \to 0} P(CS) \geq P^*$ for all $\underset{\sim}{\mu} \in \Omega(\delta^*)$.

(f) Determine q as a function of a^*, ρ^* and k such that $P(CS) \geq P^* + o(\delta^{*2})$ for all $\underset{\sim}{\mu} \in \Omega(\delta^*)$.

[*Hints*: In (c), use Theorem 2.5.3 (i). Use Corollary 2.5.1 in (d).]

6.8 Refer to section 6.4.1.1. Propose three–stage and accelerated sequential procedures for this set up along the lines of Problems 6.6 – 6.7 and verify the associated results (a) – (f). Note, however, that under the LFC, one now has "equality" instead of "inequality".

6.9 Let $S^{*2}(n) = n(n-1)^{-1}S^2(n)$ as in Problem 6.7. Suppose that one uses $S^{*2}(n)$, instead of $S^2(n)$, in order to estimate σ^{*2} in the purely sequential procedure (6.4.1). For the new stopping variable N and selection procedure $\mathscr{P}_N^{**}$, verify the results given in Problem 6.5.

6.10 Suppose one uses $S^{*2}(n)$, instead of $S^2(n)$, in the three–stage procedure considered in Problem 6.6. For such new three–stage procedure, verify the results given in Problem 6.5 (a), (b), (c) and (d).

6.11 Prove Theorem 6.4.3.

REFERENCES

Bechhofer (1954), Csenki (1980), Dudewicz and Dalal (1975), Dudewicz and Taneja (1981), Frischtak (1973), Gibbons, Olkin and Sobel (1977), Gupta (1963), Gupta, Nagel and Panchapakesan (1973), Gupta and Panchapakesan (1979), Gupta and Sobel (1957), Krishnaiah and Armitage (1965), Mukhopadhyay (1979, 1981, 1983a,b, 1985b), Mukhopadhyay and Chou (1984), Mukhopadhyay and Vik (1985), Rinott (1978), Robbins, Sobel and Starr (1968), Slepian (1962), Smirnov (1961).

7

ESTIMATION AFTER SELECTION AND RANKING

7.1 Introduction

In Chapters 3 and 4, the problems of selecting the *best population* under the *indifference zone* formulation were extensively discussed for normal as well as two–parameter exponential populations. Various types of sequential and multistage selection methodologies were proposed in order to select the best population such that P(CS) is at least or approximately at least the nominal preassigned value P^* whenever the parameters came from the preference zone.

In the present chapter, we explore a few other ideas and directions. Suppose that one of the original selection procedures had stopped with $X_{i1},\dots,X_{iN}$ from π_i, $i = 1,\dots,k$ and certain selection and ranking decisions had already been made with the P^*–requirement. But, then the experimenter turns around and wishes to estimate a few important parametric functions, say, of $\underset{\sim}{\mu}$ or $\underset{\sim}{\theta}$ or σ. In other words, such estimation problems were not in the forefront in the early stages of formulation and experimentation and thus the stopping rules giving rise to the appropriate stopping variable N were not geared toward solving the estimation problems as such. The materials discussed in Chapter 5 are indeed fundamentally different from those that we present here. In the present scenario, we assume that we already have available the random samples $X_{i1},\dots,X_{iN}$ as well as N that correspond to a particular selection methodology discussed in Chapters 3 and 4, and we provide various interesting characteristics of associated estimators of some useful parametric functions. There is considerable amount of literature on estimating certain parameter of the "selected" population. Some references are furnished in section 7.4. It is clear that we are handling a different type of problem. The new results presented here will perhaps lead to other types of interesting developments having more breadth in the future.

7.2 THE NORMAL PROBLEM

Suppose that the ith population's distribution is $N(\mu_i, \sigma^2)$, where all the parameters are unknown. We write $\underset{\sim}{\mu} = (\mu_1,...,\mu_k)$ and assume that we are working under the general framework of Chapter 3 and one of the specific selection methodologies has ended up with the random samples $X_{i1},...,X_{iN}$ and N from the population π_i, $i = 1,...,k$. At this stage, the experimenter may wish to estimate (a) μ_i for any fixed i, (b) σ^2, the population variance, (c) a few linear parametric functions of $\underset{\sim}{\mu}$ simultaneously, etc. In the case of (a), the experimenter will perhaps estimate μ_i by $\overline{X}_{iN}$ and in the case of (b), the experimenter will perhaps estimate σ^2 by means of $U(N)$, the randomly stopped pooled sample variance defined in (3.1.1). We shall give specifics about (c) in section 7.2.3. But then, what are the associated properties of such natural estimators? It is shown that in some circumstances one can derive even *asymptotic second-order characteristics.*

7.2.1 ESTIMATION OF MEANS

For a fixed i, one wishes to estimate μ_i by means of $\overline{X}_{iN}$, $i = 1,...,k$. Suppose that for an integrable function $g_2(\cdot)$, one has

$$E[g_2(\overline{X}_{in})] = g_2^*(n, \underset{\sim}{\mu}, \sigma^2), \tag{7.2.1}$$

with a certain functional $g_2^*(\cdot)$. Then, assume the sufficient conditions listed in Theorem 3.2.1, and additionally assume that $E[g_2^*(N, \underset{\sim}{\mu}, \sigma^2)]$ is finite. Then one obtains

$$E[g_2(\overline{X}_{iN})] = \Sigma_{n=m_1}^{\infty} E[g_2(\overline{X}_{iN}) \big| N = n]P(N = n)$$

$$= \Sigma_{n=m_1}^{\infty} E[g_2(\overline{X}_{in}) \big| N = n]P(N = n)$$

$$= \Sigma_{n=m_1}^{\infty} E[g_2(\overline{X}_{in})]P(N = n)$$

$$= \Sigma_{n=m_1}^{\infty} g_2^*(n, \underset{\sim}{\mu}, \sigma^2)P(N = n)$$

$$= E[g_2^*(N, \underset{\sim}{\mu}, \sigma^2)]. \tag{7.2.2}$$

Suppose that $g_2(x) = x$, then $g_2^*(n, \underset{\sim}{\mu}, \sigma^2) = \mu_i$. Also, suppose that $g_2(x) = (x - \mu_i)^2$, then $g_2^*(n, \underset{\sim}{\mu}, \sigma^2) = \sigma^2 n^{-1}$. Thus, for each selection methodology described in sections 3.3 – 3.7, one obtains from (7.2.2),

$$E[\overline{X}_{iN}] = \mu_i \text{ and } E[(\overline{X}_{iN} - \mu_i)^2] = \sigma^2 E(N^{-1}). \tag{7.2.3}$$

In other words, $\overline{X}_{iN}$ is an unbiased estimator of μ_i and one also has an exact expression of the variance of such an estimator.

Theorem 7.2.1. *For the modified two–stage procedure (3.4.1) – (3.4.2), purely sequential procedure (3.5.1), accelerated sequential procedure (3.6.4) – (3.6.6), and the three–stage procedure (3.7.2) – (3.7.3), we have as* $\delta^* \to 0$:

$$\text{Var}(\overline{X}_{iN}) = \sigma^2 C^{-1} + o(C^{-1})$$

for every fixed $i = 1,...,k$ *where* $C = h^2\sigma^2/\delta^{*2}$ *and* h *comes from (3.2.4). In the case of purely sequential and accelerated sequential methodologies, the starting sample size* m_1 *has to satisfy the conditions* (a) $m_1 \geq 3$ *when* $k = 2$, (b) $m_1 \geq 2$ *when* $k \geq 3$.

Proof. From (7.2.3), one has

$$\mathrm{Var}(\overline{X}_{iN}) = \sigma^2 C^{-1} E(C/N) \tag{7.2.4}$$

and thus it will suffice to show that $E(C/N) \to 1$ as $\delta^* \to 0$. For each procedure under consideration, $N/C \to 1$ w.p. 1 as $\delta^* \to 0$, and hence liminf $E(C/N) \geq 1$ as $\delta^* \to 0$ in view of Fatou's Lemma.

In the case of modified two–stage procedure, as $\delta^* \to 0$, one has from (3.4.3),

$$\limsup E(C/N) \leq \limsup E[(h^2/\tau^2)\{\sigma^2/U(m_1)\}]$$

$$= \limsup (h^2/\tau^2)k(m_1 - 1)E[\{\chi^2_{k(m_1-1)}\}^{-1}]$$

$$= \limsup \tfrac{1}{2}(h^2/\tau^2)k(m_1 - 1)\,\Gamma\{\frac{k(m_1-1)-2}{2}\}/\Gamma\{\frac{k(m_1-1)}{2}\} = 1,$$

by the properties of gamma functions, since $m_1 \to \infty$ and $\tau \to h$, as $\delta^* \to 0$. Hence, the result in the case of modified two–stage procedure.

We consider the purely sequential procedure (3.5.1). Recall that $N = Q + 1$ where Q was defined in (3.5.6). Thus, $P\{N \leq \epsilon C\} = 0(C^{-\frac{1}{2}k(m_1-1)})$ for all $0 < \epsilon < 1$, in view of Theorem 2.4.8 (i). Hence, $E(C/N) = E\{(C/N)I(N > \epsilon C)\} + E\{(C/N)I(N \leq \epsilon C)\} = E(I) + E(II)$, say. Now, $I \leq \epsilon^{-1}$, and hence I is uniformly integrable. Also, $I \xrightarrow{P} 1$, thus $E(I) = 1 + o(1)$ for $m_1 \geq 2$. On the other hand, $E(II) \leq m_1^{-1}CP(N \leq \epsilon C) =$

$0(C)0(C^{-\frac{1}{2}k(m_1-1)}) = o(1)$ if $m_1 > 1 + 2k^{-1}$. Hence, the result in the case of purely sequential procedure.

For the accelerated sequential procedure (3.6.4) – (3.6.6), Theorem 2.5.3 (i) leads to $P\{N \le \epsilon C\} = 0(C^{-\frac{1}{2}k(m_1-1)})$ for $0 < \epsilon < \rho$. The rest of the proof is similar to the one for the purely sequential situation.

In the case of the three–stage procedure, from Theorem 2.3.3 (i), one immediately gets $P\{N < \epsilon C\} = 0(C^{-\frac{1}{2}p/r})$ for $0 < \epsilon < \rho$, where $p(>0)$ can be arbitrarily large and r comes from (3.7.1). Then, $E\{(C/N)I(N > \epsilon C)\}$ can be shown to be $1 + o(1)$ as in the case of the purely sequential situation. However, $E\{(C/N)I(N \le \epsilon C)\} = 0(C)0(C^{-\frac{1}{2}p/r}) = o(1)$ if we choose $p > 2r$. Hence the result. □

For the purely sequential procedure (3.5.1), accelerated sequential procedure (3.6.4) – (3.6.6) and the three–stage procedure (3.7.2) – (3.7.3), we now attempt at strengthening the preceding result. Recall that

$$E\{(\overline{X}_{iN} - \mu_i)^2\} = \sigma^2 E(N^{-1}) = \sigma^2 C^{-2} E\{(N - C)^2 N^{-1} - N + 2C\}. \tag{7.2.5}$$

For the purely sequential procedure, from Theorem 3.5.1 (iii) one concludes that $(N - C)^2/C \xrightarrow{\mathscr{L}} 2k^{-1}\chi_1^2$ as $\delta^* \to 0$. Also, $(N - C)^2/C$ is uniformly integrable if $m_1 > 1 + 2k^{-1}$ in view of (3.5.8). Now, for $0 < \epsilon < 1$, $(N - C)^2 N^{-1} I(N > \epsilon C) \le \epsilon^{-1}(N - C)^2 C^{-1}$ and hence obviously $(N - C)^2 N^{-1} I(N > \epsilon C)$ is uniformly integrable if $m_1 > 1 + 2k^{-1}$, that is one has $E\{(N - C)^2 N^{-1} I(N > \epsilon C)\} = 2k^{-1} + o(1)$ if $m_1 > 1 + 2k^{-1}$. On the other hand, $E\{(N - C)^2 N^{-1} I(N \le \epsilon C)\} \le (1 + \epsilon^2) m_1^{-1} C^2 P(N \le \epsilon C) = 0(C^2)0(C^{-\frac{1}{2}k(m_1-1)}) = o(1)$ if $m_1 > 1 + 4k^{-1}$. Thus, one has

$$E\{(N - C)^2 N^{-1}\} = 2k^{-1} + o(1), \tag{7.2.6}$$

if $m_1 > 1 + 4k^{-1}$. Now, combine (7.2.5), (7.2.6) with Theorem 3.5.2 (i) to conclude that

$$\mathrm{Var}(\overline{X}_{iN}) = \sigma^2 C^{-1} + (2k^{-1} - \beta)\sigma^2 C^{-2} + o(C^{-2}), \tag{7.2.7}$$

in the purely sequential case provided that (a) $m_1 \geq 4$ when $k = 2$, (b) $m_1 \geq 3$ when $k = 3, 4$, and (c) $m_1 \geq 2$ when $k \geq 5$.

In the case of the accelerated sequential procedure, Theorem 3.6.1 (iii) shows that $(N - C)^2/C \xrightarrow{\mathscr{L}} 2(k\rho)^{-1}\chi_1^2$ as $\delta^* \to 0$. Also, $(N - C)^2/C$ is uniformly integrable if $m_1 > 1 + 2k^{-1}$ in view of Theorem 2.5.2 (ii). Again, for $0 < \epsilon < \rho$, we have $P\{N \leq \epsilon C\} = 0(C^{-\frac{1}{2}k(m_1-1)})$. That is, for $m_1 > 1 + 4k^{-1}$, one can show that $E\{(N - C)^2/N\} = 2(k\rho)^{-1} + o(1)$, as before. Thus, combine this with (7.2.5) and Theorem 3.6.2 (i) to obtain

$$\sigma^2 C^{-1} + \sigma^2\{4(k\rho)^{-1} - q - 1\}C^{-2} + o(C^{-2}) \leq \mathrm{Var}(\overline{X}_{iN})$$

$$\leq \sigma^2 C^{-1} + \sigma^2\{4(k\rho)^{-1} - q\}C^{-2} + o(C^{-2}), \tag{7.2.8}$$

if $m_1 \geq 4$.

In the case of the three–stage procedure, Theorem 3.7.1 (iii) shows that $(N - C)^2/C \xrightarrow{\mathscr{L}} 2(k\rho)^{-1}\chi_1^2$ as $\delta^* \to 0$. From Theorem 2.3.3 (ii), we also know that $(N - C)^2/C$ is uniformly integrable. Hence, as before, one can claim that $E\{(N - C)^2/N\} = 2(k\rho)^{-1} + o(1)$ since one has $P\{N \leq \epsilon C\} = 0(C^{-\frac{1}{2}p/r})$ for arbitrary $p(>0)$. Now, combine this with (7.2.5) and Theorem 3.7.2 (i) to obtain

$$\mathrm{Var}(\overline{X}_{iN}) = \sigma^2 C^{-1} + \sigma^2\{2(k\rho)^{-1} - \beta'\}C^{-2} + o(C^{-2}), \tag{7.2.9}$$

as $\delta^* \to 0$.

7.2.2 ESTIMATION OF VARIANCE

Here, *only* the *purely sequential procedure* (3.5.1) is considered and the problem is estimation of the common unknown variance σ^2. When the sequential sampling process stops, one has $U(N)$, defined in (3.1.1), available at this stopped stage. It will be quite natural then to estimate σ^2 by means of $U(N)$. We now aim at finding an expansion of $E\{U(N)\}$. From (3.5.3), one gets

$$E\{U(N)\} = E\{(N-1)^{-1}\Sigma_{i=1}^{N-1} W_i'\} = \sigma^2 k^{-1} E[Q^{-1}\Sigma_{i=1}^{Q} W_i], \qquad (7.2.10)$$

where Q and W's come from (3.5.6). Let us write $S_Q = \Sigma_{i=1}^{Q} W_i$. Then, the excess over the boundary for the stopping rule (3.5.6) is denoted by $\mathscr{R}^* = kQ(Q+1)C^{-1} - S_Q$. From (7.2.10), one gets

$$\begin{aligned} E\{U(N)\} &= \sigma^2 k^{-1} E[k(Q+1)C^{-1} - \mathscr{R}^* Q^{-1}] \\ &= \sigma^2 k^{-1}\{k(C+\beta+o(1))C^{-1}\} - \sigma^2 k^{-1} E(\mathscr{R}^* Q^{-1}) \\ &= \sigma^2 + \beta\sigma^2 C^{-1} + o(C^{-1}) - \sigma^2 k^{-1} E(\mathscr{R}^* Q^{-1}), \end{aligned} \qquad (7.2.11)$$

from Theorem 3.5.2 (i) if (a) $m_1 \geq 3$ when $k = 2$, (b) $m_1 \geq 2$ when $k \geq 3$. Lemma 2.1 of Woodroofe (1977) shows that $\mathscr{R}^{*r}$ is dominated for $m_1 \geq 2$, with $r > 0$. Now, obviously $\mathscr{R}^*(C/Q)I(Q > \frac{1}{2}C) \leq 2\mathscr{R}^*$ which shows that $\mathscr{R}^*(C/Q)I(Q > \frac{1}{2}C)$ is uniformly integrable and thus $E[\mathscr{R}^*(C/Q)I(Q > \frac{1}{2}C)] = \nu + o(1)$ if $m_1 \geq 2$. Recall that ν was defined in (3.5.7). However, with $r > 1$, $s > 1$, $r^{-1} + s^{-1} = 1$, and using Holder's inequality, one then gets $E[\mathscr{R}^*(C/Q)I(Q \leq \frac{1}{2}C)] = O(C)E^{1/r}[\mathscr{R}^{*r}]P^{1/s}(Q \leq \frac{1}{2}C)$ which becomes $O(C)O(1)O(C^{-\frac{1}{2}(m_1-1)k/s}) = o(1)$ if $m_1 > 1 + 2sk^{-1}$, that is if (a) $m_1 \geq 3$ when $k = 2$, (b) $m_1 \geq 2$ when $k \geq 3$, where one chooses $s = 1.01$, say.

Hence, from (7.2.11), we obtain

$$E\{U(N)\} = \sigma^2 + \sigma^2 C^{-1}(\beta - \nu k^{-1}) + o(C^{-1})$$

$$= \sigma^2 - 2\sigma^2(kC)^{-1} + o(C^{-1}), \tag{7.2.12}$$

if (a) $m_1 \geq 3$ when $k = 2$, (b) $m_1 \geq 2$ when $k \geq 3$, since $\beta = (\nu - 2)/k$.

The next query would perhaps be regarding the mean square error of $U(N)$. Recall (3.5.3) and note that $U(N)$ is the average of i.i.d. random variables, that is $n^{\frac{1}{2}}\{U(n) - \sigma^2\} \xrightarrow{\mathscr{L}} N(0, 2k^{-1}\sigma^4)$ as $n \to \infty$. Now, Anscombe's (1952) theorem will imply that $C^{\frac{1}{2}}\{U(N) - \sigma^2\} \xrightarrow{\mathscr{L}} N(0, 2k^{-1}\sigma^4)$, that is one has $C\{U(N) - \sigma^2\}^2 \xrightarrow{\mathscr{L}} 2k^{-1}\sigma^4\chi_1^2$ as $\delta^* \to 0$. From Theorem 2 of Chow et al. (1979), one will now conclude that

$$C\{U(N) - \sigma^2\}^2 \text{ is uniformly integrable,} \tag{7.2.13}$$

if

$$(N/C)^2 \text{ is uniformly integrable.} \tag{7.2.14}$$

Now, in order to prove (7.2.14), we go back to the proof of Theorem 3.5.1 (ii). From the basic inequality (3.5.2), one easily obtains

$$N^2 \leq m_1^2 + 2h^2 m_1 U(N-1)\delta^{*-2} + h^4 U^2(N-1)\delta^{*-4}$$

$$\leq m_1^2 + 2h^2\delta^{*-2}W^* + h^4\delta^{*-4}W^{*2},$$

with $W^* = \sup_{n \geq 2} \{(n-1)^{-1}\Sigma_{i=1}^{n-1} W_i'\}$. Thus,

$$(N/C)^2 \leq m_1^2 + 2h^2 W^* + h^4 W^{*2},$$

for sufficiently small δ^*. Now, Wiener's (1939) dominated ergodic theorem coupled with the dominated convergence theorem lead to the conclusion that $E[(N/C)^2] \to 1$ as $\delta^* \to 0$. Hence, $(N/C)^2$ is uniformly integrable since

$(N/C)^2 \xrightarrow{P} 1$ as $\delta^* \to 0$. So, one has (7.2.14), and hence from (7.2.13) one immediately obtains,

$$E[\{U(N) - \sigma^2\}^2] = 2\sigma^4(kC)^{-1} + o(C^{-1}) = 2\sigma^2\delta^{*2}(kh^2)^{-1} + o(\delta^{*2}), \quad (7.2.15)$$

as $\delta^* \to 0$, if $m_1 \geq 2$.

7.2.3 MULTIPARAMETER ESTIMATION

After implementing one of the selection methodologies of Chapter 3, the experimenter may then wish to estimate certain linear functions of $\underset{\sim}{\mu}$ by means of a certain fixed–size simultaneous confidence region. For example, one may wish to estimate $\mu_1 - \mu_2$ and $\mu_1 + \mu_2 - 2\mu_3$ simultaneously when $k \geq 3$. So, having available $X_{i1},\ldots,X_{iN}$ from π_i, $i = 1,\ldots,k$, we suppose that we wish to estimate $\underset{\sim}{\nu}^*$, say, where $\underset{\sim}{\nu}^* = A\underset{\sim}{\mu}$ and A is a $r \times k$ known real matrix with Rank $(A) = r(\leq k)$. In other words, we estimate r independent linear functions of $\underset{\sim}{\mu}$ simultaneously. Let us write $\overline{\underset{\sim}{X}}_n' = (\overline{X}_{1n},\ldots,\overline{X}_{kn})$, then $\overline{\underset{\sim}{Y}}_n = A\overline{\underset{\sim}{X}}_n \sim N_r(\underset{\sim}{\nu}^*, \sigma^2 n^{-1}H)$ for fixed n and $H = AA'$. The following fixed–size ellipsoidal confidence region for $\underset{\sim}{\nu}^*$ is proposed:

$$\mathscr{R}_N = \{\underset{\sim}{\nu}^* : (\overline{\underset{\sim}{Y}}_N - \underset{\sim}{\nu}^*)' H^{-1}(\overline{\underset{\sim}{Y}}_N - \nu^*) \leq a_0^2\delta^{*2}\} \quad (7.2.16)$$

for a fixed number $a_0 \epsilon R^+$. The volume of this ellipsoid is fixed because both a_0 and δ^* have been fixed in advance. For each of the procedures considered in Chapter 3, the confidence coefficient associated with the region $\mathscr{R}_N$ is given by

$$P\{\underset{\sim}{\nu}^* \epsilon \mathscr{R}_N\} = E[F_1(N\delta^{*2}a_0^2/\sigma^2)], \quad (7.2.17)$$

where $F_1(x) = P\{\chi_r^2 \le x\}$ for $x > 0$. For the modified two–stage, purely sequential, accelerated sequential and three–stage selection methodologies, we know that $N/C \to 1$ w.p. 1, that is $N\delta^{*2}/\sigma^2 \to h^2$ w.p. 1 as $\delta^* \to 0$. Here, h comes from (3.2.4). Hence, by the dominated convergence theorem, we obtain from (7.2.17)

$$\lim P\{\underset{\sim}{\nu}^* \in \mathcal{R}_N\} = F_1(h^2 a_0^2) \text{ as } \delta^* \to 0. \tag{7.2.18}$$

Thus, one gets the *asymptotic first–order approximation* of the confidence coefficient associated with the simultaneous fixed–size confidence region $\mathcal{R}_N$ given by (7.2.16). In what follows, we show how one can obtain the asymptotic second–order expansion of $P\{\underset{\sim}{\nu}^* \in \mathcal{R}_N\}$ when "N", the stopping time, comes from the *purely sequential selection procedure (3.5.1).* If N is determined by the corresponding three–stage or accelerated sequential procedure, then such second–order expansions can be derived by exploiting techniques that led to Theorem 2.3.4 and equation (3.6) in Mukhopadhyay and Solanky (1991) respectively. The reader is also referred to Mukhopadhyay and Al–Mousawi (1986). Other relevant references can be obtained from this latter paper and Sinha (1991).

We write $f_1(x) = \{2^{\frac{1}{2}r}\Gamma(\frac{1}{2}r)\}^{-1} x^{\frac{1}{2}r-1} \exp(-\frac{1}{2}x)$ for $x > 0$, that is $F_1(x) = \int_0^x f_1(y)dy$. Now, we prove the following result.

Theorem 7.2.2. *For the purely sequential procedure (3.5.1), we have as* $\delta^* \to 0$:

$$P\{\underset{\sim}{\nu}^* \in \mathcal{R}_N\} = F_1(h^2a_0^2) + a_0^2\delta^{*2}\sigma^{-2}\{\beta f_1(h^2a_0^2) + k^{-1}a_0^2h^2f_1'(h^2a_0^2)\} + o(\delta^{*2}),$$

if (a) $m_1 > 1 + 2k^{-1}$ *when* $r \ge 4$ *or* $r = 2$, (b) $m_1 > 1 + 3k^{-1}$ *when* $r = 3$, (c) $m_1 > 1 + 5k^{-1}$ *when* $r = 1$; *where* $k \ge r$ *and* h *comes from (3.2.4). Also,* $f_1(x) = F_1'(x)$ *and* $f_1'(x) = F_1''(x)$ *for* $x > 0$, *and* $\mathcal{R}_N$ *comes from (7.2.16).*

Proof. The steps follow the lines of proof of Theorem 4 of Mukhopadhyay and Al–Mousawi (1986). We give only a few basic arguments for completeness. From (7.2.17), one obtains

$$P\{\underset{\sim}{\nu}^*\epsilon\, \mathscr{R}_N\} = F_1(h^2 a_0^2) + a_0^2 \delta^{*2} \sigma^{-2} F_1'(h^2 a_0^2) E(N - C)$$

$$+ \tfrac{1}{2} a_0^4 h^2 \delta^{*2} \sigma^{-2} E\{(N - C)^2 C^{-1} F_1''(a_0^2 Z)\}, \qquad (7.2.19)$$

where Z is a suitable random variable between h^2 and $N\delta^{*2}/\sigma^2$. Of course, $E(N - C) = \beta + o(1)$ if $m_1 > 1 + 2k^{-1}$ in view of Theorem 3.5.2 (i). The term $E\{(N - C)^2 C^{-1} F_1''(a_0^2 Z)\}$ has to be evaluated in the same way as in the various cases considered in the proof of Theorem 4 of Mukhopadhyay and Al–Mousawi (1986). Eventually, one can prove that

$$E\{(N - C)^2 C^{-1} F_1''(a_0^2 Z)\} = 2k^{-1} F_1''(a_0^2 h^2) + o(1), \qquad (7.2.20)$$

if (a) $m_1 > 1 + 2k^{-1}$ when $r \geq 4$ or $r = 2$, (b) $m_1 > 1 + 3k^{-1}$ when $r = 3$, (c) $m_1 > 1 + 5k^{-1}$ when $r = 1$, and $k \geq r$. Now, combining (7.2.19) – (7.2.20), one obtains the desired result. □

7.2.3.1 Point Estimation

Suppose that one proposes $\underset{\sim}{\overline{Y}}_N$ as the point estimator of $\underset{\sim}{\nu}^*$ and considers measuring the loss in estimation by means of

$$L_N = (\underset{\sim}{\overline{Y}}_N - \underset{\sim}{\nu}^*)' \, G \, (\underset{\sim}{\overline{Y}}_N - \underset{\sim}{\nu}^*), \qquad (7.2.21)$$

where G is an $r \times r$ known p.d. matrix. For each procedure proposed in Chapter 3, the risk associated with (7.2.21) simplifies to

$$E(L_N) = E\{tr(\underset{\sim}{\overline{Y}}_N - \underset{\sim}{\nu}^*)' \; G \; (\underset{\sim}{\overline{Y}}_N - \underset{\sim}{\nu}^*)\}$$

$$= \Sigma_{n=m_1}^{\infty} E\{tr(\underset{\sim}{\overline{Y}}_n - \underset{\sim}{\nu}^*)' \; G \; (\underset{\sim}{\overline{Y}}_n - \underset{\sim}{\nu}^*)\}P(N = n)$$

$$= E\{tr(GH)\sigma^2/N\}, \tag{7.2.22}$$

where $tr(GH)$ stands for the trace of the matrix GH. Under the same sufficient conditions given in Theorem 7.2.1, it follows immediately that

$$E(L_N) = tr(GH)\sigma^2 C^{-1} + o(C^{-1}), \tag{7.2.23}$$

for the modified two–stage, purely sequential, accelerated sequential and three–stage selection methods considered in Chapter 3. Also, one can put (7.2.22) in the perspective of (7.2.5) and immediately obtain asymptotic second–order expansions of $E(L_N)$ along the similar lines of (7.2.7), (7.2.8) and (7.2.9), under appropriate conditions on m_1, respectively for the purely sequential, accelerated sequential and three–stage methodologies. See also Ghosh et al. (1976) and Sinha (1991). Further details are omitted for brevity.

7.3 THE NEGATIVE EXPONENTIAL PROBLEM

Suppose that the ith population's distribution is given by $f(x; \theta_i, \sigma)$, defined in (2.2.5), where all the parameters are unknown. We write $\underset{\sim}{\theta} = (\theta_1,...,\theta_k)$ and assume that we are working under the general framework of Chapter 4 and one of the specific selection methodologies has ended up with the random samples $X_{i1},...,X_{iN}$ and N from the population π_i, $i = 1,...,k$. At this stage, the experimenter may like to estimate the θ_i's or σ. One may consider the natural estimators $\tilde{X}_{iN}$ and $V(N)$, defined in (4.1.1), for θ_i and σ respectively. In this section, we derive some of the asymptotic properties associated with such natural estimators.

7.3.1 ESTIMATION OF LOCATIONS

For a fixed i, one wishes to estimate θ_i by $\tilde{X}_{iN}$, $i = 1,...,k$. Suppose that for an integrable function $g_3(\cdot)$, one has

$$E[g_3(\tilde{X}_{iN})] = g_3^*(n, \underset{\sim}{\theta}, \sigma), \qquad (7.3.1)$$

with a certain functional $g_3^*(\cdot)$. Then, assume the sufficient conditions listed in Theorem 4.2.1, and additionally assume that $E[g_3^*(N, \underset{\sim}{\theta}, \sigma)]$ is finite. Then one obtains

$$\begin{aligned} E[g_3(\tilde{X}_{iN})] &= \Sigma_{n=m_1}^{\infty} E[g_3(\tilde{X}_{iN}) \big| N = n]P(N = n) \\ &= \Sigma_{n=m_1}^{\infty} E[g_3(\tilde{X}_{in}) \big| N = n]P(N = n) \\ &= \Sigma_{n=m_1}^{\infty} E[g_3(\tilde{X}_{in})]P(N = n) \\ &= E[g_3^*(N, \underset{\sim}{\theta}, \sigma)]. \end{aligned} \qquad (7.3.2)$$

Suppose that $g_3(x) = x$, then $g_3^*(n, \underset{\sim}{\theta}, \sigma) = \mu_i + \sigma n^{-1}$. Also, if $g_3(x) = (x - \mu_i)^2$, then $g_3^*(n, \underset{\sim}{\theta}, \sigma) = 2\sigma^2 n^{-2}$. Thus, for each selection procedure described in Chapter 4, one obtains from (7.3.2)

$$E[\tilde{X}_{iN}] = \theta_i + \sigma E(N^{-1}) \text{ and } E[(\tilde{X}_{iN} - \theta_i)^2] = 2\sigma^2 E(N^{-2}). \qquad (7.3.3)$$

In other words, one has exact expressions for the bias and the mean square error of the estimator $\tilde{X}_{iN}$ for θ_i.

Theorem 7.3.1. *For the modified two–stage procedure (4.4.1) – (4.4.2), purely sequential procedure (4.5.1), accelerated sequential procedure (4.6.4) – (4.6.6), and the three–stage procedure (4.7.2) – (4.7.3), we have as* $\delta^* \to 0$:

(i) $E[\tilde{X}_{iN}] = \theta_i + \sigma C^{-1} + o(C^{-1})$;

(ii) $E[(\tilde{X}_{iN} - \theta_i)^2] = 2\sigma^2 C^{-2} + o(C^{-2})$;

for every fixed $i = 1,...,k$ *where* $C = a\sigma/\delta^*$ *and* a *comes from (4.2.4). In the case of purely sequential and accelerated sequential methodologies, the starting sample size* m_1 *has to satisfy the conditions* (a) $m_1 \geq 3$ *when* $k = 2$, (b) $m_1 \geq 2$ *when* $k \geq 3$ *for part (ii). The condition* $m_1 \geq 2$ *suffices for part (i).*

Proof. Under appropriate conditions on m_1, one needs to show that

$$E(C/N) \to 1,\ E(C^2/N^2) \to 1 \ \text{as} \ \delta^* \to 0. \tag{7.3.4}$$

First, we turn to the modified two–stage procedure. From (4.4.4), note that $E\{(C/N)^s\} \leq \{2ak(m_1 - 1)/b(m_1)\}^s E[\{\chi^2_{2k(m_1-1)}\}^{-s}]$ for $s = 1, 2$. The rest of the proof in this case follows along Theorem 7.2.1. The proofs in the case of the three–stage procedure, purely sequential procedure and the accelerated sequential procedure are also similar to those in Theorem 7.2.1. Details are omitted. □

Next, one can improvise on the techniques that led to (7.2.7) – (7.2.9) in order to obtain refinements of the results given in Theorem 7.3.1 by providing expansions of $E(\tilde{X}_{iN})$ and $E[(\tilde{X}_{iN} - \theta_i)^2]$ up to the terms of order $o(C^{-2})$ and $o(C^{-4})$ respectively. These may be considered as fairly routine also.

7.3.2 ESTIMATION OF SCALE

Only the purely sequential procedure (4.5.1) is considered here and the problem is estimation of the common unknown scale parameter σ. When the sequential sampling process stops, one has $V(N)$ available as the estimator of σ. Use the embedding techniques of Lombard and Swanepoel (1978) as in section 4.5 and note that $N = Q + 1$ where

$$Q = \inf\{n \geq m_1 - 1 : \Sigma_{i=1}^{n} W_i \leq 2kn(n+1)C^{-1}\}, \tag{7.3.5}$$

W's being i.i.d. χ^2_{2k}. Along the lines of Mukhopadhyay and Ekwo (1987), one can verify that $V(N)$ and $\frac{1}{2}\sigma\{kQ\}^{-1}\Sigma_{i=1}^{Q} W_i$ have identical distributions. Thus,

$$E[V(N)] = \tfrac{1}{2}\sigma k^{-1} E\{Q^{-1}\Sigma_{i=1}^{Q} W_i\}. \tag{7.3.6}$$

We write $S_Q = \Sigma_{i=1}^{Q} W_i$ and denote the excess over the boundary in (7.3.4) by $\mathcal{R}^{**} = 2kQ(Q+1)C^{-1} - S_Q$. Thus, from (7.3.6), one obtains

$$E[V(N)] = \tfrac{1}{2}\sigma k^{-1} E[2k(Q+1)C^{-1} - \mathcal{R}^{**}Q^{-1}]$$

$$= \sigma\{(C + \beta'' + o(1))C^{-1}\} - \tfrac{1}{2}\sigma k^{-1} E(\mathcal{R}^{**}Q^{-1}), \tag{7.3.7}$$

for $m_1 \geq 2$, where β'' comes from (5.2.49). The term $E\{\mathcal{R}^{**}Q^{-1})$ can be tackled along the similar lines that led to (7.2.12) and one obtains

$$E\{\mathcal{R}^{**}(C/Q)\} = \nu + o(1),$$

if $m_1 \geq 2$, where ν comes from (4.5.8). Hence, (7.3.7) simplifies to

$$E[V(N)] = \sigma - \sigma(kC)^{-1} + o(C^{-1}), \tag{7.3.8}$$

for $m_1 \geq 2$. This is fairly similar to (7.2.12) obtained in the normal case.

Next, $n^{\frac{1}{2}}\{V(n) - \sigma\} \xrightarrow{\mathscr{L}} N(0, k^{-1}\sigma^2)$ as $n \to \infty$. Hence, via Anscombe's (1952) theorem, we can conclude that $C^{\frac{1}{2}}\{V(N) - \sigma\} \xrightarrow{\mathscr{L}} N(0, k^{-1}\sigma^2)$ as $\delta^* \to 0$. Also, in view of Theorem 2 of Chow et al. (1979), one obtains uniform integrability of $C\{V(N) - \sigma\}^2$ for $m_1 \geq 2$, along the lines of (7.2.13). Hence, the mean square error associated with $V(N)$ is given by

$$E[\{V(N) - \sigma\}^2] = \sigma^2(kC)^{-1} + o(C^{-1}) = \sigma(ka)^{-1}\delta^* + o(\delta^*), \quad (7.3.9)$$

as $\delta^* \to 0$, if $m_1 \geq 2$.

7.3.3 MULTIPARAMETER ESTIMATION

We now discuss simultaneous fixed–size confidence region estimation of $\underset{\sim}{\theta}$ along the lines of Mukhopadhyay et al. (1988). Having recorded $X_{i1},\ldots,X_{iN}$ from π_i, $i = 1,\ldots,k$, where N is the stopping time arising from one of the methodologies given in Chapter 4, we propose

$$\mathscr{R}_N^* = [\tilde{X}_{1N} - a_0\delta^*, \tilde{X}_{1N}] \times \cdots \times [\tilde{X}_{kN} - a_0\delta^*, \tilde{X}_{kN}] \quad (7.3.10)$$

as the simultaneous confidence region of $\underset{\sim}{\theta}$. Here, $a_0 \in R^+$ is a fixed number. The region $\mathscr{R}_N^*$ has a fixed size in the sense that the volume of this k–dimensional cube has been fixed at a certain preassigned level. Now,

$$P\{\underset{\sim}{\theta} \in \mathscr{R}_N^*\} = E[\{1 - \exp(-N\delta^* a_0/\sigma)\}^k]. \quad (7.3.11)$$

This is exactly $E[\psi_2(N\delta^* a_0/\sigma)]$ where $\psi_2(x)$ was defined in (5.2.40). Hence, the first– and second–order asymptotic properties of this simultaneous estimation methodology can be easily derived along the lines of section 5.2.2. Details are omitted for brevity.

7.4 NOTES AND OVERVIEW

This is a rich area. The types of problems we have addressed here can be easily adapted in the contexts of Chapters 5 and 6. Also, the general remarks made in sections 3.10, 4.10 and 5.4 remain equally valid in this situation. Thus, one can very well imagine that the problems for second–order asymptotics in this present context are quite plentiful and the area is fairly vast and wide open.

We point out that there is considerable amount of literature on estimating certain parameter of the "selected" population. One will find several other related references from Vellaisamy et al. (1988), Venter (1988), and Hsieh (1981). There are available recent contributions for handling the selection and estimation problems simultaneously. One should refer to Gupta and Miescke (1990, 1993) for such problems and other references. The directions and approaches presented in this chapter provide interesting supplements to the existing literature on estimation after selection and ranking.

Complements and Problems

7.1 (a) Suppose that $g_2(x) = |x - \mu_i|^s$ in (7.2.1) for some fixed and known $s > 0$. Derive $E[g_2(\overline{X}_{iN})]$.

(b) For the purely sequential procedure (3.5.1), accelerated sequential procedure (3.6.4) – (3.6.6), and the three–stage procedure (3.7.2) – (3.7.3), derive the asymptotic second–order expansions of $E[g_2(\overline{X}_{iN})]$. Give particular attention to sufficient conditions involving m_1.

7.2 Obtain an expansion for $E[U^*(N)]$ along the lines of (7.2.12) in the purely sequential procedure (3.5.1).

7.3 Obtain the results given in Theorem 7.2.1 and (7.2.7) when $U(n)$ is replaced by $U^*(n)$ in the purely sequential procedure (3.5.1).

7.4 (a) Suppose $k = 3$ and $A = \begin{pmatrix} 1 & -1 & 0 \\ 1 & 1 & -2 \end{pmatrix}$. Then, summarize the result given in Theorem 7.2.2 for this special case.

(b) If one asks for 95% coverage probability for the simultaneous confidence region $\mathscr{R}_N$ given by (7.2.16), then how should one choose "a_0"? If a_0 goes a little over or under this "choice", then what will happen to the resulting confidence coefficient in relation with the nominal value 95%?

7.5 Obtain the second–order expansion given in Theorem 7.2.2 for the three–stage methodology of Chapter 3.

7.6 (a) Suppose that $g_3(x) = (x - \theta_i)^s$ in (7.3.1) for some fixed and known $s > 0$. Derive $E[g_3(\tilde{X}_{iN})]$.

(b) For the purely sequential procedure (4.5.1), accelerated sequential procedure (4.6.4) – (4.6.6), and the three–stage procedure (4.7.2) – (4.7.3), derive the asymptotic second–order expansions of $E[g_3(\tilde{X}_{iN})]$. Give particular attention to sufficient conditions involving m_1.

7.7 Obtain an expression for $E[V^*(N)]$ along the lines of (7.3.7) for the purely sequential procedure (4.5.1).

7.8 Obtain the result given in (7.3.9) when $V(n)$ is replaced by $V^*(n)$ in the purely sequential procedure (4.5.1).

7.9 (a) Suppose $k \geq 2$ and the purely sequential procedure (4.5.1), accelerated sequential procedure (4.6.4) – (4.6.6) or the three–stage procedure (4.7.2) – (4.7.3) is being used. Let $\Delta = \theta_1 - \theta_2$ and consider $I_N = [\tilde{X}_{1N} - \tilde{X}_{2N} \pm a_0\delta^*]$ as the fixed–width confidence interval for Δ. Derive the asymptotic second–order expansion of $P\{\Delta \in I_N\}$. Give particular attention to sufficient conditions involving m_1.

[*Hints*: See Mukhopadhyay and Hamdy (1984b)].

(b) If $k \geq 3$, $\Delta = \theta_1 + \theta_2 - 2\theta_3$ and $I_N = [\tilde{X}_{1N} + \tilde{X}_{2N} - 2\tilde{X}_{3N} \pm a_0\delta^*]$, then derive the asymptotic first–order property of the fixed–width confidence interval I_N.

(c) Is it possible to obtain the asymptotic second–order expansion of $P\{\Delta \in I_N\}$ in (b)? Give particular attention to sufficient conditions involving m_1.

7.10 (a) Suppose $k \geq 2$ and the purely sequential procedure (4.5.1) is being used and $\Delta = \theta_1 - \theta_2$ is estimated by $\tilde{X}_{1N} - \tilde{X}_{2N}$. Derive the asymptotic second–order expansion of $E[\{(\tilde{X}_{1N} - \tilde{X}_{2N}) - \Delta\}^2]$. Give particular attention to sufficient conditions involving m_1.

(b) Is it possible to obtain a similar result when $k \geq 3$, $\Delta = \theta_1 + \theta_2 - 2\theta_3$, and Δ is estimated by $\tilde{X}_{1N} + \tilde{X}_{2N} - 2\tilde{X}_{3N}$?

7.11 Derive asymptotic second–order expansions of $P\{\underset{\sim}{\nu}^* \in \mathscr{R}_N\}$, given in (7.2.18), when N corresponds to the accelerated sequential selection procedure (3.6.4) – (3.6.6) or the three–stage selection procedure (3.7.2) – (3.7.3). Give particular attention to sufficient conditions involving m_1.

[*Hints*: See Mukhopadhyay and Al–Mousawi (1986).]

7.12 Consider (7.2.22), and derive asymptotic second–order expansions of $E(L_N)$ up to the order $o(C^{-2})$ when N corresponds to the purely sequential procedure (3.5.1), or accelerated sequential procedure (3.6.4) – (3.6.6), or the three–stage procedure (3.7.2) – (3.7.3). Give particular attention to sufficient conditions involving m_1.

7.13 Complete the proof of Theorem 7.3.1.

7.14 Obtain asymptotic second–order expansions of the results given in Theorem 7.3.1 (i) and (ii) up to the orders $o(C^{-2})$ and $o(C^{-4})$ respectively when N corresponds to the accelerated sequential procedure (4.6.4) – (4.6.6), or the three–stage procedure (4.7.2) – (4.7.3), or the purely sequential procedure (4.5.1). Give particular attention to sufficient conditions involving m_1.

7.15 Refer to (7.3.6). Show that $V(N)$ and $\frac{1}{2}\sigma\{kQ\}^{-1}\Sigma_{i=1}^{Q}W_i$ have identical distributions.

[*Hints*: See Mukhopadhyay and Ekwo (1987).]

REFERENCES

Anscombe (1952), Chow, Hsiung and Lai (1979), Ghosh, Sinha and Mukhopadhyay (1976), Gupta and Miescke (1990, 1993), Hsieh (1981), Lombard and Swanepoel (1978), Mukhopadhyay and Al–Mousawi (1986), Mukhopadhyay and Ekwo (1987), Mukhopadhyay and Hamdy (1984b), Mukhopadhyay, Hamdy and Darmanto (1988), Mukhopadhyay and Solanky (1991), Sinha (1991), Vellaisamy, Kumar and Sharma (1988), Venter (1988), Wiener (1939), Woodroofe (1977).

8

ADDITIONAL TOPICS

8.1 Introduction

This chapter addresses certain additional interesting topics in selection and ranking. The body of the literature covered in Chapters 3 – 7 is quite rich indeed. However, in order to broaden the scope of the book even further, we introduce and briefly address a few types of otherwise appealing selection and ranking procedures.

In Chapters 3 – 7, we considered only non–elimination type selection and estimation methodologies, because our types of second–order asymptotics are not known yet in the case of various available elimination–type procedures. In section 8.2, we address some elimination type selection procedures, in the normal case, as developed in Paulson (1964) and recently improved upon by Hartmann (1988, 1991), among others. In this vein, one should also look into Swanepoel and Geertsema (1973). This section gives a brief summary of a practical idea that needs considerable theoretical developments in the future in the light of Chapters 3 – 7.

Section 8.3 introduces an interesting way to derive likelihood based purely sequential selection rules for normal as well as two–parameter exponential populations. We add that the present approach leads to selection methodologies that are somewhat different from and possibly little worse than the corresponding selection rules of Bechhofer et al. (1968) in some instances, but at the same time one will notice that such a simple minded approach has been fairly successfully applied in a wide variety of interesting problems.

In Chapters 3 – 4, we had considered normal and two–parameter exponential populations respectively, assuming that the populations have a common but unknown scale parameter, and hence sample sizes from each population were also taken to be all equal. In section 8.4, we summarize some of the available multistage selection techniques when the scale parameters are unknown and unequal and the sample sizes are also unequal. We start with the fundamental results of Dudewicz and Dalal (1975) followed by some of the related later developments.

Section 8.5 considers a modified version of Gupta's (1956) subset selection problem where the second–order asymptotics have been successfully applied. The problem discussed here consists of selecting a subset from a collection of normal populations, having unknown means and a common variance, in such a way that we pick a preassigned proportion of "good" populations. In the unknown variance case, Santner (1976) proposed a two–stage subset selection rule, and we discuss other multistage and sequential subset selection rules with the associated second–order asymptotics along the lines of the recent formulation and developments given in Mukhopadhyay and Panchapakesan (1992).

In section 8.6, we summarize some nonparametric selection rules of Geertsema (1972) for identifying the population associated with the largest location parameter. A good amount of discussion related to moderate sample studies is also included here. The remarks given in section 8.6.5 show that the purely sequential selection rule (3.5.1) can indeed be used in the distribution–free case and even then the resulting methodology will remain *asymptotically second–order efficient* under very mild regularity conditions.

One readily observes that this chapter deals with some of the important gaps and concerns that have been left out from Chapters 3 – 7 for various reasons. One will also observe that the types and depths of the results covered in each topic are somewhat uneven. This is naturally affected by the authors' selection bias. We decided to include only a few additional topics where the general second–order theories as developed in Chapter 2 are believed to have utmost future potential for further application and research.

There are different types of selection rules that are given within the topics of any particular section. In this Chapter, the selection rules are referred to as $\mathcal{Q}_N(i, j)$ corresponding to the jth decision rule given in the section 8.i, i = 2,...,6. Such a convention is expected to enhance in the location of various selection rules as needed.

8.2 ELIMINATION TYPE SELECTION PROCEDURES

Let us go back to the original problem of selecting the "best" normal population in the indifference zone framework as introduced in Chapter 3. That is, one wishes to identify the population associated with the largest mean, $\mu_{[k]}$. Now, consider the following scenario as an example. Suppose $k = 10$ and the purely sequential sampling scheme (3.5.1) has gathered 20 observations from each population and it turns out that $20\delta^{*2} < h^2U(20)$, in other words, the sampling process (3.5.1) wants to continue further sampling. Recall that $\overline{X}_{in}$ stands for the sample mean obtained from π_i based on n random samples, $i = 1,...,10$. At this point, suppose that one has observed $\overline{X}_{1\,20} = 3.18$ while $\overline{X}_{2\,20}$ through $\overline{X}_{10\,20}$ are all around the value 30. The original purely sequential scheme (3.5.1) will of course continue by taking *one additional sample from each* of the $k = 10$ populations. But, the experimenter may think at this stage that the first population can be practically taken out of any further considerations because there may be at best some remote possibility of the first population's eventually turning out as the "best" population. Of course, such decisions at any stage should be made by looking at the $\overline{X}$ values in conjunction with the observed estimate of σ^2 at that stage. That is, one should realize that at some stage or any stage of sampling, one or more populations may be taken out of considerations for any further sampling because for some reasons, these may be thought to be "far away" from the potentially "best" population. This is the basic idea of "elimination" of inferior populations. One important point needs emphasis here. If one deletes one or more populations at some stage from further considerations based on random samples obtained from all populations thus far and then proceeds sampling only from the remaining ones according to some sampling design, then the evaluation of overall P(CS) and controlling it at some predetermined level P* will turn out to be a whole lot trickier. Paulson (1964) pioneered the idea of eliminating inferior populations early enough from further considerations and yet maintained the nominal level of P(CS).

One should recall the basic notation and formulation from sections 3.1 – 3.2. First we give a brief description of Paulson's (1964) procedure, followed by some discussions of a few recent developments in this area.

8.2.1 PAULSON'S PROCEDURE

Let us first assume that the common variance σ^2 is known and we wish to consider a selection procedure such that $P(CS) \geq P^*$ whenever $\underset{\sim}{\mu} \in \Omega(\delta^*)$, the preference zone. Paulson (1964) proposed a class of truncated sequential procedures devised as follows.

With $\lambda \epsilon (0, \delta^*)$, define $a_\lambda^* = \sigma^2(\delta^* - \lambda)^{-1} \ln\{(k-1)/(1-P^*)\}$. Let $Z_\lambda = [\lambda^{-1} a_\lambda^*]^*$. Now, one starts with X_{i1} from population π_i, $i = 1,\ldots,k$ and eliminates π_i from any further consideration if

$$X_{i1} < \max_{1 \leq j \leq k} X_{j1} - a_\lambda^* + \lambda. \tag{8.2.1}$$

We emphasize that we eliminate π_i for all indices i that satisfies (8.2.1). If all but one population are eliminated, one then stops the experiment and selects the remaining population as the "best" population. Otherwise, the experimenter takes one more observation from all the populations which are yet left for consideration. At the sth stage, one eliminates any of the remaining populations π_i if

$$\Sigma_{j=1}^{s} X_{ij} < \text{Max}\, \Sigma_{j=1}^{s} X_{rj} - a_\lambda^* + s\lambda, \tag{8.2.2}$$

$s = 2, 3,\ldots,Z_\lambda$ where the "Max" is obtained from all populations left for consideration after the $(s-1)^{th}$ stage. In case only one population is left for consideration after the sth stage, one stops the experiment and selects this population as the best one. The experimenter proceeds to the $(s+1)^{th}$ stage otherwise. If the experimenter reaches the stage $s = Z_\lambda$ and still more than one population remains for further consideration, then the experiment is terminated at the $(Z_\lambda + 1)^{th}$ stage by means of selecting the population associated with the largest cumulative sample sum at this step. This elimination type selection procedure is denoted by $\mathscr{Q}_{\underset{\sim}{N}}(2, 1)$ where one writes $\underset{\sim}{N} = (N_1,\ldots,N_k)$, N_i being the random number of random samples needed in order to eliminate the population π_i, $i = 1,\ldots,k$. For the selection procedure $\mathscr{Q}_{\underset{\sim}{N}}(2, 1)$, Paulson (1964) had shown that

$$P(CS) \geq P^* \text{ whenever } \underset{\sim}{\mu} \in \Omega(\delta^*), \tag{8.2.3}$$

$\Omega(\delta^*)$ being the preference zone defined earlier in (3.1.3). Paulson (1964) had indicated that the choice $\lambda = \delta^*/4$ seemed to work well in practice. In this vein, one should also look at Fabian (1974a,b). Various comparisons of the selection procedure $\mathscr{Q}_{\underset{\sim}{N}}(2, 1)$ with the corresponding ones suggested by Bechhofer et al. (1968) were carried out in Ramberg (1966), Ramberg and Bechhofer (1973) via computer simulations.

We add that the elimination type selection procedure $\mathscr{Q}_{\underset{\sim}{N}}(2, 1)$ satisfies the property that the expected number of total sample sizes to be drawn from all populations $\pi_1,\ldots,\pi_k$ actually decreases as the configuration of the mean vector $\underset{\sim}{\mu}(\in\Omega(\delta^*))$ becomes "more favorable". Perng (1969) also provided an asymptotic (as $P^* \to 1$) expression for $E(N_i)$, $i = 1,\ldots,k$.

In the case when the common variance σ^2 is unknown, Paulson (1964) proposed the following elimination type selection procedure which we denote by $\mathscr{Q}_{\underset{\sim}{N}}(2, 2)$. One starts with $m_1(\geq 2)$ from each population and obtains $\overline{X}_{im_1}$, $U(m_1)$ as in (3.1.1). Let $\nu^* = k(m_1 - 1)$ and

$$d_1 = \tfrac{1}{2}[\{(k-1)/(1-P^*)\}^{2/\nu^*} - 1]\nu^*,$$

$$d_2 = \tfrac{4}{3}\, U(m_1) d_1 \delta^{*-1} \text{ and } Z^* = [4d_2\delta^{*-1}]^*. \tag{8.2.4}$$

If $m_1 > Z^*$, one stops sampling and selects the population associated with the largest $\overline{X}_{im_1}$ value. If $m_1 \leq Z^*$, then one eliminates population π_i if

$$\Sigma_{j=1}^{m_1} X_{ij} < \underset{1 \leq j \leq k}{\text{Max}} \{\Sigma_{r=1}^{m_1} X_{jr}\} - d_2 + \tfrac{1}{4} m_1 \delta^*. \tag{8.2.5}$$

If all but one population are eliminated by such a process, one then stops the experiment and selects the remaining population as the best population. Otherwise, the experimenter proceeds to the $(m_1 + 1)^{th}$ step by taking one more sample from all the populations left for consideration. Next, at the $(m_1 + s)^{th}$ stage, the experimenter eliminates population π_i if for $s = 1, 2, \ldots, Z^* - m_1$,

$$\Sigma_{j=1}^{m_1+s} X_{ij} < \text{Max}\{\Sigma_{r=1}^{m_1+s} X_{jr}\} - d_2 + \tfrac{1}{4}(m_1+s)\delta^*, \tag{8.2.6}$$

where the "Max" is obtained from all populations left for consideration prior to this stage. If the experimenter is left with more than one population, then the sampling is terminated at the $(Z^* + 1)^{th}$ stage by selecting the population associated with the largest cumulative sample sum as the best one. This procedure $\mathcal{Q}_{\underset{\sim}{N}}(2, 2)$ was shown to satisfy the P^* requirement given by (8.2.3) in Paulson (1964). Appropriate choice of m_1 is very crucial and one is referred to Paulson (1964) for related comments.

8.2.2 FURTHER DEVELOPMENTS

In the usual (Koopman–Darmois) one–parameter exponential family of distributions, Hoel and Mazumder (1969) extended the elimination type procedure $\mathcal{Q}_{\underset{\sim}{N}}(2, 1)$ for selecting the best population. Swanepoel and Venter (1975) introduced the problem of selecting the population associated with the largest location parameter in the distribution–free framework via elimination type sequential procedures. The developments in Swanepoel and Geertsema (1973, 1976) and Bechhofer and Goldsman (1987, 1989) are particularly relevant.

The recent results of Hartmann (1988, 1991) are very interesting. He has provided a very clever way to improve upon Paulson's (1964) original elimination type selection procedures in the normal case. Some overview and extensions in this area can also be obtained from Hartmann (1993).

8.3 LIKELIHOOD BASED SELECTION PROCEDURES

In the indifference zone formulation, we now address a few likelihood based selection procedures. In the case of $k(\geq 2)$ normal populations having unknown means, but known equal variance σ^2, Mukhopadhyay (1983a) formulated the problem of selecting the population associated with the largest population mean via multiple hypothesis testing. This particular procedure is slightly worse than the procedure proposed in section 4.3.1 of Bechhofer et al. (1968). Yet, we include the Mukhopadhyay formulation, because the approach proposed here works for several other types of problems, for example, (a) selecting the largest normal mean when the common variance is unknown, (b) selecting the smallest normal variance, when the variances are unknown and unequal. One can also analogously handle similar problems in terms of negative exponential populations. This section is devoted to such aspects for *non-elimination type selection procedures* via *likelihood comparisons.*

8.3.1 SELECTING THE LARGEST NORMAL MEAN

Suppose first there are $k(\geq 2)$ independent normal populations $\pi_1,\ldots,\pi_k$ having the common variance σ^2, the mean of π_i being μ_i with $\mu_i \in \mathbb{R}$, $\sigma \in \mathbb{R}^+$, $i = 1,\ldots,k$. It is assumed that $\mu_1,\ldots,\mu_k$ are unknown and let $\mu_{[1]} \leq \cdots \leq \mu_{[k]}$ be the ordered values of μ's. We suppose that there is just one population associated with $\mu_{[k]}$ and our goal is to select this population which is referred to as the *best* one, as in Chapter 3. Bechhofer's (1954) indifference zone formulation is adopted and hence given $\delta^*(>0)$, one defines the *preference zone*

$$\Omega(\delta^*) = \{\underset{\sim}{\mu} = (\mu_1,\dots,\mu_k): \mu_{[k]} - \mu_{[k-1]} \geq \delta^*\}. \tag{8.3.1}$$

Given $P^* \epsilon (k^{-1}, 1)$, we also require that

$$P(CS) \geq P^* \text{ whenever } \underset{\sim}{\mu} \epsilon \Omega(\delta^*). \tag{8.3.2}$$

First, purely sequential procedures are proposed for which the goal in (8.3.2) is met when σ^2 is assumed known. Then, some ad hoc procedures are proposed when σ^2 is unknown and the goal in (8.3.2) is met only approximately.

For each such procedure, it turns out that the parameter configuration $\mu_{[1]} = \cdots = \mu_{[k-1]} = \mu_{[k]} - \delta^*$ indeed minimizes $P(CS)$ in $\Omega(\delta^*)$, and thus the least favorable configuration (LFC) is precisely given by $\mu_{[1]} = \cdots = \mu_{[k-1]} = \mu_{[k]} - \delta^*$. Hence, throughout this section, we consider the selection problem *only under the LFC.*

8.3.1.1 Common Known Variance

We wish to decide among k hypotheses, namely

$$H_i : \mu_i = \mu_{[k]},\ i = 1,\dots,k \tag{8.3.3}$$

such that $P(CS \mid H_i) \geq P^*$ for *all* $i = 1,\dots,k$. We suppose we have available a sequence $X_{i1},\dots,X_{in},\dots$ of i.i.d. random variables from π_i, and we consider taking an equal number of samples from each π_i, $i = 1,\dots,k$. Having recorded $X_{i1},\dots,X_{in}$ from π_i, $i = 1,\dots,k$ we propose to work with the maximal invariant (with respect to the group of location shift by the same amount) sequence given by $\underset{\sim}{U}_j$ where

$$\underset{\sim}{U}' = (X_{2j} - X_{1j},\dots,X_{kj} - X_{1j})$$

$$= (U_{1j}, \ldots, U_{k-1j}), \text{ say}, \tag{8.3.4}$$

for $j = 1,\ldots,n$, $n = 1, 2,\ldots$. One may also invoke Hall et al. (1965) for further justifications. It is easy to see that $\underset{\sim}{U}_1,\ldots,\underset{\sim}{U}_n$ are i.i.d. $N_{k-1}(\underset{\sim}{\theta}, \sigma^2\Sigma)$ where $\underset{\sim}{\theta}' = (\mu_2 - \mu_1, \mu_3 - \mu_1,\ldots,\mu_k - \mu_1) = (\theta_1,\ldots,\theta_{k-1})$, say, and $\Sigma = (\sigma_{ij})$ having $\sigma_{ii} = 2$, $\sigma_{ij} = 1$ for all $i \neq j = 1,\ldots,k-1$. Note that Σ has a very special structure and its inverse is given by (σ^{ij}) where $\sigma^{ii} = (k-1)/k$ and $\sigma^{ij} = -1/k$ for all $i \neq j = 1,\ldots,k-1$. Plainly, the hypothesis H_i specifies a unique value of the parameter vector $\underset{\sim}{\theta}$, and more specifically, we can write our selection problem as equivalent with the testing problem among $H_1,\ldots,H_k$ where

$$H_1 : \theta_i = -\delta^* \text{ for all } i = 1,\ldots,k-1,$$

$$H_j : \theta_{j-1} = \delta^*, \ \theta_i = 0 \text{ for all } i \neq j-1, \tag{8.3.5}$$

with $j = 2,\ldots,k$. In other words, $H_1,\ldots,H_k$ given by (8.3.3) and (8.3.5) are actually equivalent.

Suppose $f^{(o)}(\underset{\sim}{u}_j | \underset{\sim}{\theta})$ is the p.d.f. of $\underset{\sim}{U}_j$ at $\underset{\sim}{u}_j$ when $\underset{\sim}{\theta}$ obtains. Then

$$f^{(o)}(\underset{\sim}{u}_j | \underset{\sim}{\theta}) = (2\pi\sigma^2)^{-\frac{1}{2}(k-1)} |\Sigma|^{-\frac{1}{2}} \exp\{-\tfrac{1}{2}g^{(o)}(\underset{\sim}{u}_j | \underset{\sim}{\theta})/\sigma^2\} \tag{8.3.6}$$

where

$$\begin{aligned} g^{(o)}(\underset{\sim}{u}_j | \underset{\sim}{\theta}) &= (\underset{\sim}{u}_j - \underset{\sim}{\theta})' \, \Sigma^{-1}(\underset{\sim}{u}_j - \underset{\sim}{\theta}) \\ &= (k-1)k^{-1}\Sigma_{i=1}^{k-1} u_{ij}^2 - 2\Sigma_{i=1}^{k-1} u_{ij}\theta_i - 2k^{-1}\Sigma\Sigma_{i<\ell}^{k-1} u_{ij}u_{\ell j} \\ &\quad + 2k^{-1}(\Sigma_{i=1}^{k-1}\theta_i)(\Sigma_{i=1}^{k-1} u_{ij}) + (k-1)k^{-1}\Sigma_{i=1}^{k-1}\theta_i^2 \\ &\quad - 2k^{-1}\Sigma\Sigma_{i<\ell}^{k-1}\theta_i\,\theta_\ell . \end{aligned} \tag{8.3.7}$$

From (8.3.6), it is straightforward to verify that

$$g^{(o)}(\underset{\sim}{u}_j|H_1) = (k-1)k^{-1}\,\Sigma_{i=1}^{k-1} u_{ij}^2 - 2k^{-1}\Sigma\Sigma_{i<\ell}^{k-1} u_{i\ell}u_{\ell j}$$

$$+\,2\delta^*k^{-1}\,\Sigma_{i=1}^{k-1} u_{ij} + (k-1)k^{-1}\,\delta^{*2},$$

$$g^{(o)}(\underset{\sim}{u}_j|H_s) = (k-1)k^{-1}\,\Sigma_{i=1}^{k-1} u_{ij}^2 - 2\delta^* u_{s-1j} - 2k^{-1}\Sigma\Sigma_{i<\ell}^{k-1} u_{i\ell}u_{\ell j}$$

$$+\,2\delta^*k^{-1}\,\Sigma_{i=1}^{k-1} u_{ij} + (k-1)k^{-1}\,\delta^{*2}, \qquad (8.3.8)$$

for $s = 2,\ldots,k$. Thus,

$$f(\underset{\sim}{u}_\ell|H_i)/f(\underset{\sim}{u}_\ell|H_j)$$

$$= \exp\{-\tfrac{1}{2}[h^{(o)}(\underset{\sim}{u}_\ell|H_i) - h^{(o)}(\underset{\sim}{u}_\ell|H_j)]/\sigma^2\}$$

$$= \begin{cases} \exp(-\delta^* u_{j-1\,\ell}/\sigma^2) \text{ if } i = 1,\ j\neq 1 \\ \exp\{-\delta^*(u_{j-1\ell} - u_{i-1\ell})/\sigma^2\} \text{ if } i\neq j=2,\ldots,k. \end{cases} \qquad (8.3.9)$$

We now utilize the ideas introduced by Robbins (1970) and later further developed in Khan (1973). But these techniques are also justifiable in essence from the works of Armitage (1950). In what follows, we choose a doubly indexed sequence of constants (>1) and consider $a_{ij} = (k-1)/(1-P^*)$ for all $i \neq j = 1,\ldots,k$, $k^{-1} < P^* < 1$ being preassigned. One writes

$$f_{i,n} = \text{Likelihood function of } \underset{\sim}{U}_1,\ldots,\underset{\sim}{U}_n \text{ under the hypothesis } H_i,\ i = 1,\ldots,k. \qquad (8.3.10)$$

From (8.3.9) it follows that

$$f_{j,n}/f_{i,n} = \exp\{-\delta^*\Sigma_{\ell=1}^{n}(X_{i\ell} - X_{j\ell})/\sigma^2\}, \qquad (8.3.11)$$

for all $1 \le i,\ j \le k$. We now define the stopping rule as follows. Start with one sample from each π and then continue by taking one more sample from each π at a time according to the stopping rule as follows.

$\mathcal{R}_N(3, 1)$: The stopping variable N is given by the $\inf\{n \geq 1 : 1 \geq \sup_{j \neq i} [a_{ij} f_{j,n}/f_{i,n}]$ for some $i\}$ or infinity if no such n, and when N stops with i, we decide for the hypothesis $H_i : \mu_i = \mu_{[k]}$, $i = 1,\ldots,k$. (8.3.12)

It was shown in Mukhopadhyay (1983a) that $P(N < \infty | H_i) = 1$ and $P(CS | H_i) \geq P^*$ for all $i = 1,\ldots,k$. Using (8.3.11), one can simplify (8.3.12) and write

$$N = \inf\{n \geq 1 : \delta^* \sup_{j \neq i}\{\Sigma_{\ell=1}^{n}(X_{j\ell} - X_{i\ell})\} \leq \sigma^2 \ln(\tfrac{1-P^*}{k-1}) \text{ for some } i\}$$

$$= \infty \text{ if no such } n.$$

If N stops with i, the selection procedure $\mathcal{R}_N(3, 1)$ picks π_i as the *best* population. It is easy to check that if $\mathcal{R}_N(3, 1)$ selects π_i as the best population then

$$\overline{X}_{iN} = \max_{1 \leq j \leq k} \overline{X}_{jN}, \tag{8.3.13}$$

where recall from Chapter 3 that $\overline{X}_{in} = n^{-1}\Sigma_{j=1}^{n} X_{ij}$.

Mukhopadhyay (1983a) had shown that the selection procedure $\mathcal{R}_N(3, 1)$ given by (8.3.11) saves sample sizes on the average significantly compared with the traditional fixed–sample size procedure discussed in section 2.3 of Gibbons et al. (1977) and in section 3.2 of Chapter 3. On the other hand, if we denote the stopping variable considered in section 4.3.1 of Bechhofer et al. (1968) by N_{BKS}, then one can easily verify that $N_{BKS} \leq N$ w.p. 1. However, the method of construction of our present selection procedure $\mathcal{R}_N(3, 1)$ given by (8.3.12) or (8.3.13) easily extends for other types of problems and distributions and hence this particular approach is included in our discussion.

Remark 8.3.1. One may actually choose $a_{ij} > 1$ arbitrarily such that $\Sigma_{i \neq j} a_{ij}^{-1} \leq 1 - P^*$ and implement the selection procedure $\mathcal{R}_N(3, 1)$. This is the reason why we opt for writing a_{ij} even though these are taken to be all equal in (8.3.10). See Khan (1973).

8.3.1.2 Common Unknown Variance

The goal is similar to what we had in section 8.3.1.1, but now it is assumed that the common variance $\sigma^2 \epsilon R^+$ is unknown. As in (3.1.1), let us estimate σ^2 by means of $U(n) = \{k(n-1)\}^{-1}\Sigma_{i=1}^{k}\Sigma_{j=1}^{n}(X_{ij} - \overline{X}_{in})^2$ with $n \geq 2$. The ad hoc procedures discussed here were introduced in Mukhopadhyay and Chou (1982a). We also write $Z_{ij\ell} = X_{j\ell} - X_{i\ell}$ and $\overline{Z}_{ijn} = n^{-1}\Sigma_{\ell=1}^{n} Z_{ij\ell}$, $\ell = 1,\ldots,n$, $i \neq j = 1,\ldots,k$. One looks at the stopping rule (8.3.12) and defines the following adaptive sequential selection procedure.

$\mathcal{R}_N(3, 2)$: Let $m_1 = \max\left\{[\{-\ \delta^{*-2}\ell n\ (\frac{1-P^*}{k-1})\}^{1/(\gamma+1)}]^* + 1,\ 2\right\}$ for an arbitrary but fixed $\gamma \epsilon R^+$. One starts with m_1 samples from each π and then proceeds by taking one sample at a time from each population according to the stopping time $N = \inf\{n \geq m_1 : \delta^* \sup_{j \neq i} [n\overline{Z}_{ijn}]/U(n) \leq \ell n\ (\frac{1-P^*}{k-1})$ for some $i\}$, or ∞ if no such n. When N stops with i, we decide for the hypothesis $H_i : \mu_i = \mu_{[k]}$. (8.3.14)

The choice of m_1 is motivated by the modified two–stage procedure (3.4.1) –

(3.4.2) and the *asymptotic* (as $P^* \to 1$) expression of $E(N|H_i)$, namely $-\delta^{*-2}\sigma^2 \ln\left(\frac{1-P^*}{k-1}\right)$, for the selection methodology $\mathcal{R}_N(3,1)$ in (8.3.12). See equation (2.5) in Mukhopadhyay (1983a).

One may also utilize the ideas from Baker (1950) and Hall (1962) in order to construct other appropriate test procedures for H_i, $i = 1,\ldots,k$. With fixed $m_1(\geq 2)$, let

$$r_n(U(m_1)) = \delta^* \Sigma_{\ell=1}^{n} Z_{ij\ell}/U(m_1),$$

$$a_{om_1} = \tfrac{1}{2}\nu^*(\alpha^{-2/\nu^*} - 1),$$

$$\alpha = (1-P^*)/(k-1),\ \nu^* = k(m_1-1), \tag{8.3.15}$$

and one may consider the following selection procedure.

$\mathcal{R}_N(3,3)$: Start with $\{X_{i1},\ldots,X_{im_1} : i = 1,\ldots,k\}$ and then obtain $Z_{ij(m_1+1)}, Z_{ij(m_1+2)},\ldots$ successively. Then, for each $n(\geq m_1)$, after observing Z_{ijn}, we stop sampling and decide for H_i, $i = 1,\ldots,k$, if $\sup_{j\neq i} r_n(U(m_1)) \leq -a_{om_1}$ for some i, having obtained m_1 from (8.3.14). (8.3.16)

In the sequential procedure (8.3.16), one estimated σ^2 only once, by means of $U(m_1)$. Modify (8.3.15) in obvious fashions to define $r_n(U(n))$, a_{on}, α and $\nu^* = k(n-1)$ for $n \geq m_1$. Now, the following selection methodology is in order.

$\mathcal{R}_N(3,4)$: Observe $\{X_{i1}, X_{i2},\ldots ; i = 1,\ldots,k\}$ and thus obtain Z_{ij1}, $Z_{ij2},\ldots$ successively. For each $n(\geq 2)$, after observing Z_{ijn}, we stop sampling and decide for H_i if $\sup_{j\neq i} r_n(U(n)) \leq -a_{on}$ for some i. (8.3.17)

Remark 8.3.2. In the special cases $k = 2$, several *asymptotic* (as $P^* \to 1$) *first-order properties* of the selection methodologies $\mathcal{Q}_N(3, 1)$, $\mathcal{Q}_N(3, 2)$, $\mathcal{Q}_N(3, 3)$ and $\mathcal{Q}_N(3, 4)$ were developed in Mukhopadhyay and Chou (1982a).

Remark 8.3.3. Through extensive sets of simulations, it was shown in Mukhopadhyay and Chou (1982a) that moderate sample performances of these selection procedures were quite satisfactory. We also mention that, in an unpublished technical report, Ray Chaudhuri (1986) considered the case of unknown σ^2 and defined the *preference zone* as

$$\Omega(\delta^*) = \{(\underset{\sim}{\mu}, \sigma) : \mu_{[k]} - \mu_{[k-1]} \geq \delta^* \sigma\}, \tag{8.3.18}$$

for a preassigned $\delta^*(>0)$. And then, she essentially followed the formulation considered in section 8.3.1.1 via $(\overline{X}_{2n} - \overline{X}_{1n},\ldots,\overline{X}_{kn} - \overline{X}_{1n})/U(n)$ and multivariate t–distributions in order to control $P(CS)$ at the P^* level in $\Omega(\delta^*)$ given by (8.3.18).

8.3.2 SELECTING THE SMALLEST NORMAL VARIANCE

Suppose there are $k(\geq 2)$ independent normal populations $\pi_1,\ldots,\pi_k$ where π_i is assumed to have the mean $\mu_i \in R$ and unknown variance $\sigma_i^2 \in R^+$, $i = 1,\ldots,k$. Let $\sigma_{[1]}^2 \leq \cdots \leq \sigma_{[k]}^2$ be the ordered variances. We assume that there is just one population associated with $\sigma_{[1]}^2$, and this is referred to as the *best* population. We wish to select the population associated with $\sigma_{[1]}^2$ under the *indifference zone* formulation. This problem was addressed in Mukhopadhyay and Chou (1982b) via likelihood comparisons. Given two preassigned numbers $\delta^* \in (0, 1)$ and $P^* \in (k^{-1}, 1)$, one defines the *preference zone* as

$$\Omega(\delta^*) = \{\underset{\sim}{\theta} = (\sigma_1^2,\ldots,\sigma_k^2, \mu_1,\ldots,\mu_k) : \sigma_{[1]}^2 \leq \delta^* \sigma_{[2]}^2\}, \tag{8.3.19}$$

with the convention that in $\Omega(\delta^*)$ we consider only those μ's which are unknown. Our goal is to consider sequential selection procedures such that

$$P(CS) \geq P^* \text{ whenever } \underset{\sim}{\theta} \in \Omega(\delta^*).$$

For each type of procedure discussed here, it turns out that the parameter configuration $\delta^{*-1}\sigma^2_{[1]} = \sigma^2_{[2]} = \cdots = \sigma^2_{[k]}$ indeed minimizes $P(CS)$ in $\Omega(\delta^*)$, whatever be μ's. Hence, the least favorable configuration (LFC) is precisely given by $\delta^{*-1}\sigma^2_{[1]} = \sigma^2_{[2]} = \cdots = \sigma^2_{[k]}$, and thus we consider the selection problem *only under the LFC.*

8.3.2.1 All Means Unknown

We assume that $\mu_1,\ldots,\mu_k$ are *all* unknown and we wish to decide among k hypotheses, namely

$$H_i : \sigma^2_i = \sigma^2_{[1]},\ i = 1,\ldots,k, \tag{8.3.20}$$

such that $P(CS \mid H_i) \geq P^*$ for all $i = 1,\ldots,k$. We suppose we have available a sequence $X_{i1},\ldots,X_{in},\ldots$ of i.i.d. random variables from π_i, and we consider taking an equal number of samples from each π. Having recorded $X_{i1},\ldots,X_{in}$ from π_i with $n \geq 2$, we write $\overline{X}_{in}$ and S^2_{in} as in (3.1.1). Let us denote

$$\underset{\sim}{T}_n = \left(\frac{S^2_{2n}}{S^2_{1n}},\ \ldots,\ \frac{S^2_{kn}}{S^2_{1n}}\right)$$
$$= (T_{1n},\ldots,T_{k-1n}), \text{ say.} \tag{8.3.21}$$

We define $\theta_i = \sigma^2_1/\sigma^2_{i+1}$, $i = 1,\ldots,k-1$ and write $\underset{\sim}{\theta}' = (\theta_1,\ldots,\theta_{k-1})$. Using the results of Hall et al. (1965), T_{in} can be easily seen to be invariantly sufficient for θ_i, $i = 1,\ldots,k-1$ with respect to the group of equal positive scale transformations. It is easy to see that $H_1,\ldots,H_k$ can be equivalently rephrased in the following way.

$$H_1 : \theta_i = \delta^* \text{ for all } i = 1,\ldots,k-1,$$

$$H_j : \theta_{j-1} = \delta^{*-1},\ \theta_i = 1 \text{ for all } i \neq j-1, \tag{8.3.22}$$

with $j = 2,\ldots,k$. Let $a_{ij} = (k-1)/(1-P^*)$ for all $i \neq j = 1,\ldots,k$.

Now, one uses Kimbal's (1951) multivariate F–distribution (see page 240 of Johnson and Kotz (1972)) and obtains the p.d.f. of $\underset{\sim}{T}_n$ at $\underset{\sim}{t}_n$ as

$$f^{(1)}(\underset{\sim}{t}_n \mid \underset{\sim}{\theta})$$
$$= C^*(n, k) \prod_{i=1}^{k-1} t_{in}^{\frac{1}{2}(n-3)} \prod_{i=1}^{k-1} \theta_i^{\frac{1}{2}(n-1)} (1+ \Sigma_{i=1}^{k-1}\theta_i t_{in})^{-\frac{1}{2}k(n-1)} \tag{8.3.23}$$

where $C^*(n, k) = \{\Gamma\ \tfrac{1}{2}(n-1)\}^{-k}\ \{\Gamma\ \tfrac{1}{2}(kn-k)\}$. We follow the convention that $f^{(1)}(\underset{\sim}{t}_n \mid \underset{\sim}{\theta})$ is of the specified form when $\underset{\sim}{t}_n \in R^{+k-1}$. As in (8.3.10), we write

$$f_{i,n}^{(1)} = \text{Likelihood of } \underset{\sim}{T}_n \text{ under the hypothesis } H_i,\ i = 1,\ldots,k. \tag{8.3.24}$$

Now, for all $1 \leq i,\ j \leq k$,

$$f_{j,n}^{(1)}/f_{i,n}^{(1)} = Y_{ijn}^{-\frac{1}{2}k(n-1)} \tag{8.3.25}$$

where

$$Y_{ijn} = \frac{c_{j1}+c_{j2}T_{1n}+\cdots+c_{jk}T_{k-1n}}{c_{i1}+c_{i2}T_{1n}+\cdots+c_{ik}T_{k-1n}}, \tag{8.3.26}$$

with $c_{rs} = 1$ or δ^* according as $r = s$ or $r \neq s$. Along the lines of $\mathscr{Q}_N(3, 1)$, we propose the following selection methodology. Start with two samples from each π and then proceed by taking one more sample from each according to the stopping rule

$\mathcal{R}_N(3,5)$: $N = \inf\{n \geq 2 : \sup_{j \neq i} [a_{ij} f_{j,n}/f_{i,n}] \leq 1$ for some $i\}$ or infinity if no such n, and when N stops with i, we decide for the hypothesis $H_i : \sigma_i = \sigma_{[1]}$, $i = 1,\ldots,k$. (8.3.27)

It was shown in Mukhopadhyay and Chou (1982b) that $P(N < \infty | H_i) = 1$ and $P(CS | H_i) \geq P^*$ for all $i = 1,\ldots,k$.

In the special case of two populations, Mukhopadhyay and Chou (1982b) derived various *first-order asymptotic* (as $P^* \to 1$) properties of the selection methodology $\mathcal{R}_N(3, 5)$. That paper also included extensive discussions on moderate sample performances of the selection procedure $\mathcal{R}_N(3, 5)$, particularly in comparison with the fixed sample size rule as given in Gibbons et al. (1977, Chapter 5). We add that Hoel's (1971) selection rule agrees with ours in (8.3.27) when $k = 2$.

Remark 8.3.4. The case when all the means are known is similar in principle to the case when all the means are unknown, since we can define $S_{in}^2 = n^{-1}\Sigma_{j=1}^n (X_{ij} - \mu_i)^2$ and $\underset{\sim}{T}_n = (T_{1n},\ldots,T_{k-1n})$ where $T_{in} = S_{i+1}^2/S_{1n}^2$, for $i = 1,\ldots,k-1$. One will observe an obvious difference in the expression of $f^{(1)}(\underset{\sim}{t}_n | \underset{\sim}{\theta})$ from that given in (8.3.23), since the degrees of freedom of each component χ^2 is now n instead $(n-1)$. The procedure $\mathcal{R}_N(3, 5)$ in (8.3.27) will be appropriately modified.

Remark 8.3.5. In the case of a bivariate normal population with unknown means μ_1, μ_2 unknown variances σ_1^2, σ_2^2 and unknown correlation coefficient ρ, the selection procedure $\mathcal{R}_N(3, 5)$ was further generalized in order to identify the component associated with the smaller variance $\sigma_{[1]}^2$ via likelihood comparisons in Mukhopadhyay and Chou (1985). Some of the *first–order asymptotics* were also included in Mukhopadhyay and Chou (1985).

8.3.2.2 Few Known Means

To get the idea across, we consider $k = 2$ first and suppose that μ_1 is known, but μ_2 is unknown. Define

$$S_{1n}^2 = n^{-1}\Sigma_{j=1}^{n}(X_{1j} - \mu_1)^2,\ S_{2n}^2 = (n-1)^{-1}\Sigma_{j=1}^{n}(X_{2j} - \overline{X}_{2n})^2$$

$$T_{1n} = S_{2n}^2/S_{1n}^2,\quad \theta_1 = \sigma_1^2/\sigma_2^2.$$

Then

$$f^{(1)}(t_{1n} \mid \theta_1) = a^*(n) t_{1n}^{\frac{1}{2}(n-3)} \theta_1^{\frac{1}{2}(n-1)} / \{n + (n-1)\theta_1 t_{1n}\}^{\frac{1}{2}(2n-1)}, \tag{8.3.28}$$

where one has $a^*(n) = \{\Gamma\ \frac{1}{2}(2n-1)\}\ \{\Gamma(\frac{1}{2}n)\ \Gamma\frac{1}{2}(n-1)\}^{-1} n^{-\frac{1}{2}n}(n-1)^{\frac{1}{2}(n-1)}$. From (8.3.28), it follows that

$$f_{1,n}^{(1)}/f_{2,n}^{(1)} = \{\delta^* + (\tfrac{n-1}{n})T_{1n}\}^{\frac{1}{2}(2n-1)}\{1 + (\tfrac{n-1}{n})\delta^* T_{1n}\}^{\frac{1}{2}(-2n+1)}\delta^{*-\frac{1}{2}}. \tag{8.3.29}$$

Now, one can naturally propose a selection rule along the lines of $\mathcal{R}_N(3, 5)$ in (8.3.27). This was considered in section 3.3 of Mukhopadhyay and Chou (1982b).

In the case $k = 3$, if μ_1 is known but μ_2 and μ_3 are unknown, then also one can modify $\mathcal{R}_N(3, 5)$ appropriately. Define S_{1n}^2, S_{2n}^2 as we had in this subsection, and let $S_{3n}^2 = (n-1)^{-1}\Sigma_{j=1}^{n}(X_{3j} - \overline{X}_{3n})^2$. Consider $\underset{\sim}{T}_n = (T_{1n}, T_{2n})$ where $T_{1n} = S_{2n}^2/S_{1n}^2$ and $T_{2n} = S_{3n}^2/S_{1n}^2$. Now,

$$f^{(1)}(\underset{\sim}{t}_n \mid \underset{\sim}{\theta}) = b^*(n)(t_{1n}t_{2n})^{\frac{1}{2}(n-3)}(\theta_1\theta_2)^{\frac{1}{2}(n-1)}\ \{1 + (\tfrac{n-1}{n})\theta_1 t_{1n}$$

$$+ (\frac{n-1}{n})\theta_2 t_{2n}\}^{-\frac{1}{2}(3n-2)}, \qquad (8.3.30)$$

where $b^*(n) = (\frac{n-1}{n})^{n-1} \Gamma\frac{1}{2}(3n-2)/\{\Gamma(\frac{1}{2}n)\ [\Gamma\ \frac{1}{2}(n-1)]^2\}$, and one obtains the expression of $f^{(1)}_{i,n}/f^{(1)}_{j,n}$. Hence, the selection rule $\mathcal{R}_N(3, 5)$ can be improvised properly to suit the present situation. The situation when μ_1, μ_2 are known, but μ_3 is unknown can again be handled in a fairly similar fashion. Moderate sample size performances of such techniques were commented upon in Mukhopadhyay and Chou (1982b).

8.3.3 SELECTING THE LARGEST EXPONENTIAL LOCATION WHEN THE COMMON SCALE PARAMETER IS KNOWN

Suppose that there are $k(\geq 2)$ independent exponential populations $\pi_1,\ldots,\pi_k$, the p.d.f. associated with π_i being given by $f(x;\ \theta_i,\ \sigma)$, defined in (2.2.5) with $\theta_i \in R$ and $\sigma \in R^+$, $i = 1,\ldots,k$. It is assumed that $\theta_1,\ldots,\theta_k$ are unknown and σ is known. Let $\theta_{[1]} \leq \cdots \leq \theta_{[k]}$ be the ordered θ–values and we suppose that there is just one population associated with $\theta_{[k]}$. Our goal is to select this population associated with $\theta_{[k]}$ which is referred to as the *best* one, as in Chapter 4. Bechhofer's (1954) indifference zone formulation is adopted and hence given $\delta^*(>0)$, one defines the *preference zone*

$$\Omega(\delta^*) = \{\underset{\sim}{\theta} = (\theta_1,\ldots,\theta_k) : \theta_{[k]} - \theta_{[k-1]} \geq \delta^*\}. \qquad (8.3.31)$$

Given $P^* \epsilon (k^{-1}, 1)$, we also require that

$$P(CS) \geq P^* \text{ whenever } \underset{\sim}{\theta} \epsilon \Omega(\delta^*). \qquad (8.3.32)$$

Purely sequential procedures are proposed for which the goal in (8.3.32) is met. The present discussion is a brief summary of the results obtained in Miller and Mukhopadhyay (1990).

One should note that for the proposed methodology, the parameter configuration $\theta_{[1]} = \cdots = \theta_{[k-1]} = \theta_{[k]} - \delta^*$ indeed minimizes P(CS) in $\Omega(\delta^*)$, and hence the least favorable configuration (LFC) is precisely given by $\theta_{[1]} = \cdots = \theta_{[k-1]} = \theta_{[k]} - \delta^*$. Thus, throughout this section, we consider the selection *problem only under the LFC.* We wish to decide among k hypotheses

$$H_i : \theta_i = \theta_{[k]},\ i = 1,\ldots,k \tag{8.3.33}$$

such that $P(CS \mid H_i) \geq P^*$ for all $i = 1,\ldots,k$. We suppose we have available a sequence $X_{i1},\ldots,X_{in},\ldots$ of i.i.d. random variables from π_i and we consider taking equal number of samples from each π_i, $i = 1,\ldots,k$. Having recorded $X_{i1},\ldots,X_{in}$ from π_i, let us denote $\tilde{X}_{in} = \min\{X_{i1},\ldots,X_{in}\}$ as in (4.1.1). Now the likelihood function of $X_{i1},\ldots,X_{in}$, $i = 1,\ldots,k$ is given by

$$\sigma^{-kn} \exp[-\{\Sigma_{i=1}^{k} \Sigma_{j=1}^{n}(X_{ij} - \theta_i)\}/\sigma] \prod_{i=1}^{k} I(\tilde{X}_{in} > \theta_i). \tag{8.3.34}$$

Observe that $\underset{\sim}{U}_n$ is the maximal invariant (with respect to equal location shifts) where $\underset{\sim}{U}'_n = (U_{1n},\ldots,U_{k-1n})$ with $U_{in} = \tilde{X}_{i+1n} - \tilde{X}_{in}$, $i = 1,\ldots,k-1$. The p.d.f. of $\underset{\sim}{U}_n$ can be found from the p.d.f. of $\underset{\sim}{U}^{*\prime}_n = (\tilde{X}_{1n}, \tilde{X}_{2n} - \tilde{X}_{1n},\ldots, \tilde{X}_{kn} - \tilde{X}_{1n}) = (U_{on}, U_{1n},\ldots,U_{k-1n})$, say, where

$$f^{(2)}(\underset{\sim}{u}^*_n \mid \underset{\sim}{\theta})$$

$$= (n/\sigma)^k \exp[-n\{ku_{on} + \Sigma_{i=1}^{k-1} u_{in} - \Sigma_{i=1}^{k} \theta_i\}/\sigma], \tag{8.3.35}$$

with its support being $u_{on} > \theta_1$ and $u_{in} + u_{on} > \theta_{i+1}$, $i = 1,\ldots,k-1$. From (8.3.35), the p.d.f of $\underset{\sim}{U}_n$ can be written as

$$f^{(3)}(\underset{\sim}{u}_n \mid \underset{\sim}{\theta}) = \int f^{(2)}(\underset{\sim}{u}_n^* \mid \theta) du_{on}, \tag{8.3.36}$$

and this integral depends only on the difference of location parameters. So, in principle, under $H_i : \theta_i = \theta_{[k]}$, we can write down the expression of $f_{i,n}$, the p.d.f. of $U_{1n},...,U_{k-1n}$ and obtain $f_{i,n}/f_{j,n}$ for all $1 \leq i, \; j \leq k$. Define $a_{ij} = (k-1)/(1-P^*)$ and let us propose the following selection methodology along the lines of $\mathcal{Q}_N(3, 1)$ given by (8.3.12). We take one sample at a time from each π.

$\mathcal{Q}_N(3,6)$: Let $N = \inf\{n \geq 1 : \sup_{j \neq i} [a_{ij} f_{j,n}/f_{i,n}] \leq 1$ for some $i\}$ or infinity if no such n, and when N stops with i, we decide for the hypothesis $H_i : \theta_i = \theta_{[k]}$, $i = 1,...,k$. (8.3.37)

However, in this situation, it is not possible to explicitly simplify $f_{i,n}/f_{j,n}$ along the lines of (8.3.11) for arbitrary $k(\geq 2)$ since the support in (8.3.35) depends on $\theta_1,...,\theta_k$. Thus, Miller and Mukhopadhyay (1990) considered the special cases $k = 2$ and $k = 3$ in detail in order to point out this major difficulty as mentioned above. In that paper, certain *asymptotic* (as $P^* \to 1$) *first-order characteristics* for the proposed methodologies were also investigated in conjunction with a study of moderate sample performances of such selection rules.

Remark 8.3.6. If one assumes that π_i has unknown location parameter $\mu_i \in R$ and unknown scale parameter $\sigma_i \in R^+$, $i = 1,...,k$ and the problem is to select that population associated with $\sigma_{[1]}$, the smallest scale, under Bechhofer's (1954) indifference zone formulation, then such a problem can be formulated and solved in a fairly straightforward fashion along the lines of section 8.3.2.1. This was the subject matter in the paper of Mukhopadhyay and Miller (1990).

8.4 Selection procedures when variances are unknown and unequal

In Chapters 3 and 4, we had discussed problems associated with the selection of the largest location parameter under the indifference zone formulation. The specific parameteric models considered were normal and two–parameter exponential with equal but unknown variance. The sample sizes considered from each population were also all equal. Now, we briefly address analogous problems when the *population variances* are *unknown and unequal* and the *sample sizes are also unequal*. A few available *two–stage* procedures are summarized with associated *exact* as well as *first–order asymptotic properties*. Certain types of *purely sequential* selection procedures are also included together with *first–order asymptotics*. In the latter development some indications are given for possible *second–order asymptotics*.

8.4.1 SELECTING THE BEST NORMAL POPULATION

Suppose we have π_i, $i = 1,\ldots,k(\geq 2)$ independent populations and we have available a sequence $X_{j1},\ldots,X_{jn_j},\ldots$ of i.i.d. $N(\mu_j, \sigma_j^2)$ random variables from π_j where we assume that $\mu_j \in R$ and $\sigma_j \in R^+$ are all unknown parameters, $j = 1,\ldots,k$. As before, we write $\mu_{[k]}$ for the largest μ–value and the population associated with $\mu_{[k]}$ is defined as the *best* population. It is assumed that there is just one population associated with $\mu_{[k]}$.

Along the lines of (3.1.3), given $\delta^* > 0$, we define Bechhofer's (1954) *preference zone*,

$$\Omega(\delta^*) = \{\underset{\sim}{\mu} = (\mu_1,\ldots,\mu_k) : \mu_{[k]} - \mu_{[k-1]} \geq \delta^*\}. \tag{8.4.1}$$

For a given $P^* \in (k^{-1}, 1)$, we are interested in selection procedures such that P(CS) is *at least* P^* or *asymptotically* (as $\delta^* \to 0$) at least P^* whenever the true parameters $\underset{\sim}{\mu} \in \Omega(\delta^*)$, whatever the unknown variances $\sigma_1^2,\ldots,\sigma_k^2$.

Dudewicz (1971) had shown the nonexistence of any single stage procedure for such a selection problem. In the paper of Dudewicz and Dalal (1975), two types of two–stage selection procedures called P_E and P_R were proposed. The procedure P_R is based on sample means and variances from each population and we first summarize this along with some of its properties obtained in Rinott (1978). Section 6.2.1 is very relevant in this case. A few observations obtained in Mukhopadhyay (1979) are also included. Then, certain purely sequential procedures are introduced with possible asymptotic second–order characteristics.

8.4.1.1 Preliminaries

Having recorded $X_{i1},\dots,X_{in_i}$ from π_i with $n_i \geq 2$, we write

$$\overline{X}_i(n_i) = n_i^{-1}\Sigma_{j=1}^{n_i} X_{ij},\ \ S_i^2(n_i) = (n_i-1)^{-1}\Sigma_{j=1}^{n_i}(X_{ij}-\overline{X}_i(n_i))^2,$$

$$T(\underset{\sim}{n}) = \max_{1\leq i\leq k} \overline{X}_i(n_i), \tag{8.4.2}$$

where $\underset{\sim}{n} = (n_1,\dots,n_k)$, $i = 1,\dots,k$. These notations are slightly different from those employed in (3.1.1). A natural selection procedure is simply to choose the population associated with the largest sample mean, that is, one implements the following selection procedure.

$$\mathcal{R}_{\underset{\sim}{n}}(4,1): \text{Select } \pi_j \text{ as the best population}$$

$$\text{if } \overline{X}_j(n_j) = T(\underset{\sim}{n}). \tag{8.4.3}$$

Let $\pi_{(i)}$ stand for the population associated with $\mu_{[i]}$ and suppose that the corresponding sample size, sample mean and sample variance are respectively written as $n_{(i)}$, $\overline{X}_{(i)}(n_{(i)})$ and $S_{(i)}^2(n_{(i)})$. For the moment *pretend* that $\sigma_1^2,\dots,\sigma_k^2$ are known and as per Bechhofer's (1954) suggestion, suppose one chooses $n_1,\ldots,n_k$ in such a way that these are smallest numbers and

$$\sigma_1^2/n_1 = \cdots = \sigma_k^2/n_k = h_3^2/\delta^{*2}, \tag{8.4.4}$$

with some appropriate $h_3 = h_3(k, P^*)$, to be defined shortly. Now, one implements the selection procedure $\mathcal{Q}_{\underset{\sim}{n}}(4, 1)$ from (8.4.3), and then the associated probability of correct selection is given by

$$P_{\underset{\sim}{n}}(CS) = P\{\overline{X}_{(i)}(n_{(i)}) < \overline{X}_{(k)}(n_{(k)}), i = 1,\ldots,k-1\}$$

$$= P\{Z_{(i)} < Z_{(k)} + \delta_i(\delta^*/h_3)^{-1}, i = 1,\ldots,k-1\}, \tag{8.4.5}$$

where $\delta_i = \mu_{[k]} - \mu_{[i]} \geq \delta^*$ when $\underset{\sim}{\mu}\in\Omega(\delta^*)$, $i = 1,\ldots,k-1$ and $Z_{(1)},\ldots,Z_{(k)}$ are i.i.d. $N(0, 1)$ random variables.

From (8.4.5) one observes that

$$\inf_{\underset{\sim}{\mu}\in\Omega(\delta^*)} P_{\underset{\sim}{n}}(CS) = P\{Z_{(i)} < Z_{(k)} + h_3, i = 1,\ldots,k-1\}, \tag{8.4.6}$$

and this infimum is attained when $\delta_i = \delta^*$ for all $i = 1,\ldots,k-1$, that is, if $\mu_{[1]} = \ldots = \mu_{[k-1]} = \mu_{[k]} - \delta^*$, a parameter configuration that is referred to as the least favorable configuration (LFC). Now, comparing (8.4.6) with (3.2.3) and (3.8.1), we immediately note that if we replace h_3 by means of $h = h(k, P^*)$ obtained from (3.2.4), then

$$\inf_{\underset{\sim}{\mu}\in\Omega(\delta^*)} P_{\underset{\sim}{n}}(CS) = P^*, \tag{8.4.7}$$

and the infimum is attained at the LFC. Hence, from (8.4.4), the choices for sample sizes are then given by

$$n_i^* = h\sigma_i^2/\delta^{*2}, \quad i = 1,...,k. \tag{8.4.8}$$

8.4.1.2 Two–Stage Procedure

In reality, $\sigma_1^2,...,\sigma_k^2$ are of course unknown and in this situation, Dudewicz and Dalal (1975) proposed the following *two–stage procedure* based on $\overline{X}_i$ and S_i^2, $i = 1,...,k$. In the notation of Dudewicz and Dalal (1975) and Rinott (1978), this procedure is referred as P_R.

Let $m_1(\geq 2)$ be the starting sample size from each π and we write $h_4 = h_4(m_1, k, P^*)$ be a positive constant, to be determined specifically a little later. Define

$$N_i = \max\{m_1, [h_4^2 S_i^2(m_1)\delta^{*-2}]^* + 1\}, \tag{8.4.9}$$

$i = 1,...,k$. If $N_i = m_1$, we do not take any more samples from π_i. However, if $N_i > m_1$, then we sample the difference $(N_i - m_1)$ from π_i. Based on the totality of all samples $X_{i1},...,X_{iN_i}$ from π_i, we now implement the selection rule $\mathscr{R}_{\underset{\sim}{N}}(4, 1)$ defined in (8.4.3) where one writes $\underset{\sim}{N} = (N_1,...,N_k)$. Before we proceed any further, one should note that N_i's given by (8.4.9) actually mimics the expressions of n_i^*'s defined in (8.4.8).

The constant h_4 has to satisfy a specific integral equation, obtained by Rinott (1978). In what follows we essentially provide the derivations from Rinott (1978). This derivation uses a very elegant probability inequality, called Slepian's (1962) inequality. For completeness, we first state Slepian's (1962) inequality.

Slepian's Inequality: Let $\underset{\sim}{X} = (X_1,...,X_s)$ be a random variable distributed as $N_s(\underset{\sim}{0}, \Sigma)$ where $\Sigma = (\sigma_{ij})$ is a p.d. correlation matrix. Then, for real numbers

$c_1,\ldots,c_s$, we have the following:

(i) $P\{X_i \le c_i,\ i = 1,\ldots,s\}$ is strictly increasing as a function of σ_{ij}, $i \ne j = 1,\ldots,s$; (8.4.10)

(ii) In particular if $\sigma_{ij} > 0$, $i \ne j = 1,\ldots,s$, then

$$P\{X_i \le c_i,\ i = 1,\ldots,s\} > \prod_{i=1}^{s} P(X_i \le c_i). \tag{8.4.11}$$

Theorem 8.4.1. *For the two–stage procedure (8.4.9), with* h_4 *defined in (8.4.16), we have* $P(CS) > P^*$ *for all* $\underset{\sim}{\mu} \in \Omega(\delta^*)$ *and all unknown* $\sigma_1^2,\ldots,\sigma_k^2$.

Proof. The following proof is adapted from Rinott (1978). One first observes that the LFC associated with $\mathscr{Q}_{\underset{\sim}{N}}(4, 1)$ is again given by $\mu_{[1]} = \cdots = \mu_{[k-1]} = \mu_{[k]} - \delta^*$. In what follows we work *only under this LFC.* One obtains

$$\begin{aligned} P(CS) &= P\{\overline{X}_{(i)}(N_{(i)}) < \overline{X}_{(k)}(N_{(k)}),\ i = 1,\ldots,k-1\} \\ &= P\{Y_i < B_i,\ i = 1,\ldots,k-1\}, \end{aligned} \tag{8.4.12}$$

where

$$Y_i = \{\overline{X}_{(i)}(N_{(i)}) - \overline{X}_{(k)}(N_{(k)}) + \delta^*\}\,\{[\sigma_{(i)}^2/N_{(i)}] + [\sigma_{(k)}^2/N_{(k)}]\}^{-\frac{1}{2}},$$

$$B_i = \delta^*\{[\sigma_{(i)}^2/N_{(i)}] + [\sigma_{(k)}^2/N_{(k)}]\}^{-\frac{1}{2}},$$

and also let us define

$$B_i^* = h_4\{[\sigma_{(i)}^2/S_{(i)}^2(m_1)] + [\sigma_{(k)}^2/S_{(k)}^2(m_1)]\}^{-\frac{1}{2}},$$

$i = 1,\ldots,k-1$. From (8.4.9), we get $N_i \ge h_4^2\, S_i^2(m_1)\delta^{*-2}$ for all $i = 1,\ldots,k$ and hence $B_i \ge B_i^*$ for all $i = 1,\ldots,k-1$ so that one can write from (8.4.12)

$$P(CS) \geq P\{Y_i \leq B_i^*, i = 1,...,k-1\}. \qquad (8.4.13)$$

Now, given $S_1^2(m_1),...,S_k^2(m_1)$, one observes that $(Y_1,...,Y_{k-1})$ is distributed as a $(k-1)$ dimensional normal random variable having mean vector $\underset{\sim}{0}$ and the dispersion matrix $\Sigma = (\sigma_{ij})$ where Σ is the $(k-1) \times (k-1)$ p.d. matrix with

$$\sigma_{ii} = 1 \text{ and } \sigma_{ij} = \sigma_{(k)}^2 B_i B_j \{\delta^{*2} N_{(k)}\}^{-1},$$

for $i \neq j = 1,...,k-1$. In this conditional distribution, $\sigma_{ij} > 0$ for $i \neq j = 1,...,k-1$. Thus,

$$P\{Y_i \leq B_i^*, i = 1,...,k-1 \mid S_1^2(m_1),...,S_k^2(m_1)\}$$

$$> \prod_{i=1}^{k-1} P\{Y_i \leq B_i^* \mid S_1^2(m_1),...,S_k^2(m_1)\}, \qquad (8.4.14)$$

by virtue of Slepian's (1962) inequality given in (8.4.11). Now, one combines (8.4.13) – (8.4.14) to obtain

$$P(CS) > E\{\prod_{i=1}^{k-1} \Phi(B_i^*)\}$$

$$= E\{\prod_{i=1}^{k-1} \Phi(h_4[(m_1-1)(W_i^{-1} + W_k^{-1})]^{-\frac{1}{2}})\}, \qquad (8.4.15)$$

where $W_1,...,W_k$ are i.i.d. $\chi^2_{m_1-1}$ random variables. Now, let us choose $h_4 = h_4(m_1, k, P^*)$ in order to satisfy

$$\int_0^\infty \int_0^\infty \Phi(h_4[(m_1-1)(x^{-1} + y^{-1})]^{-\frac{1}{2}}) f^0(x) f^0(y) dx dy = P^*, \qquad (8.4.16)$$

$f^0(x)$ being the p.d.f. of $\chi^2_{m_1-1}$, $x > 0$. Now, the proof is complete. □

Remark 8.4.1. Rinott (1978) also includes interesting comparisons of the original two–stage procedures P_E and P_R of Dudewicz and Dalal (1975). Unfortunately, Rinott's (1978) paper does not include a table to determine h_4 values. For $k = 2$, h_4 coincides with h_5 where

$$\int_{-\infty}^{\infty} G_4(t + h_5)\, g_4(t)dt = P^*, \tag{8.4.17}$$

with $G_4(\cdot)$ and $g_4(\cdot)$ respectively being the d.f. and p.d.f. of a t–distribution with $(m_1 - 1)$ degrees of freedom. In other words, for $k = 2$, h_4–values can be obtained from tables of Dudewicz and Dalal (1975). The h_4–values for $k = 3, 4, 5$ are available in Tables 4 – 6 in Mukhopadhyay and Chou (1984), for various values of P^*.

8.4.1.3 The Case of Two Populations

We now reconsider the two–stage procedure $\mathcal{Q}_{\underset{\sim}{N}}(4, 1)$ obtained from (8.4.9) in the special case $k = 2$, and summarize some of the comments made in Mukhopadhyay (1979). From Remark 8.4.1, recall that h_4 satisfies the equation (8.4.17). It is reasonable to believe that $h_4 = h_4(m_1)$ is a decreasing function of m_1, if P^* is held fixed, and this fact was proved in Lemma 1 of Mukhopadhyay (1979). Essentially, what this means is that if one starts with a larger pilot sample size, then one may expect to take fewer samples at the second stage. Also, one has $\lim h_4(m_1) = h$ as $m_1 \to \infty$, for fixed P^* where

$$\int_{-\infty}^{\infty} \Phi(x + h)\, \phi(x)dx = P^*, \tag{8.4.18}$$

the integral equation being the same as in (3.2.4) with $k = 2$. In other words, one can claim that

$$\lim E(N_i/n_i^*) > 1 \text{ as } \delta^* \to 0, \tag{8.4.19}$$

$i = 1, 2$. That is, the two–stage procedure (8.4.9) oversamples compared with n_i^*'s given by (8.4.8), even asymptotically.

Along the lines of section 3.4, one can let m_1 depend on δ^* and grow at a slower rate than that of n_i^*'s and obtain a *modified* version of the *two–stage* procedure (8.4.9) in order to conclude *asymptotic first–order efficiency* in comparison with n_i^*'s, that is then

$$\lim E(N_i/n_i^*) = 1 \text{ as } \delta^* \to 0, \tag{8.4.20}$$

$i = 1, 2$. Such a procedure and an appropriate purely sequential procedure were investigated in Mukhopadhyay (1979).

In this special case, $P_{\underset{\sim}{n}}(CS)$ given by (8.4.5) is at least P^* provided that one has

$$\sigma_1^2 n_1^{-1} + \sigma_2^2 n_2^{-1} \le \delta^{*2}/a^2, \tag{8.4.21}$$

where $\Phi(a) = P^*$. Obviously "a" is positive and $2^{\frac{1}{2}}a = h$, having obtained "h" from (8.4.18). Minimizing the total sample size $(n_1 + n_2)$ subjected to the restriction (8.4.21) leads to *optimal* solutions

$$n_1^{**} = a^2\sigma_1(\sigma_1+\sigma_2)/\delta^{*2},\ n_2^{**} = a^2\sigma_2(\sigma_1+\sigma_2)/\delta^{*2} \tag{8.4.22}$$

and

$$n_1^{**}/n_2^{**} = \sigma_1/\sigma_2, \tag{8.4.23}$$

had σ_1, σ_2 been known. In the case $k = 2$, one should truly mimic the expressions given in (8.4.22) and not those considered in (8.4.8). Mukhopadhyay (1979) proposed the following *asymptotically first-order efficient selection procedure* which is *purely sequential* in nature.

One starts with $m_1(\geq 2)$ samples from each π. Then, at any subsequent step, one has to decide which population to sample from if one more sample is needed. So, the following sampling scheme is proposed along the lines of Robbins et al. (1967).

Sampling Scheme: If at any stage we have taken i samples from π_1 and j samples from π_2, then we take the next sample from π_1 or π_2 according as

$$i/j < S_1(i)/S_2(j) \text{ or } i/j \geq S_1(i)/S_2(j). \tag{8.4.24}$$

Stopping Rule: One uses the sampling scheme (8.4.24) and stops for the first integers $N_1 = n_1(\geq m_1)$, $N_2 = n_2(\geq m_2)$ for which one has

$$n_i \geq a^2 S_i(n_i)\{S_1(n_1) + S_2(n_2)\}/\delta^{*2}, \tag{8.4.25}$$

simultaneously, $i = 1, 2$.

Once the purely sequential procedure (8.4.24) – (8.4.25) stops, we use the selection procedure $\mathcal{R}_{\tilde{N}}(4, 1)$ given in (8.4.3). One should observe that the sampling scheme (8.4.24) mimics the optimal ratio given by (8.4.23). In fact, one can show that $\lim(N_1/N_2) = \sigma_1/\sigma_2$ w.p. 1 as $\delta^* \to 0$. On the other hand, the stopping rule takes into account the actual expressions of n_i^{**}'s given in (8.4.23). *First-order asymptotic properties* of this procedure were proven in Mukhopadhyay's (1979) Theorem 5. *Asymptotic second-order characteristics* of such a procedure are yet unknown.

8.4.2 SELECTING THE BEST NEGATIVE EXPONENTIAL POPULATION

Suppose we have independent populations π_i, $i = 1,\ldots,k(\geq 2)$ and we have available a sequence $X_{i1},\ldots,X_{in_i},\ldots$ of i.i.d. random variables having the p.d.f. $f(x; \theta_i, \sigma_i)$ given by (2.2.5) where we assume that $\theta_i \in R$ and $\sigma_i \in R^+$ are all unknown parameters, $i = 1,\ldots,k$. As in Chapter 4, we write $\theta_{[k]}$ for the largest θ–value and the population associated with $\theta_{[k]}$ is defined as the *best* population. It is assumed that there is just one population associated with $\theta_{[k]}$.

Along the lines of (4.1.2), given $\delta^* > 0$, we define Bechhofer's (1954) *preference zone,*

$$\Omega(\delta^*) = \{\underset{\sim}{\theta} = (\theta_1,\ldots,\theta_k) : \theta_{[k]} - \theta_{[k-1]} \geq \delta^*\}. \tag{8.4.26}$$

For a given $P^* \in (k^{-1}, 1)$, we are interested in selection procedures such that P(CS) is *at least* P^* or *asymptotically* (as $\delta^* \to 0$) at least P^* whenever the true parameters $\underset{\sim}{\theta} \in \Omega(\delta^*)$, whatever the unknown scale parameters $\sigma_1,\ldots,\sigma_k$.

Mukhopadhyay (1984) first proposed a two–stage procedure in the case of two populations. It provided a clever application of the Studentized smaller χ^2 distribution and its appropriate percentiles. Later, Mukhopadhyay and Hamdy (1984a) came up with a different two–stage procedure for arbitrary $k \geq 2$, and this procedure was shown to be superior to the earlier procedure. In what follows, we summarize the methodologies from Mukhopadhyay and Hamdy (1984a). The results of Lam and Ng (1990) are also quite relevant here.

8.4.2.1 Preliminaries

Having recorded $X_{i1},\ldots,X_{in_i}$ from π_i with $n_i \geq 2$, we write

$$\tilde{X}_i(n_i) = \min\{X_{i1},\ldots,X_{in_i}\},$$

$$V_i(n_i) = (n_i - 1)^{-1} \Sigma_{j=1}^{n_i} (X_{ij} - \tilde{X}_i(n_i)),$$

$$T^*(\mathbf{n}) = \max_{1 \le i \le k} \tilde{X}_i(n_i), \tag{8.4.27}$$

where $\mathbf{n} = (n_1,...,n_k)$, $i = 1,...,k$. These notations are slightly different from those employed in (4.1.1). A natural selection procedure is simply to choose the population associated with the largest among minimum order statistics, that is, one implements the following selection procedure.

$$\mathcal{R}_{\mathbf{n}}(4, 2): \text{Select } \pi_j \text{ as the best population}$$

$$\text{if } \tilde{X}_j(n_j) = T^*(\mathbf{n}). \tag{8.4.28}$$

Let $\pi_{(i)}$ stand for the population associated with $\theta_{[i]}$ and suppose that the corresponding sample size, minimum order statistics and the estimator of scale parameter are respectively written as $n_{(i)}$, $\tilde{X}_{(i)}(n_{(i)})$ and $V_{(i)}(n_{(i)})$.

8.4.2.2 Two–Stage Procedure

Since $\sigma_1,...,\sigma_k$ are all unknown, we need to carry out sampling in at least two stages. The following methodology and associated results were put forth in Mukhopadhyay and Hamdy (1984a).

Let $m_1(\ge 2)$ be the starting sample size from each π and we write $h_6 = h_6(m_1, k, P^*)$ be a positive constant, to be defined specifically a little later. Define

$$N_i = \max\{m_1, [h_6 V_i(m_1)\delta^{*-1}]^* + 1\}, \tag{8.4.29}$$

$i = 1,...,k$. If $N_i = m_1$, we do not take any more samples from π_i. However, if $N_i > m_1$, then we sample the difference $(N_i - m_1)$ from π_i. Based on the totality of all samples $X_{i1},...,X_{in_i}$ from π_i, we now implement the selection rule $\mathcal{Q}_{\underset{\sim}{N}}(4, 2)$ defined in (8.4.28) where $\underset{\sim}{N} = (N_1,...,N_k)$.

Theorem 8.4.2. *For the two-stage procedure (8.4.29), with* h_6 *defined in (8.4.37), we have* $P(CS) \geq P^*$ *for all* $\underset{\sim}{\theta}\epsilon\Omega(\delta^*)$ *and all unknown* $\sigma_1,...,\sigma_k$.

Proof. Obviously one can write (with $\delta_i = \theta_{[k]} - \theta_{[i]}$)

$$P(CS) = P\{\tilde{X}_{(i)}(n_{(i)}) < \tilde{X}_{(k)}(n_{(k)}), i = 1,...,k-1\}$$

$$= P\{[\tilde{X}_{(i)}(n_{(i)}) - \theta_{[i]}] < [\tilde{X}_{(k)}(n_{(k)}) - \theta_{[k]}] + \delta_i, \quad i = 1,...,k-1\}. \tag{8.4.30}$$

Now, one has $\delta_i \geq \delta^*$ for all $i = 1,...,k-1$ whenever $\underset{\sim}{\theta}\epsilon\Omega(\delta^*)$, and hence from (8.4.30) one can write

$$P(CS) \geq P\{[\tilde{X}_{(i)}(n_{(i)}) - \theta_{[i]}] < [\tilde{X}_{(k)}(n_{(k)}) - \theta_{[k]}] + \delta^*, \quad i = 1,...,k-1\}, \tag{8.4.31}$$

for every $\underset{\sim}{\theta}\epsilon\Omega(\delta^*)$, and it follows that

$$\underset{\underset{\sim}{\theta}\epsilon\Omega(\delta^*)}{\text{Inf}} P(CS) = P\{[\tilde{X}_{(k)}(n_{(k)}) - \theta_{[k]}] + \delta^* > [\tilde{X}_{(i)}(n_{(i)}) - \theta_{[i]}], \quad i = 1,...,k-1\}, \tag{8.4.32}$$

while this infimum is attained when $\delta_i = \delta^*$ for all $i = 1,...,k-1$, that is if $\theta_{[1]} = \cdots = \theta_{[k-1]} = \theta_{[k]} - \delta^*$, a parameter configuration that is referred to as the least favorable configuration (LFC). From this point onwards, we work only under the LFC.

Let A_i^* be the event that $[\tilde{X}_{(k)}(n_{(k)}) - \theta_{[k]}] + \delta^* > [\tilde{X}_{(i)}(n_{(i)}) - \theta_{[i]}]$ for $i = 1,\ldots,k-1$. Next, note that $I(\bigcap_{i=1}^{k} N_i = n_i)$ and $(\tilde{X}_1(n_1),\ldots,\tilde{X}_k(n_k))$ are independent for all fixed $\underset{\sim}{n}$. Thus, for all $i = 1,\ldots,k-1$, one has

$$P(A_i^*)=E[1-N_{(k)}\sigma_{(i)}\{N_{(k)}\sigma_{(i)}+ N_{(i)}\sigma_{(k)}\}^{-1}\exp\{-N_{(i)}\delta^*/\sigma_{(i)}\}]$$
$$\geq E[1 - \exp\{-N_{(i)}\delta^*/\sigma_{(i)}\}]. \tag{8.4.33}$$

Now, utilize Bonferroni's inequality, and combine (8.4.32) – (8.4.33) to write

$$P(CS) \geq P(A_1^* \cap \cdots \cap A_{k-1}^*)$$
$$\geq \Sigma_{i=1}^{k-1} P(A_i^*) - (k-2). \tag{8.4.34}$$

Next, one combines (8.4.33) – (8.4.34) and hence

$$P(CS) \geq 1 - \Sigma_{i=1}^{k-1} E[\exp\{-h_6 V_{(i)}(m_1)/\sigma_{(i)}\}], \tag{8.4.35}$$

since $N_{(i)} \geq h_6 V_{(i)}(m_1)\delta^{*-1}$ w.p. 1 for all $i = 1,\ldots,k-1$. Now, one simply observes that $(2m_1 - 2)\, V_{(i)}(m_1)/\sigma_{(i)}$ are all i.i.d. $\chi^2_{2m_1-2}$ random variables and plug in the moment generating function of $\chi^2_{2m_1-2}$ in (8.4.35) in order to obtain

$$P(CS) \geq 1 - (k-1)\{1 + h_6(m_1-1)^{-1}\}^{-(m_1-1)}. \tag{8.4.36}$$

Now, determine $h_6 = h_6(m_1, k, P^*)$ such that

$$h_6 = (m_1-1)\,[\{(1-P^*)^{-1}(k-1)\}^{1/(m_1-1)}-1], \tag{8.4.37}$$

and the proof will be complete in view of (8.4.36). □

Remark 8.4.2. We should emphasize that the two–stage methodology (8.4.29) can be implemented without the help of any specialized table since we have an explicit expression for h_6 available in (8.4.37).

Remark 8.4.3. It is interesting to note that h_6 can also be viewed in a somewhat different way. From (8.4.35), one can write

$$P^* = (2-k) + (k-1)E[1 - \exp\{-h_6 V_{(1)}(m_1)/\sigma_{(1)}],$$

and hence h_6 can be determined from

$$P\{0 < F_{2,2m_1-2} < h_6\} = (P^*+k-2)/(k-1). \tag{8.4.38}$$

For the two–stage procedure (8.4.29) one can show that

$$\lim E(N_i/n_i^o) = h_6 a_o^{*-1} \text{ as } \delta^* \to 0, \tag{8.4.39}$$

where $n_i^o = a_o^* \sigma_i/\delta^*$ and $a_o^* = \ln\{(k-1)/(1-P^*)\}$, $i = 1,\ldots,k$. These n_i^o's can be thought of as being the optimal fixed sample sizes required from π_i had σ_i's been known. Combine (8.4.37) – (8.4.38) and conclude that $h_6 = h_6(m_1)$, for fixed k and P^*, is a strictly decreasing function of m_1 and $\lim h_6(m_1) = a_o^*$ as $m_1 \to \infty$. In other words, the limiting ratio in (8.4.39) is always larger than unity, that is the two–stage procedure (8.4.29) oversamples compared with n_i^o's, even asymptotically. An *asymptotically first–order efficient modified two–stage* procedure was also proposed in section 2.1 of Mukhopadhyay and Hamdy (1984a), followed by moderate sample comparisons of the two–stage procedure (8.4.37) with that of Mukhopadhyay (1984).

8.5 Subset Selection of Good Populations

In this section, we address a variation of Gupta's (1956) subset selection formulation. One will recall our discussions from section 1.2.1.2. As in Chapter 3, we let $\pi_1,...,\pi_k$ denote $k(\geq 2)$ independent normal populations with unknown means $\mu_1,...,\mu_k$ and a common variance σ^2 which may be known or unknown. Let $\mu_{[1]} \leq \cdots \leq \mu_{[k]}$ denote the ordered μ_i's. Any population associated with $\mu_{[k]}$ is called the *best* population. The classical subset selection procedures $\mathscr{P}_n^*$ and $\tilde{\mathscr{P}}_n^{**}$, defined in (1.2.8) and (1.2.9) respectively, of Gupta (1956) were proposed for selecting a nonempty subset of the k given populations so that the selected subset contains the best population with a guaranteed minimum probability P*. In the case of known as well as unknown σ^2, Gupta proposed a single–stage procedure (depending on the case) which is defined for samples of a common arbitrary size n taken from each population. This is in contrast with the situation using the *indifference zone* formulation of Bechhofer (1954) where a single–stage procedure exists only when σ^2 is known. In Gupta's subset formulation, the expected size of the selected subset serves as an operating characteristic for the performance of any *valid* decision rule that guarantees a specified minimum P(CS), that is, selection of any subset containing the best population. There are several variations of this basic goal. In selecting the best population, one may be interested in selecting *good* populations, namely, populations with $\mu_i \geq \mu_{[k]} - \delta^*$, where $\delta^* > 0$ is a specified constant. A *correct selection* in this case can be defined, for example, as the selection of a nonempty subset containing all *good* populations, or as that of a nonempty subset which excludes all bad (that is, not good) populations. The P(CS) is controlled at a pre–specified level P*. Such a formulation has been considered among others by Desu (1970) and Carroll et al. (1975). Good and bad populations have been defined by Lehmann (1961) in terms of a standard or control population. To specify one of several possibilities in Lehmann's formulation, we can control the probability of selecting at least one good population and minimize the expected number of bad populations included in the selected subset. In all these above formulations, a *single–stage procedure* can be defined.

Considering good and bad populations defined earlier in terms of $\mu_{[k]}$, Santner (1976) studied selecting a nonempty subset by controlling the P(CS) (CS here stands for selecting a subset that contains no bad populations) *and* also the expected proportion of good populations selected. This cannot be achieved by a single–stage procedure when σ^2 is unknown. Santner (1976) considered a *two–stage procedure.* A similar problem in the case of two–parameter exponential populations has been studied by Liang and Panchapakesan (1993).

In this section, we continue with our earlier definition of a good population with respect to the largest mean. We wish to construct a subset selection rule with a guarantee that the expected proportion of good populations selected is at least p, where $0 < p < 1$ is preassigned. It should be noted that there is at least one good population among the k given populations and that we are not concerned with correct selection just defined as such in the traditional sense. Here, the material is adapted from the recent technical report of Mukhopadhyay and Panchapakesan (1992). We first consider (section 8.5.1) the case of known σ^2. In this case, we propose a single–stage procedure. When σ^2 is unknown, it is not possible to predetermine the sample size required in order to control the expected proportion of good populations selected, and hence we consider various multistage procedures (section 8.5.2) in this situation. Specifically, we propose *purely sequential, accelerated sequential* and *three–stage methodologies* (section 8.5.2) and address different types of associated *second–order asymptotics* for several characteristics, such as the P(CS) and average sample size. Such second–order asymptotic theories are well established in the area of selection and ranking under the indifference zone formulation of Bechhofer (1954) as can be seen from Chapter 3. The thrust of the present discussions lie in formulating and studying such *non–elimination type* methodologies in the area of *subset selection* by developing appropriate notions of second–order asymptotics. It should be emphasized that similar problems do not arise in the classical formulation of Gupta (1956) because there a single–stage selection procedure can be found even in the case when σ^2 is unknown.

8.5.1 KNOWN COMMON VARIANCE CASE

Let us suppose that $X_{i1}, X_{i2}, \ldots$ is a sequence of i.i.d. random variables available from π_i, $i = 1,\ldots,k$. Having recorded $X_{i1},\ldots,X_{in}$ from π_i, recall from (3.1.1) that we write $\overline{X}_{in} = n^{-1} \Sigma_{j=1}^{n} X_{ij}$, $S_{in}^2 = (n-1)^{-1}\Sigma_{j=1}^{n} (X_{ij} - \overline{X}_{in})^2$, and $U(n) = k^{-1} \Sigma_{i=1}^{k} S_{in}^2$, for $n \geq 2$ and $i = 1,\ldots,k$. Note that $U(n)$ stands for the usual unbiased estimator of σ^2 if σ^2 is assumed unknown. Now, having $\overline{X}_{in}$ from π_i, $i = 1,\ldots,k$, we propose the subset selection rule

$$\mathcal{R}_n(5,1) : \text{Select } \pi_i \text{ if } \overline{X}_{in} \geq T_n - (\delta^* + d), \tag{8.5.1}$$

where $T_n = \max\{\overline{X}_{1n},\ldots,\overline{X}_{kn}\}$ and $d(>0)$ is to be determined so that the expected proportion of *good* populations selected is at least p, $0 < p < 1$ being preassigned, as one will recall.

Define

$$\Omega_i(\delta^*) = \begin{cases} \{\underset{\sim}{\mu} = (\mu_1,\ldots,\mu_k) \mid \mu_{[k-i+1]} \geq \mu_{[k]} - \delta^* > \mu_{[k-i]}\}, & i=1,\ldots,k-1 \\ \{\underset{\sim}{\mu} = (\mu_1,\ldots,\mu_k) \mid \mu_{[1]} \geq \mu_{[k]} - \delta^*\}, & i = k. \end{cases} \tag{8.5.2}$$

Also, let $S(n)$ denote the number of good populations included in the selected subset and let $E^{(i)}[S(n)]$ denote the expected value of $S(n)$ for $\underset{\sim}{\mu} \in \Omega_i(\delta^*)$. Finally, let $\overline{X}_{(in)}$ denote the $\overline{X}_{jn}$ associated with $\mu_{[i]}$, $i = 1,\ldots,k$. Setting $D = \delta^* + d$, it is easy to see that

$$E^{(i)}[S(n)] = \Sigma_{j=k-i+1}^{k} P\{\overline{X}_{(jn)} \geq \max_t \overline{X}_{(tn)} - D\}$$

$$= \Sigma_{j=k-i+1}^{k} P\{n^{\frac{1}{2}}(\overline{X}_{(jn)} - \mu_{[j]}) \geq n^{\frac{1}{2}}(\overline{X}_{(tn)} - \mu_{[t]})$$

$$+ n^{\frac{1}{2}}(\mu_{[t]} - \mu_{[j]}) - n^{\frac{1}{2}}D,\ t = 1,\ldots,k;\ t \neq j\}$$

$$= \Sigma_{j=k-i+1}^{k} P\{Z_j \geq Z_t + n^{\frac{1}{2}}(\mu_{[t]} - \mu_{[j]} - D)\sigma^{-1}, t \neq j\}, \tag{8.5.3}$$

where Z's are i.i.d. N(0, 1) random variables. It is obvious that $E^{(i)}[S(n)]$ is nonincreasing in $\mu_{[s]}$, $s = 1,\ldots,k-i$, when all other $\mu_{[j]}$'s are kept fixed. Thus, the infimum of $E^{(i)}[S(n)]$ over $\Omega_i(\delta^*)$ occurs when $\mu_{[1]} = \cdots = \mu_{[k-i]}$. Now, following the calculus of Santner (1976, Section 3), one can show that $dE^{(i)}[S(n)]/d\mu_{[k-i+1]} \geq 0$. By observing the symmetry of $E^{(i)}[S(n)]$ in $(\mu_{[k-i+1]},\ldots,\mu_{[k]})$, we conclude, as Santner (1976) does, that $E^{(i)}[S(n)]$ is nondecreasing in any $\mu_{[s]}$, $s = k-i+1,\ldots,k$. For any fixed $\mu_{[k]}$, it now follows that the infimum of $E^{(i)}[S(n)]$ occurs when $\mu_{[1]} = \cdots = \mu_{[k-1]} = \mu_{[k]} - \delta^*$, a configuration independent of i. Since the expression (8.5.3) for $E^{(i)}[S(n)]$ depends only on the differences between $\mu_{[i]}$'s, the infimum can be evaluated at $\underset{\sim}{\mu}^{0}(\delta^*) = (0,\ldots,0,\delta^*)$. Thus, by a straightforward evaluation, we have

$$E[S(n) \mid \underset{\sim}{\mu}^{0}(\delta^*), \sigma^2]$$

$$= (i-1)\int_{-\infty}^{\infty} \Phi^{k-2}[t + n^{\frac{1}{2}}D\sigma^{-1}]\, \Phi[t + n^{\frac{1}{2}}(D-\delta^*)\sigma^{-1}]\, \phi(t)\, dt$$

$$+ \int_{-\infty}^{\infty} \Phi^{k-1}[t + n^{\frac{1}{2}}(D+\delta^*)\sigma^{-1}]\, \phi(t)\, dt. \tag{8.5.4}$$

Denoting the two integrals in (8.5.4) by $I_1(n)$ and $I_2(n)$, respectively, one obtains

$$E[S(n) \mid \underset{\sim}{\mu}^{o}(\delta^*), \sigma^2] = (i-1) I_1(n) + I_2(n). \qquad (8.5.5)$$

Now, let $P(n)$ denote the proportion of good populations included in the selected subset. It is required that

$$E[P(n) \mid \underset{\sim}{\mu}, \sigma^2] \geq p \text{ for all } \underset{\sim}{\mu} \text{ and } \sigma^2, \qquad (8.5.6)$$

where $0 < p < 1$ is preassigned. Toward this end, note that

$$\min_{1 \leq i \leq k} \inf_{\Omega_i(\delta^*)} E[P(n) \mid \underset{\sim}{\mu}, \sigma^2]$$

$$= \min_{1 \leq i \leq k} i^{-1} \inf_{\Omega_i(\delta^*)} E^{(i)}[S(n) \mid \underset{\sim}{\mu}, \sigma^2]$$

$$= \min_{1 \leq i \leq k} \{(i-1)\, I_1(n) + I_2(n)\}/i$$

$$= \min_{1 \leq i \leq k} [I_1(n) + \{I_2(n) - I_1(n)\}/i]. \qquad (8.5.7)$$

Observe that $I_2(n) - I_1(n) \geq 0$ and that this difference converges to zero as $n \to \infty$, using the inequality

$$I_1(n) \geq \int_{-\infty}^{\infty} \Phi^{k-1}(t + n^{\frac{1}{2}} d\sigma^{-1})\, \phi(t)\, dt. \qquad (8.5.8)$$

Combining (8.5.7) and (8.5.8), we can meet the requirement (8.5.6) in a conservative fashion by choosing $d = h\sigma/n^{\frac{1}{2}}$ and $h = h(k, p)$ is given by

$$g_0(h) = \int_{-\infty}^{\infty} \Phi^{k-1}(t + h)\, \phi(t)\, dt = p. \qquad (8.5.9)$$

The function $g_0(\cdot)$ was also defined right before Theorem 3.5.2. The values of h or a known multiple of it have been tabulated for selected values of k and p (P* in the tables) by Bechhofer (1954), Gupta (1963), and Gupta et al. (1973). Note that (8.5.9) is the same as that in (3.2.4), replacing P* by p. See also Table A.1 in Gibbons et al. (1977).

In the above discussion, we note that a conservative choice of $d = d(n)$ can be found for any given sample size n. On the other hand, if we choose a value of d arbitrarily, then we can find the smallest sample size $n = n(d)$ such that

$$n \geq [h\sigma/d]^2 = n^*, \text{ say,} \tag{8.5.10}$$

and use the same decision rule given by (8.5.1) with $D = d + \delta^*$. Of course, in this case, the rule becomes

$$\mathcal{Q}_n(5, 2) : \text{Select } \pi_i \text{ if } \overline{X}_{in} \geq T_n - D. \tag{8.5.11}$$

However, when σ^2 is unknown, we cannot implement either of the subset selection rules given by (8.5.1) and (8.5.11) because we cannot determine either n(d) or d(n) (when the other is given) needed to meet the requirement (8.5.6). In the next section, we introduce several types of multistage subset selection procedures in the *unknown variance case.*

8.5.2 UNKNOWN COMMON VARIANCE CASE

Suppose that we start with $d(>0)$ and $\delta^*(>0)$, and implement the subset selection procedure $\mathcal{Q}_n(5, 2)$ given by (8.5.11) with $D = \delta^* + d$. Then, $E[P(n) \mid \underset{\sim}{\mu}, \sigma^2] \geq p$ for all $\underset{\sim}{\mu}$ and σ^2 when $0 < p < 1$ is preassigned if n is the smallest integer $\geq n^*$, given by (8.5.10). However, n^* is unknown since σ^2 is unknown.

Let us suppose in general that one devises a statistical methodology giving rise to a positive integer valued random variable N estimating n^*

and implements the rule

$$\mathcal{R}_N(5, 2) : \text{Select } \pi_i \text{ if } \overline{X}_{iN} \geq T_N - D, \tag{8.5.12}$$

where $T_N = \max\{\overline{X}_{1N},...,\overline{X}_{kN}\}$.

Theorem 8.5.1: *Suppose that the stopping variable* N *satisfies the following conditions: The distribution of* N *does not depend on* $\underset{\sim}{\mu}$; $I(N = n)$ *is independent of* $(\overline{X}_{1n},...,\overline{X}_{kn})$ *for every fixed* $n \geq 2$; *and* $P(N < \infty) = 1$. *Then*

$$E[P(N) \mid \underset{\sim}{\mu}, \sigma^2] \geq E[f(ND^2/\sigma^2)], \tag{8.5.13}$$

for all $\underset{\sim}{\mu}$ *and* σ^2, *where*

$$f(y) = g_o(y^{\frac{1}{2}}), \quad y > 0, \tag{8.5.14}$$

$g_o(\cdot)$ *is given by (8.5.9) which is same as that in Chapter 3.*

Proof. As before, $S(n)$ denotes the number of good populations in the selected subset. Let $m_1(\geq 2)$ be the starting sample size. Then

$$E^{(i)}[S(N)] = \Sigma_{n=m_1}^{\infty} E^{(i)}[S(N) \mid N = n] \; P(N = n)$$

$$= \Sigma_{n=m_1}^{\infty} E^{(i)}[S(n)] \; P(N = n),$$

in view of the assumed conditions. Now, (8.5.5) leads to

$$E[S(N) \mid \underset{\sim}{\mu}^o(\delta^*), \sigma^2] = (i-1) \; E[I_1(N)] + E[I_2(N)].$$

Hence, one has

$$\min_{1\le i\le k} \inf_{\Omega_i(\delta^*)} E[P(N)\,|\,\underset{\sim}{\mu}, \sigma^2] = \min_{1\le i\le k} i^{-1} \inf_{\Omega_i(\delta^*)} E^{(i)}[S(N)\,|\,\underset{\sim}{\mu}, \sigma^2]$$

$$= \min_{1\le i\le k} [E\{I_1(N)\} + E\{I_2(N) - I_1(N)\}/i] \ge E[I_1(N)] \ge E[f(Nd^2/\sigma^2)]. \quad \square$$

Now, we introduce a few multistage subset selection methodologies for which we can claim *asymptotic* (as $d \to 0$) *second-order properties.*

8.5.2.1 Purely Sequential Procedure

We start with $m_1(\ge 2)$ samples from each π_i, and then we take one sample at a time from each π_i according to the following stopping rule:

$$N = N(d) = \inf\{n \ge m_1 : n \ge h^2 U(n)/d^2\}. \tag{8.5.15}$$

Obviously, conditions of Theorem 8.5.1 hold and one implements the subset selection procedure $\mathcal{Q}_N(5, 2)$ given by (8.5.12).

Now, let

$$\nu = (k+2)/2 - \Sigma_{n=1}^{\infty} n^{-1} E[\max(0, \chi^2_{nk} - 2nk)], \; \beta = k^{-1}(\nu - 2), \tag{8.5.16}$$

as in (3.5.7) and Theorem 3.5.2.

Theorem 8.5.2. *For the purely sequential subset selection procedure with the stopping rule (8.5.15), we have as* $d \to 0$:

(i) $E(N) = n^* + \beta + o(1)$ *if* (a) $m_1 \ge 3$ *when* $k = 2$ *and* (b) $m_1 \ge 2$ *when* $k \ge 3$;

(ii) $E[P(N)\,|\,\underset{\sim}{\mu}, \sigma^2] \ge p + h^2 n^{*-1}\{\beta f'(h^2) + h^2 f''(h^2) k^{-1}\} + o(d^2)$ *if*

(a) $m_1 \geq 4$ *when* $k = 2$, (b) $m_1 \geq 3$ *when* $k = 3, 4, 5$, *and* (c) $m_1 \geq 2$ *when* $k \geq 6$;

for all $\underset{\sim}{\mu}$ *and* σ^2, *where* β *is given in (8.5.16).*

Proof. First observe that one can write

$$N = \inf\{n \geq m_1 \ : \ kn(n-1)/n^* \geq \Sigma_{i=1}^{n-1} W_i\}$$

$$= 1 + \inf\{n \geq m_1-1 \ : \ kn^{*-1}n^2(1 + n^{-1}) \geq \Sigma_{i=1}^{n-1} W_i\}, \qquad (8.5.17)$$

where W's are i.i.d. χ_k^2. The representation (8.5.17) is similar to that considered in the proof of Theorem 3.5.2. The rest is omitted for brevity. □

8.5.2.2 Accelerated Sequential Procedure

In terms of implementation, the purely sequential procedure (8.5.15) can turn out to be fairly inconvenient. Let us however attempt to make the purely sequential procedure operationally more convenient by accelerating the sampling process. In this methodology, we use $U^*(n) = n(n-1)^{-1}U(n)$ as the estimator of σ^2.

Let us choose and fix $q^*(\geq 0)$ and $0 < \rho < 1$. In order to cut sampling operations, we estimate only a fraction of n^* first by means of purely sequential sampling. We start with $m_1(\geq 2)$ samples from each π_i. Then, we take one sample at a time from each π_i according to the following stopping rule:

$$t = t(d) = \inf\{n \geq m_1 : n \geq \rho h^2 U^*(n)/d^2\}, \qquad (8.5.18)$$

and end up with $X_{i1},...,X_{it}$ from π_i, $i = 1,...,k$. Note that $t(d)$ estimates ρn^*, a fraction of the "optimal", although conservative, fixed sample size n^*. Next, we define

$$N_1 = N_1(d) = [h^2 \upsilon^*(t)\, d^{-2} + q^*]^* + 1, \tag{8.5.19}$$

and

$$N = N(d) = \max\{t(d), N_1(d)\}. \tag{8.5.20}$$

Note that $N_1(d)$ estimates n^* based on t samples from each π_i. If $N = t$, we do not take any more samples. If $N > t$, then we sample the difference $(N - t)$ from each π_i in a single batch at this stage. Now, based on $X_{i1},\ldots,X_{iN}$ from π_i, $i = 1,\ldots,k$, we implement the subset selection procedure $\mathcal{Q}_N(5, 2)$ given by (8.5.12). Such methodologies, one will recall, are known as accelerated sequential procedures and the associated unified theory has been developed in section 2.5. These types of procedures have earlier been introduced in section 3.6 for the indifference zone formulation.

Theorem 8.5.3. *For the accelerated sequential subset selection procedure with stopping rule given by (8.5.18) – (8.5.20), we have as* $d \to 0$:

(i) $q^* - 2(k\rho)^{-1} \le \liminf E(N - n^*) \le \limsup E(N - n^*) \le q^* - 2(k\rho)^{-1} + 1$ *if* $m_1 \ge 4$;

(ii) $E[P(N) \mid \underset{\sim}{\mu}, \sigma^2] \ge p + o(d^2)$ *if* $q^* = \max\{0, 2(k\rho)^{-1} - h^2 f''(h^2) \times [k^2 \rho f'(h^2)]^{-1}\}$ *with* (a) $m_1 \ge 5$ *when* $k = 2$, *and* (b) $m_1 \ge 4$ *when* $k \ge 3$;

for all $\underset{\sim}{\mu}$ *and* σ^2.

Its proof is similar to that of Theorem 3.6.2 and hence it is omitted.

8.5.2.3 Three–Stage Procedure

In this methodology, we attempt to cut sampling operations almost to a minimum and yet maintain the appropriate second–order asymptotic properties. Towards this end, we estimate a fraction of n^* first and then n^* successively by means of batch sampling. We choose and fix $0 < \rho < 1$ and let

$$\beta' = \tfrac{1}{2} - 2(k\rho)^{-1},\ m_1^* = -\beta' - h^2\rho^{-1}\{f''(h^2)/f'(h^2)\}.$$

We start with $m_1(\geq 2)$ samples from each π_i, assuming that $m_1 = m_1(d) = 0(n^{*1/r})$ for some $r > 1$. Define

$$T = T(d) = \max\{m_1, [\rho h^2 U(m_1)d^{-2}]^* + 1\},$$

and

$$N = N(d) = \max\{T, [h^2 U(T)d^{-2} + m_1^*]^* + 1\}. \tag{8.5.21}$$

If $T = m_1$, then we do not sample any more from any population at the second stage. However, if $T > m_1$, then we sample the difference at the second stage to extend $X_{i1},\ldots,X_{im_1}$ to $X_{i1},\ldots,X_{iT}$ from π_i, $i = 1,\ldots,k$. Next, based on T samples from each population, we determine $U(T)$ and thus determine N. Note that T and N, respectively, estimate ρn^* and n^*. If $N = T$, we do not sample any more from any population at the third stage. However, if $N > T$, then we sample the difference at the third stage in order to extend $X_{i1},\ldots,X_{iT}$ to $X_{i1},\ldots,X_{iN}$ from π_i, $i = 1,\ldots,k$. Now, based on $X_{i1},\ldots,X_{iN}$ from π_i, one implements the subset selection procedure $\mathcal{Q}_N(5, 2)$ given by (8.5.12). Such three–stage procedures have earlier been introduced in section 3.7 for the indifference zone formulation. The unified theory of three–stage sampling techniques and associated second–order asymptotics were developed in section 2.3.

Theorem 8.5.4. *For the three–stage subset selection procedure with the stopping rule given by (8.5.21), we have as* $d \to 0$:

(i) $E(N) = n^* - h^2\rho^{-1}\{f''(h^2)/f'(h^2)\} + o(1)$;

(ii) $E[P(N)\,|\,\underset{\sim}{\mu}, \sigma^2] \geq p + o(d^2)$;

for all $\underset{\sim}{\mu}$ *and* σ^2.

Its proof is similar to that of Theorem 3.7.2(i) and Theorem 3.7.3, and hence omitted.

8.6 NONPARAMETRIC SELECTION PROCEDURES

In Chapters 3 and 4, we had discussed problems associated with the selection of the largest location parameter under the indifference zone formulation, but under the specific parametric models of normal and two–parameter negative exponential distributions. Because of certain pivotal and independence properties of relevant estimators, in both these problems we could write P(CS) as the expectation of certain known but complicated functions of the appropriate stopping variables. This was really the key. Now, we relax the special distributional assumptions and briefly describe some nonparametric selection procedures for the *largest center of symmetry.* The methodologies cover quite some ground and these are also fairly general in nature, but one should note that at this time, *second–order asymptotic characteristics* of selection procedures considered here are yet *unknown.* We review only a few available techniques and present appropriate *first–order asymptotics* of such methodologies. Some of the issues related to moderate sample size performances are briefly addressed in section 8.6.4.

8.6.1 PRELIMINARIES

Let $\pi_1,\ldots,\pi_k$ denote $k(\geq 2)$ independent populations with distribution functions $F_1^0(x - \Delta_1),\ldots,F_1^0(x - \Delta_k)$ respectively where $F_1^0(x)$ is continuous and *symmetric* about $x = 0$. Thus, $\Delta_1,\ldots,\Delta_k$ are the *centers of symmetry* of the respective distributions. We assume that the function F_1^0 and the location parameters $\Delta_1,\ldots,\Delta_k$ are unknown and we denote by $\Delta_{[1]} \leq \cdots \leq \Delta_{[k]}$ the ordered Δ_i values. The population associated with $\Delta_{[k]}$ is referred to as the *best* population and we assume that there is only one like that.

In Chapter 3, we had $F_1^0(x) = \Phi(x)$ and $\Delta_i = \mu_i$, $i = 1,\ldots,k$, if in addition one also assumes $\sigma = 1$ in that formulation. The distributions considered in Chapter 4 do not satisfy the present requirements. The problem considered here is to select the best population in such a way that P(CS) is approximately P* whenever $\Delta_{[k]} - \Delta_{[k-1]} \geq \delta^*$, where one has $\delta^*(>0)$ and $k^{-1} < P^* < 1$, two *preassigned* constants. Let us write the preference zone as

$$\Omega(\delta^*) = \{\underset{\sim}{\theta} = (\Delta_1,\ldots,\Delta_k) : \Delta_{[k]} - \Delta_{[k-1]} \geq \delta^*\}, \tag{8.6.1}$$

and let $\mathscr{F}$ denote a subclass of continuous and symmetric (around zero) distributions satisfying certain regularity conditions to be specified in the sequel. Purely sequential and accelerated sequential selection procedures are described here for which we can conclude that

$$\liminf P(CS) \geq P^* \text{ whenever } \underset{\sim}{\theta} \in \Omega(\delta^*) \text{ and } F_1^0 \in \mathscr{F}, \tag{8.6.2}$$

as $\delta^* \to 0$.

Having recorded i.i.d. random variables $X_{i1},\ldots,X_{in}$ from π_i, we propose to estimate Δ_i by a suitable statistic $L_i(n)$, say, where it is assumed that $n^{\frac{1}{2}}\{L_i(n) - \Delta_i\} \xrightarrow{\mathscr{L}} N(0, A_1^{-2})$ as $n \to \infty$, for $i = 1,\ldots,k$, $F_1^0 \in \mathscr{F}$. Here $A_1 = A_1(F_1^0)$ is assumed to be a finite and positive functional of F_1^0. A natural decision rule will then be the following.

$\mathscr{R}_n(6, 1)$: Select π_j as the population associated with

$$\Delta_{[k]} \text{ if } L_{jn} = \max_{1 \leq i \leq k} L_{in}. \tag{8.6.3}$$

For such a selection procedure, P(CS) is asymptotically (as $n \to \infty$) at least P* if

$$n \geq h^2(A_1\delta^*)^{-2} = n^*, \text{ say} \tag{8.6.4}$$

where "h" is the solution of the integral equation

$$\int_{-\infty}^{\infty} \Phi^{k-1}(y + h)\, \phi(y)\, dy = P^*. \tag{8.6.5}$$

One will immediately recall that $h = h(k, P^*)$ is exactly the same as that obtained from (3.2.4). The infimum of the asymptotic value of P(CS) is then attained and equals the target, namely P^*, in $\Omega(\delta^*)$ when $\Delta_{[1]} = \cdots = \Delta_{[k-1]} = \Delta_{[k]} - \delta^*$. The parameter configuration $\Delta_{[1]} = \cdots = \Delta_{[k-1]} = \Delta_{[k]} - \delta^*$ is *tentatively* referred to as the least favorable configuration and it is clear that this particular configuration does not depend on any specific choice for F_1^o. That is, this least favorable configuration remains the same whatever $F_1^o \in \mathcal{F}$. In this sense, the present methodologies are nonparametric in nature.

Actually, n^* given by (8.6.4) is unknown because it depends on an unknown functional of $F_1^o \in \mathcal{F}$. In the spirit of Robbins et al. (1968), a purely sequential selection procedure was introduced in Geertsema (1972). See also Swanepoel (1977). Mukhopadhyay and Solanky (1993) have recently devised an accelerated sequential selection procedure and made the original Geertsema (1972) procedure operationally more attractive. Before we elaborate on these methodologies and describe various *asymptotic* (as $\delta^* \to 0$) *first-order properties*, we list a few additional regularity conditions that $\mathcal{F}$ must satisfy, as specified in Geertsema (1972).

Assumptions: For all $\underset{\sim}{\theta} \in \Omega(\delta^*)$ and $F_1^o \in \mathcal{F}$, we assume that

(i) $n^{\frac{1}{2}}(L_i(n) - \Delta_i) = A_1^{-1} Z_i(n) + o(1)$ a.s. as $n \to \infty$ where $Z_i(n)$ is a standardized average of i.i.d. random variables having finite second moment and $0 < A_1 = A_1(F_1^o) < \infty$. (8.6.6)

(ii) S_n^2 is a suitable estimator of A_1^{-2} based on $X_{i1},\ldots,X_{in}$, $i = 1,\ldots,k$ such that $\lim S_n^2 = A_1^{-2}$ w.p. 1 as $n \to \infty$. (8.6.7)

(iii) The set $\{\delta^{*2}N(\delta^*) : \delta^* > 0\}$ is uniformly integrable where $N = N(\delta^*)$ is an appropriate stopping variable and the random sample size from each population π. (8.6.8)

Throughout this section, we work under these assumptions (i) – (iii) given by (8.6.6) – (8.6.8).

8.6.1.1 Purely Sequential Procedure

Geertsema (1972) proposed the following purely sequential selection procedure. We have mentioned in (8.6.7) that S_n^2 is a suitable strongly consistent estimator of A_1^{-2} based on $X_{i1},...,X_{in}$, $i = 1,...,k$. Along the lines of Srivastava (1966) and Robbins et al. (1968), Geertsema (1972) had proposed the following purely sequential selection methodology.

One starts with $m_1(\geq 2)$ samples from each π. Then, we keep taking one more sample at a time from each π according to the following stopping rule:

$$N = N(\delta^*) = \inf\{n \geq m_1 : n \geq h^2 S_n^2 \delta^{*-2}\}. \tag{8.6.9}$$

Under assumptions (8.6.6) – (8.6.7), it is easily seen that $P(N < \infty) = 1$ for fixed $\underset{\sim}{\theta}$, A_1, m_1 and δ^*, in other words, the purely sequential procedure (8.6.9) terminates with probability one. Based on the totality of all samples, that is having $X_{i1},...,X_{iN}$ from π_i, $i = 1,...,k$, one now implements the selection rule $\mathcal{Q}_N(6, 1)$ given by (8.6.3). Some of the basic first-order asymptotics are described in Theorem 8.6.1 in Section 8.6.2.

8.6.1.2 Accelerated Sequential Procedure

Since one by one purely sequential sampling rule (8.6.9) of Geertsema (1972) is often operationally inconvenient, Mukhopadhyay and Solanky (1993) set out to devise an accelerated version of the same. We still wish to have our final sample sizes from each π close to the unknown n^*, but we wish to reach there in fewer steps. Towards this end, we first choose and fix $0 < \rho < 1$ and let

$$t = t(\delta^*) = \inf\{n \geq m_1 : n \geq \rho h^2 S_n^2 \delta^{*-2}\}, \tag{8.6.10}$$

$$N_1 = N_1(\delta^*) = [h^2 S_t^2 \delta^{*-2}]^* + 1, \tag{8.6.11}$$

$$N = N(\delta^*) = \max\{t(\delta^*), N_1(\delta^*)\}. \tag{8.6.12}$$

We start with $m_1(\geq 2)$ samples from each π. Then, we take one sample at a time from each π sequentially and end up with $X_{i1},...,X_{it}$ from π_i, $i = 1,...,k$. Note that $t(\delta^*)$ estimates ρn^*, a fraction of n^*. Now, $N_1(\delta^*)$ estimates n^* based on all t samples from each π. If $N = t$, then we do not take any more samples. If $N > t$, we sample the difference $(N - t)$ from each π in *one single batch* at this stage. This way, we save approximately $100(1 - \rho)\%$ of the sampling operations needed by the purely sequential rule (8.6.9). Once we determine N by means of (8.6.10) – (8.6.12), based on the totality of all samples, that is having $X_{i1},...,X_{iN}$ from π_i, $i = 1,...,k$, one now implements the selection rule $\mathcal{Q}_N(6, 1)$ given by (8.6.3). In the following section, some of the associated basic *first–order asymptotic properties* are summarized.

8.6.2 FIRST–ORDER ASYMPTOTICS

The first–order asymptotic characteristics of the purely sequential and accelerated sequential selection methodologies are summarized as follows.

Theorem 8.6.1. *Assume (i) – (iii) given by (8.6.6) – (8.6.8). For both purely sequential procedure (8.6.9) and the accelerated sequential procedure (8.6.10) – (8.6.12), for all* $\underset{\sim}{\theta} \in \Omega(\delta^*)$ *and* $F_1^0 \in \mathcal{F}$, *we have as* $\delta^* \to 0$:

(i) $N \to \infty$ *w.p.* 1 *and* $E(N) \to \infty$;

(ii) $N/n^* \to 1$ *w.p.* 1;

(iii) $E(N/n^*) \to 1$;

(iv) $\liminf P(CS) \geq P^*$;

where $n^* = h^2(A_1\delta^*)^{-2}$ *given by (8.6.4).*

Proof. <u>The Purely Sequential Case</u>

First note that $N(\delta_1^*) \geq N(\delta_2^*)$ w.p. 1 if $0 < \delta_1^* < \delta_2^*$, that is $N(\delta^*)$ is nondecreasing in δ^*. Now, (8.6.6) – (8.6.7) will lead to first assertion in part (i). The second assertion in part (i) now follows by applying the monotone convergence theorem.

The basic inequality (3.5.2) in the present situation reduces to

$$h^2 S_N^2 \delta^{*-2} \leq N \leq m_1 + h^2 S_{N-1}^2 \delta^{*-2}, \tag{8.6.13}$$

and now part (ii) follows by dividing (8.6.13) throughout by means of n^* and taking limits.

In view of assumption (iii) given by (8.6.8), part (iii) is immediate from part (ii). Now, we turn to prove part (iv). For notational convenience, let us write $\pi_{(i)}$ for the population associated with $\Delta_{[i]}$ and suppose $L_{(i)}(N)$ is the corresponding statistic. Now, for $\underset{\sim}{\theta}\in\Omega(\delta^*)$, we have

$$\begin{aligned} P(CS) &= P\{L_{(i)}(N) \leq L_{(k)}(N),\ i=1,\ldots,k-1\} \\ &= P\{n^{*\frac{1}{2}}(L_{(i)}(N) - \Delta_{[i]}) \leq n^{*\frac{1}{2}}(L_{(k)}(N) - \Delta_{[k]}) \\ &\qquad + n^{*\frac{1}{2}}(\Delta_{[k]} - \Delta_{[i]}),\ i = 1,\ldots,k-1\} \end{aligned}$$

$$\geq P\{n^{*\frac{1}{2}}(L_{(i)}(N)-\Delta_{[i]}) \leq n^{*\frac{1}{2}}(L_{(k)}(N)-\Delta_{[k]}) + n^{*\frac{1}{2}}\delta^*,$$
$$i = 1,\ldots,k-1\}. \tag{8.6.14}$$

Equation (8.6.6) in conjunction with part (ii) and Anscombe's (1952) Theorem lead to the conclusion that

$$(A_1 n^{*\frac{1}{2}}(L_{(1)}(N) - \Delta_{[1]}),\ldots,A_1 n^{*\frac{1}{2}}(L_{(k)}(N) - \Delta_{[k]})) \xrightarrow{\mathscr{L}} N_k(\underset{\sim}{0}, I_{k \times k}).$$

Hence from (8.6.14), with $Y_1,\ldots,Y_k$ i.i.d. $N(0, 1)$, we obtain

$$\liminf P(CS) \geq P\{Y_i < Y_k + h,\ i = 1,\ldots,k-1\} = P^*.$$

This is part (iv).

The Accelerated Sequential Case

Part (i) follows immediately from the definition of N given in (8.6.12). For part (ii), note that $t/(\rho n^*) \to 1$ w.p. 1 which can be proved along the lines of proof given above in the "purely sequential case". Also, $t \to \infty$ w.p. 1 as $\delta^* \to 0$, hence $N_1/n^* \to 1$ w.p. 1 and immediately from (8.6.12) one can conclude part (ii), since one has $0 < \rho < 1$.

Now, the stopping variable t is of the same form as N given by (8.6.9), in other words, $\{\delta^{*2} t : \delta^* > 0\}$ is uniformly integrable by virtue of assumption (iii) in (8.6.8). Consider, for $\delta_0^*(>0)$,

$$P\{N = N_1\ \forall \delta^* < \delta_0^*\} = P\{N_1 \geq t\ \forall \delta^* < \delta_0^*\} \to 1 \text{ as } \delta_0^* \to 0$$

and

$$t\, I(t > N_1)/n^* \leq t/n^*.$$

Hence, $t\, I(t > N_1)/n^*$ is uniformly integrable which immediately implies that $E\{t\, I(t > N_1)/n^*\} = o(1)$. We now write

$$N = t\, I(t > N_1) + N_1 I(N_1 \geq t)$$

and thus, in order to verify part (iii), we only need to show that

$$\limsup E\{N_1 I(N_1 \geq t)/n^*\} \leq 1. \tag{8.6.15}$$

Now, (8.6.11) implies

$$N_1 \leq h^2 S_t^2 \delta^{*-2} + 1,$$

and we obtain

$$E\{N_1 I(N_1 \geq t)/n^*\} \leq E\{A_1^2 S_t^2\} + o(1). \tag{8.6.16}$$

But, we know from the stopped stage in (8.6.10) that $A_1^2 S_t^2 \leq t/(\rho n^*)$ holds , hence, $\limsup E\{A_1^2 S_t^2\} \leq \limsup E\{t/(\rho n^*)\} = 1$. Now, (8.6.15) follows from (8.6.16). This gives part (iii).

Proof of part (iv) is very similar to that given in the case of "purely sequential" sampling. □

8.6.3 AN EXAMPLE

Let $L_i(n)$ be the sample median obtained from $X_{i1},\ldots,X_{in}$ for $i = 1,\ldots,k$. It is well known that $A_1^2 = 4f_1^{o2}(0)$ where $f_1^o(x) = dF_1^o(x)/dx$ under assumptions (i) – (ii). Now, in order to arrive at a specific reasonable estimator of A_1^2, we follow Geertsema's (1972) approach implementing Lehmann's (1963) suggestion in this context.

Having $\frac{1}{2} < \alpha < 1$, define $\Phi(z_\alpha) = \alpha$ and write

$$a(n) = n - b(n) + 1, \tag{8.6.17}$$

$$b(n) = \max\{1, [\tfrac{1}{2}(n - z_\alpha n^{\frac{1}{2}})]^* + 1\}, \tag{8.6.18}$$

for $n = m_1, m_1 + 1, \ldots$. Let $X_{n,1}(i) \leq \cdots \leq X_{n,n}(i)$ be the order statistics obtained from the ith population π_i, $i = 1,\ldots,k$. Then, Lehmann (1963) suggests that we utilize the following consistent estimator of A_1^{-2}:

$$S_n^2 = \tfrac{1}{4} n(kz_\alpha^2)^{-1} \sum_{i=1}^{k} \{X_{n,a(n)}(i) - X_{n,b(n)}(i)\}^2. \tag{8.6.19}$$

Under such setups, one may implement Geertsema's (1972) purely sequential selection procedure (8.6.9). On the other hand, one may like to implement the accelerated sequential selection procedure (8.6.10) – (8.6.12). One can verify assumptions (i) – (iii) for this choice of estimator of A_1^{-2} and one readily notes that the first–order asymptotic properties given by Theorem 8.6.1 hold.

We remark here that S_n^2 defined in (8.6.19) is quite slow in converging towards A_1^{-2}. In simulated experiments we have noted such unsuspecting behavior of significantly underestimating A_1^{-2} by means of S_n^2 for various kinds of d.f.'s $F_1^o(\cdot)$, for example, normal, Cauchy, double exponential or certain mixtures of these. We add that severe underestimation occurs even if n is moderate or large in size. Given this observation, it is not hard to guess that the actual amount of underestimation naturally depends on how heavy the "tail" of the d.f. F_1^o really is. In other words, any ad hoc adjustment in S_n^2 should take into account both the sizes of n and S_n^2 as well as F_1^o. But F_1^o is unknown, and hence we took an empirical approach. In the case of normal, Cauchy, double exponential or certain mixtures of these, through large–scale simulations we find that the following adjusted, and yet consistent, estimator of A_1^{-2} works satisfactorily. We consider

$$S_n^{*2} = 2n\{2n + 3(nS_n^2)^{\frac{1}{2}} - 4\}^{-1} S_n^2 \tag{8.6.20}$$

as the revised estimator of A_1^{-2} for $n \geq m_1$.

Table 8.6.1. Normal Populations: Case (a)
$k=5, P^*=.95, m_1=10$

ρ	n^*	δ^*	$\bar{n}$	$s(\bar{n})$	$\bar{p}$	$s(\bar{p})$	$\bar{n}-n^*$
			Accelerated Sequential (8.6.10)–(8.6.12)				
.5	10	1.211	15.03	0.078	.986	.004	5.03
	25	0.766	28.65	0.162	.964	.006	3.65
	50	0.542	52.82	0.321	.948	.007	2.82
	100	0.383	100.63	0.592	.951	.008	0.63
	150	0.313	148.55	0.860	.935	.008	–1.45
	200	0.271	197.55	1.113	.949	.007	–2.45
.6	10	1.211	14.32	0.060	.983	.004	4.32
	25	0.766	28.23	0.168	.961	.006	3.23
	50	0.542	52.07	0.308	.955	.007	2.07
	100	0.383	100.23	0.546	.938	.008	0.23
	150	0.313	148.67	0.871	.947	.007	–1.33
	200	0.271	200.18	1.046	.957	.006	0.18
			Purely Sequential (8.6.9)				
	10	1.211	13.70	0.067	.978	.005	3.70
	25	0.766	28.70	0.157	.965	.006	3.70
	50	0.542	54.16	0.297	.968	.006	4.16
	100	0.383	102.66	0.590	.943	.007	2.66
	150	0.313	152.19	0.770	.950	.007	2.19
	200	0.271	199.97	0.990	.951	.007	–0.03

Remark 8.6.1. Instead of the simple sample median, one can certainly use Hodges–Lehmann estimator in place of $L_i(n)$. Geertsema (1972) provides further details. Also, one should refer to Carroll (1974) for several interesting generalizations via purely sequential sampling.

8.6.4 MODERATE SAMPLE PERFORMANCES AND COMMENTS

In this section, we attempt to address the following questions and associated issues:

Table 8.6.2. Double Exponential Populations: Case (b)
$k=5$, $P^*=.95$, $m_1=10$

ρ	n^*	δ^*	$\bar{n}$	$s(\bar{n})$	$\bar{p}$	$s(\bar{p})$	$\bar{n}-n^*$
			Accelerated Sequential (8.6.10)–(8.6.12)				
.5	10	.966	17.10	0.128	.986	.004	7.10
	25	.611	32.22	0.229	.966	.006	7.22
	50	.432	57.63	0.379	.965	.006	7.63
	100	.306	106.12	0.711	.952	.007	6.12
	150	.249	156.72	0.974	.951	.007	6.72
	200	.216	206.43	1.252	.944	.007	6.43
.6	10	.966	15.98	0.097	.983	.004	5.98
	25	.611	31.76	0.222	.969	.005	6.76
	50	.432	57.06	0.385	.956	.006	7.06
	100	.306	106.42	0.688	.948	.007	6.42
	150	.249	155.53	0.969	.959	.006	5.53
	200	.216	204.63	1.131	.950	.007	4.63
			Purely Sequential (8.6.9)				
	10	.966	15.43	0.097	.984	.004	5.43
	25	.611	31.42	0.205	.971	.005	6.42
	50	.432	56.78	0.355	.967	.006	6.78
	100	.306	107.13	0.656	.955	.007	7.13
	150	.249	156.52	0.881	.946	.007	6.52
	200	.216	207.86	1.061	.956	.006	7.86

1. How nonparametric are the sequential selection procedure (8.6.9) and accelerated sequential selection procedure (8.6.10) – (8.6.12)? In other words, if we use such nonparametric methodologies in complicated parametric situations, what would then be the expected characteristics?
2. How large n^* would have to be so that the first–order asymptotics can be used as guidelines and reasonable approximations in reality?
3. What is the moderate sample behavior of Geertsema's (1972) purely sequential rule? Where does our new accelerated sequential methodology stand in comparison with the Geertsema procedure for moderate values of n^*?

Table 8.6.3. Cauchy Populations: Case (c)

$k=5$, $P^*=.95$, $m_1=10$

ρ	n^*	δ^*	$\bar{n}$	$s(\bar{n})$	$\bar{p}$	$s(\bar{p})$	$\bar{n}-n^*$
			Accelerated Sequential (8.6.10)–(8.6.12)				
.5	10	1.518	17.20	0.560	.980	.020	7.20
	25	0.960	33.96	0.912	.980	.020	8.96
	50	0.679	57.70	0.635	.960	.028	7.70
	100	0.480	104.24	1.354	.962	.006	4.24
	150	0.392	154.52	1.459	.971	.007	4.52
	200	0.339	198.24	1.725	.968	.020	–2.76
.6	10	1.518	15.72	0.480	.981	.007	5.72
	25	0.960	31.96	1.008	.962	.006	6.96
	50	0.679	57.14	1.125	.937	.008	7.14
	100	0.048	102.38	1.202	.960	.006	2.38
	150	0.392	153.66	1.706	.971	.006	3.66
	200	0.339	203.44	1.359	.948	.007	3.44
			Purely Sequential (8.6.9)				
	10	1.518	18.27	0.098	.983	.004	8.27
	25	0.960	32.80	0.181	.962	.006	7.80
	50	0.679	56.72	0.313	.958	.006	6.72
	100	0.480	104.67	0.584	.957	.006	4.67
	150	0.392	153.47	0.789	.944	.007	3.47
	200	0.339	199.84	1.002	.945	.007	–0.16

4. What is the extent of "saving" in operational cost if one does indeed implement the accelerated version instead of the purely sequential one?

We address such issues via fairly exhaustive computer simulation exercises. The IBM 3090 mainframe computer and our own programs written in Fortran were used for the purpose of simulating the Geertsema procedure (8.6.9) and the accelerated methodology (8.6.10) – (8.6.12). We consider $k=5$, $P^*=0.95$ and $m_1=10$. The appropriate "h" satisfying (8.6.5) is obtained from Table A.1 of Gibbons et al. (1977). We fixed $n^* = 10, 25, 50, 100, 150$ and 200, and computed $\delta^* = \{h^2(A_1^2 n^*)^{-1}\}^{\frac{1}{2}}$ where $A_1^2 = 4f_1^{o2}(0)$, for the

Table 8.6.4. Mixture of Normal and Cauchy: Case (d)
$k=5,\ P^*=.95,\ m_1=10$

		$a = .75$		$a = .90$	
ρ	n^*	$\bar{p}$	$\bar{n}-n^*$	$\bar{p}$	$\bar{n}-n^*$
		Accelerated Sequential (8.6.10)–(8.6.12)			
.5	10	.986	8.12	.978	6.84
	25	.967	8.02	.962	7.41
	50	.963	8.28	.971	7.30
	100	.948	6.92	.947	8.75
	150	.943	7.66	.946	12.10
	200	.955	7.74	.956	9.56
.6	10	.988	9.93	.982	6.16
	25	.967	7.32	.977	7.29
	50	.962	7.81	.954	7.98
	100	.942	7.78	.955	8.39
	150	.945	9.07	.951	8.89
	200	.958	12.13	.938	10.19
		Purely Sequential (8.6.9)			
	10	.973	5.72	.980	5.23
	25	.961	6.38	.965	5.93
	50	.963	6.76	.962	8.15
	100	.936	7.94	.939	9.08
	150	.948	11.79	.955	13.99
	200	.941	13.67	.940	14.70

example considered in section 8.6.3. We write $f_1^*(y) = (2\pi)^{-\frac{1}{2}} \exp(-\frac{1}{2} y^2)$, $f_2^*(y) = \frac{1}{2} \exp(-|y|)$ and $f_3^*(y) = \{\pi(1+y^2)\}^{-1}$ for $-\infty < y < \infty$. Recall that the population π_i has the p.d.f. $f_1^o(x-\Delta_i)$ where various choices of $f_1^o(x)$ have been made as follows (with $\Delta_1 = \cdots = \Delta_{k-1} = 1$ and $\Delta_k = 1 + \delta^*$, that is π_k is the "best" population by construction): (a) $f_1^o(x) = f_1^*(x)$; (b) $f_1^o(x) = f_2^*(x)$; (c) $f_1^o(x) = f_3^*(x)$; (d) $f_1^o(x) = af_1^*(x) + (1-a)f_3^*(x)$ with $a =$ .25, .50, .75 and .90; (e) $f_1^o(x) = af_1^*(x) + (1-a)f_2^*(x)$ with $a = .01, .03, .05,$

Table 8.6.5. Mixture of Normal and Double Exponential: Case (e)
$k=5,\ P^*=.95,\ m_1=10$

ρ	n^*	$\alpha = .10$ $\bar{p}$	$\bar{n}-n^*$	$\alpha = .15$ $\bar{p}$	$\bar{n}-n^*$
		Accelerated Sequential (8.6.10)–(8.6.12)			
.5	10	.990	6.20	.994	5.53
	25	.981	5.43	.979	3.70
	50	.958	5.01	.954	1.71
	100	.945	3.19	.944	–3.04
	150	.947	1.63	.946	–10.01
	200	.958	0.63	.952	–14.93
.6	10	.983	5.31	.984	4.69
	25	.977	5.11	.977	3.33
	50	.953	4.70	.961	1.27
	100	.962	1.51	.958	–4.72
	150	.957	–1.38	.959	–12.98
	200	.953	0.13	.949	–12.50
		Purely Sequential (8.6.9)			
	10	.988	4.96	.987	4.25
	25	.960	4.77	.972	3.34
	50	.961	4.04	.955	0.77
	100	.959	2.37	.970	–4.30
	150	.960	1.55	.958	–9.80
	200	.951	–0.94	.960	–14.21

.10, .13, .15, .20, .25, .50 and .75; (f) $f_1^0(x) = \alpha f_2^*(x) + (1-\alpha)f_3^*(x)$ with $\alpha =$ .25, .50, .75 and .90. The IMSL (1987) subroutines RNNOR, RNEXP and RNCHY are used to generate pseudo random samples from normal, double exponential and Cauchy distributions respectively. Having observed $X_{i1},\ldots,X_{in}$ from π_i, we define the sample median as follows along the lines of Geertsema (1972):

$$L_i(n) = \begin{cases} X_{n,\frac{1}{2}(n+1)}^{(i)} & \text{when } n \text{ is odd} \\ X_{n,\frac{1}{2}(n+2)}^{(i)} & \text{when } n \text{ is even} \end{cases} \tag{8.6.21}$$

Table 8.6.6. Mixture of Double Exponential and Cauchy: Case (f)
$k=5,\ P^*=.95,\ m_1=10$

ρ	n^*	$a=.75$ $\bar{p}$	$a=.75$ $\bar{n}-n^*$	$a=.90$ $\bar{p}$	$a=.90$ $\bar{n}-n^*$
		Accelerated Sequential (8.6.10)–(8.6.12)			
.5	10	.987	10.34	.991	9.88
	25	.968	13.55	.975	12.98
	50	.957	19.84	.962	18.60
	100	.962	27.09	.939	30.65
	150	.953	39.15	.953	40.29
	200	.944	35.28	.953	42.11
.6	10	.991	9.13	.988	8.58
	25	.965	13.02	.969	13.50
	50	.969	18.64	.962	18.09
	100	.936	26.07	.949	28.13
	150	.952	32.07	.949	36.82
	200	.968	46.01	.950	52.49
		Purely Sequential (8.6.9)			
	10	.985	8.54	.984	8.53
	25	.967	11.43	.967	11.82
	50	.959	15.62	.949	15.62
	100	.954	24.90	.949	27.13
	150	.959	34.46	.945	36.26
	200	.957	38.27	.951	41.07

$i=1,\ldots,k$. In other words, for odd n, the sample median is the order statistic in the middle, whereas for even n, the median can be taken to be any number in between the two order statistics in the middle. For even n, however, we pick the larger of the two order statistics in the middle as our median. Also, note that *while implementing* either the purely sequential scheme or its accelerated version, in the stopping rules *we always use* S_n^* *(given by (8.6.20)) instead of* S_n.

The accelerated sequential procedure (8.6.10) – (8.6.12) has been tried with $\rho = .3, .4, .5, .6, .7$ and $.8$. Let us now elaborate on how simulations have been carried out for a particular set of values of n^*, ρ etc. regardless of what specifically $f_1^0(x)$ is under consideration. We implemented a "sampling procedure" by repeating it independently 1000 times (=J, say). Let n_j be the observed value of N and $p_j = 1(0)$ if the population having the largest center of symmetry is (is not) selected at the end of the jth iteration, $j=1,\ldots,J$. We write $\bar{n} = J^{-1} \sum_{j=1}^{J} n_j$, $s^2(\bar{n}) = (J^2-J)^{-1} \sum_{j=1}^{J} (n_j-\bar{n})^2$, $\bar{p} = J^{-1} \sum_{j=1}^{J} p_j$, $s^2(\bar{p}) = \bar{p}(1-\bar{p})/J$. In a particular situation, the expected sample size required is estimated by $\bar{n}$, having the associated estimated standard error $s(\bar{n})$. On the other hand, the estimated probability of correct selection $\bar{p}$ is to be compared with the target value, P^*. In Tables 8.6.1 – 8.6.6, we summarize some of our findings for immediate comparisons in cases (a) – (f) respectively.

While implementing the accelerated sequential procedure (8.6.10) – (8.6.12), we found that the values $\rho = .5$ and $.6$ seemed to have worked best. In Tables 8.6.1 – 8.6.3, note that $s(\bar{n})$ (and $s(\bar{p})$) remains quite comparable whether $\rho = .5$ and $.6$, however, $s(\bar{n})$ and $|\bar{n} - n^*|$ are slightly larger in the accelerated case compared with what we found in the purely sequential case. Such features are intuitively expected and Tables 8.6.1 – 8.6.3 generally validate this feeling.

Table 8.6.4 summarizes our findings partially in case (d) where f_1^0 is a mixture of $100\alpha\%$ normal and $100(1-\alpha)\%$ Cauchy. We report very briefly on the "situation" when $\alpha = .75$ and $.90$. The overall picture is quite similar when $\alpha = .25$ and $.50$. The values of $s(\bar{n})$ and $s(\bar{p})$ in these situations are found to be only slightly larger than those obtained in the two corresponding non–mixed distribution cases. Hence, in Table 8.6.4, we refrain from giving the values of $s(\bar{n})$ and $s(\bar{p})$.

Table 8.6.5 is similar in essence in terms of presentation of brief summaries in case (e). This is a very difficult mixture distribution indeed. The uniformly most powerful test for H_0: $X \sim f_1^*(x)$ against the alternative H_a: $X \sim f_2^*(x)$ rejects H_0 not only for large values of $|X|$, but also for values of $|X|$ near zero. This awkward feature gets to be even nastier in case of mixing f_1^* with f_2^*. For $a = .01, .03, .05$ and $.10$ we noted excellent match between $\bar{n}$ and n^* as well as between $\bar{p}$ and P^*. For $a = .15$ we began noticing that $\bar{n}$ started to underestimate n^* significantly and this picture kept getting worse as a became larger within our range. In all these cases, $\bar{p}$ came out to be quite close to P^*. We considered $a = .15$ as sort of the "boundary", after which moderate sample size performance, in terms of *both* $\bar{n}$ and $\bar{p}$ being in close proximity of n^* and P^* respectively, would not remain very attractive. So, we ran additional simulations for $a = .15$ and $n^* = 500, 800$ and 1100 with $J = 500$ replications. For such large n^* values, slowly but surely, we observed "asymptotics" to take over. The point is that for *all* a, asymptotics are still valid, however to notice that behavior uniformly, we do need to consider much larger values of n^* when $a \geq .15$. Table 8.6.6 is similar to Tables 8.6.4 – 8.6.5 in spirit. Now, we give a few specific pertinent remarks addressing the issues raised in questions #1–4 given in the top of this section.

Remark 8.6.2: $\bar{p}$ is very close to P^* in every single case. One should also note that $|\bar{n} - n^*|$ values in the purely sequential situation and in the accelerated sequential situation agree reasonably well, yet one saves sampling operations significantly by implementing accelerated versions. Such substantial operational improvement obtained via accelerated versions shows up quite naturally in terms of requiring significantly less computational expenses incurred during simulation exercises. For example, we add that in Tables 8.6.5 and 8.6.6, the accelerated sequential procedure for the "whole block" when $\rho = .5$ or $.6$ costs us (in computer time) approximately 50% of what we had spent in order to obtain the sequential part of these tables given in the last block.

Remark 8.6.3: The first–order asymptotics, that is $\bar{n} \approx n^*$ and $\bar{p} \approx P^*$, hold up fairly well even for moderate values of n^* in the range of 10 to 200. For larger values of n^*, we can certainly expect that $\bar{n}$ and $\bar{p}$ would be closer to n^* and P^* respectively.

Remark 8.6.4: In the simulations, we have considered a wide variety of symmetric distributions, F_1^o. *Across the board*, we note that both the purely sequential as well as the accelerated sequential methodology take the intrinsic nonparametric nature into account remarkably well. In other words, these nonparametric methodologies remain very robust even for complicated parametric situations. Again, we emphasize that the accelerated sequential techniques do a remarkable overall job and yet these are operationally more convenient than the Geertsema procedure.

8.6.5 OTHER RELATED IDEAS

The purely sequential selection procedure (3.5.1) can easily be thought of as a nonparametric procedure if π's are characterized by unknown *continuous* p.d.f.'s having the same unknown variance σ^2, $0 < \sigma^2 < \infty$. Indeed, Robbins et al. (1968) remarked about this nonparametric aspect of the methodology given in (3.5.1). In this situation, parts (i), (ii) and (iv) of Theorem 3.5.1 hold. Part (iv) of Theorem 3.5.1 holds with proper modification. Let us define $\mu_{4i} = E(X_{i1} - \mu_i)^4$ and *assume* that

(i) $\mu_{41} = \cdots = \mu_{4k} = \mu_4$, say;

(ii) μ_4 is finite and $\mu_4 > \sigma^4$.

Then, one can conclude that

$$C^{-\frac{1}{2}}(N - C) \xrightarrow{\mathscr{L}} N(0, k^{-1}(\mu_4\sigma^{-4} - 1)), \tag{8.6.22}$$

as $\delta^* \to 0$, in view of Theorem 2.4.1. Observe that $\mu_4\sigma^{-4} = 3$ in the normal case.

On the other hand, if π's are characterized by unknown p.d.f.'s having the same unknown variance, but these p.d.f.'s are either *continuous or discrete* in nature, even so the selection procedure (3.5.1) "essentially" makes sense. That is, we properly modify (3.5.1) to suit the present scenario and then implement it. First choose the starting sample size $m_1(\geq 2)$ to satisfy (3.7.1) and then use the selection procedure (3.5.1). Now, if m_1 grows at a certain rate, that is under certain range of values of r in (3.7.1), one may use the machineries from Chang and Hsiung (1979) as well as Martinsek (1983) in order to verify that

$$\lim_{\delta^* \to 0} E(N - C) \text{ is bounded,} \tag{8.6.23}$$

under appropriate moment conditions on the X_i's. That is, the purely sequential procedure (3.5.1) with the above modification, is *asymptotically second-order efficient* in the sense of (3.4.8) *without the normality assumptions* about the populations π's, as long as certain moments of π's are finite. Hence, the purely sequential methodology (3.5.1) does really cover a lot more ground than what appears on the surface. One may also study the properties associated with P(CS) via simulation exercises for different types of distributions.

8.7 Notes and Overview

The developments of appropriate asymptotic second–order theories for the problem areas considered in sections 8.2 – 8.4 and 8.6 will prove to be very challenging. One should also investigate other distributions for which the multistage methodologies introduced in section 8.5 can be extended and associated second–order characteristics can be obtained. These problem areas are wide open and hence immediate attention is expected to provide very many research directions.

Complements and Problems

8.1 (a) Verify (8.2.3).

(b) Show that the same property holds for the elimination type selection procedure $\mathscr{Q}_{\underset{\sim}{N}}(2, 2)$.

[Refer to Paulson (1964).]

8.2 For the elimination type selection procedure $\mathscr{Q}_{\underset{\sim}{N}}(2, 1)$, show that $P\{N_1 < N_2 < \cdots < N_{k-1}\} \to 1$ as $P^* \to 1$ if $\mu_1 < \mu_2 < \cdots < \mu_k$.

[Refer to Perng (1969).]

8.3 Consider the selection procedure $\mathscr{Q}_N(3, 1)$. Show that $P(N < \infty | H_i) = 1$ and $P(CS | H_i) \geq P^*$ for each $i = 1, \ldots, k$.

[Refer to Mukhopadhyay (1983a).]

8.4 Denote the stopping variable given in section 4.3.1 of Bechhofer et al. (1968) by N_{BKS}. Show that $N_{BKS} \leq N$ w.p. 1 where N corresponds to the sample size required by the selection procedure (8.3.12).

8.5 (a) For the selection procedure $\mathscr{Q}_N(3, 1)$, show that $E(N/C^*) \to 1$ as $P^* \to 1$, where $C^* = -\delta^{*-2}\sigma^2 \ln(\frac{1-P^*}{k-1})$, under each H_i, $i = 1, \ldots, k$.

[Refer to Robbins (1970) and Khan (1973).]

(b) Consider $\mathscr{Q}_N(3, 1)$ in the case $k = 2$. Show that $N/C^* \to 1$ w.p. 1 and $(N - C^*)/D^* \xrightarrow{\mathscr{L}} N(0, 1)$ as $P^* \to 1$ where one defines $D^* = \{-2 \ln(1-P^*)\}^{\frac{1}{2}} \sigma^2/\delta^{*2}$, under both H_1, H_2.

[Refer to Mukhopadhyay and Chou (1982a).]

8.6 Consider $\mathscr{R}_N(3, 2)$ in the case $k = 2$. Verify the properties in Problem 8.5 (b), under both H_1, H_2.

[Refer to Mukhopadhyay and Chou (1982a).]

8.7 Consider $\mathscr{R}_N(3, 1)$ in the case $k = 2$. Show that $N/C^* \to 1$ w.p. 1 as $P^* \to 1$ where $C^* = -\sigma^2 \delta^{*-2} \ln\{\frac{1}{2}(1-P^*)\}$, under both H_1, H_2.

8.8 Consider $\mathscr{R}_N(3, 5)$ in the case $k = 2$. Show that $N/C_1^* \to 1$ w.p. 1 and $(N - C_1^*)/D_1^* \xrightarrow{\mathscr{L}} N(0, 1)$ as $P^* \to 1$, under both H_1, H_2 where $C_1^* = -\{\ln(1 - P^*)\}/\ln\{\frac{1}{2}(1 + \delta^{*2})/\delta^*\}$ and $D_1^{*2} = \{-\frac{1}{2} \ln(1 - P^*)\} \{\ln[\frac{1}{2}(1 + \delta^{*2})/\delta^*]\}^{-3}$.

[Refer to Theorem 1 in Mukhopadhyay and Chou (1982b).]

8.9 Verify (8.3.25), (8.3.29) and (8.3.30).

8.10 Derive results like those in the first part of Problem 8.5 (b) in the context of $\mathscr{R}_N(3, 6)$ when $k = 2$.

[Refer to Theorem 2.1 of Miller and Mukhopadhyay (1990).]

8.11 In the case $k = 2$, show that h_4 satisfying (8.4.16) coincides with h_5 satisfying (8.4.17).

8.12 Consider h_4 as a function of m_1 when one holds P^* fixed. Show that $h_4(m_1)$ is a decreasing function of m_1, and $h_4(m_1) \to h$, satisfying (8.4.18), as $m_1 \to \infty$ in the case $k = 2$.

[Refer to Lemma 1 of Mukhopadhyay (1979).]

8.13 When $k = 2$, obtain a modified version of the two–stage procedure $\mathscr{Q}_{\underset{\sim}{N}}(4, 1)$ from (8.4.9) where one can claim (8.4.20).

[Refer to Mukhopadhyay (1979).]

8.14 Verify (8.4.22) – (8.4.23).

8.15 In the case $k = 2$, show that $N_1/N_2 \to \sigma_1/\sigma_2$ w.p. 1 as $\delta^* \to 0$ for the selection procedure $\mathscr{Q}_{\underset{\sim}{N}}(4, 1)$ obtained from (8.4.24) – (8.4.25). Also show that $E(N_i/n_i^{**}) \to 1$ as $\delta^* \to 0, i = 1, 2$.

[Refer to Robbins et al. (1967) and Theorem 5 of Mukhopadhyay (1979).]

8.16 Verify (8.4.31) – (8.4.32).

8.17 Verify the claims made in (8.4.38) and (8.4.39).

8.18 Obtain a modified version of the two–stage selection procedure $\mathscr{Q}_{\underset{\sim}{N}}(4, 2)$ from (8.4.29) and derive associated first–order asymptotics.

[Refer to section 2.1 of Mukhopadhyay and Hamdy (1984a).]

8.19 Prove Theorems 8.5.2, 8.5.3 and 8.5.4.

8.20 Derive the first–order asymptotics associated with the selection procedure $\mathscr{Q}_{\underset{\sim}{N}}(5, 2)$ in the case of the purely sequential procedure (8.5.15), accelerated sequential procedure (8.5.18) – (8.5.20), and the three–stage procedure (8.5.21).

8.21 Choose $m_1(\geq 2)$ to satisfy* (3.7.1) and then use the selection procedure (3.5.1). For "appropriate" choices of "r", show that lim $E(N - C)$

is bounded as $\delta^* \to 0$ in the light of section 8.6.5.

[Refer to Martinsek (1983).]

REFERENCES

Anscombe (1952), Armitage (1950), Baker (1950), Bechhofer (1954), Bechhofer and Goldsman (1987, 1989), Bechhofer, Kiefer and Sobel (1968), Carroll (1974), Carroll, Gupta and Huang (1975), Chang and Hsiung (1979), Desu (1970), Dudewicz (1971), Dudewicz and Dalal (1975), Fabian (1974a,b), Geertsema (1972), Gibbons, Olkin and Sobel (1977), Gupta (1956, 1963), Gupta, Nagel and Panchapakesan (1973), Hall (1962), Hall, Wijsman and Ghosh (1965), Hartmann (1988, 1991, 1993), Hoel (1971), Hoel and Mazumder (1969), Johnson and Kotz (1972), Khan (1973), Kimbal (1951), Lam and Ng (1990), Lehmann (1961, 1963), Liang and Panchapakesan (1993), Martinsek (1983), Miller and Mukhopadhyay (1990), Mukhopadhyay (1979, 1983a, 1984), Mukhopadhyay and Chou (1982a,b, 1984, 1985), Mukhopadhyay and Hamdy (1984a), Mukhopadhyay and Miller (1990), Mukhopadhyay and Panchapakesan (1992), Mukhopadhyay and Solanky (1993), Paulson (1964), Perng (1969), Ramberg (1966), Ramberg and Bechhofer (1973), Ray Chaudhuri (1986), Rinott (1978), Robbins (1970), Robbins, Simons and Starr (1967), Robbins, Sobel and Starr (1968), Santner (1976), Slepian (1962), Srivastava (1966), Swanepoel (1977), Swanepoel and Geertsema (1973, 1976), Swanepoel and Venter (1975).

NOTATION AND ABBREVIATIONS

i.i.d.	Independent and identically distributed
p.d.f.	Probability density function
d.f.	Distribution function
$\forall$	For all
$\square$	Indicates the end of a proof
p.d.	Positive definite
I(A)	Indicator function of set A
[x]*	Largest integer < x
$\phi(x)$	$(2\pi)^{-\frac{1}{2}} \exp(-\frac{1}{2} x^2)$
$\Phi(x)$	$\int_{-\infty}^{x} \phi(y)dy$
R; R^s	Real line; corresponding s–dimensional space
R^+; R^{+s}	Positive half of real line; corresponding s–dimensional space
$\sup_x$, $\sup_x$	$\sup_{x \in R}$, $\sup_{x \in R}$
LFC	Least favorable configuration
$h = h(k,P^*)$: Chapter 3	$\int_{-\infty}^{\infty} \{\Phi(y+h)\}^{k-1} \phi(y)dy = P^*$
$\mathrm{U}(n)$, $\mathrm{U}^*(n)$: Chapters 3, 5, 7 and 8	Pooled estimators of σ^2 defined in (3.1.1)
$\tau = \tau(m_1,k,P^*)$	Defined in (3.3.4)

$g_0(y), f(y)$: Chapter 3	$g_0(y)=\int_{-\infty}^{\infty}\{\Phi(x+y)\}^{k-1}\phi(x)dx$, $f(y) = g_0(y^{\frac{1}{2}})$, $y > 0$
$\nu = \nu(k)$: Chapter 3	Defined in (3.5.7)
$\beta = \beta(k) = k^{-1}\{\nu(k)-2\}$	Defined in Theorem 3.5.2
$\beta' = \beta'(k) = \frac{1}{2}-2(k\rho)^{-1}$	Defined in Theorem 3.7.2
$a = a(k,P^*)$: Chapter 4	$\int_0^{\infty}\{1-\exp(-z-a)\}^{k-1}\exp(-z)dz = P^*$
$V(n)$, $V^*(n)$: Chapters 4 5, 7 and 8	Pooled estimators of σ defined in (4.1.1)
$\tau_0 = \tau_0(m_1,k,P^*)$	Defined in (4.3.3)
$b^*=b^*(m_1) = 2\tau_0 k(m_1-1)$	Defined in (4.3.1)
$\nu = \nu(k)$: Chapter 4	Defined in (4.5.8)
$\beta''=\beta''(k) = k^{-1}(\frac{1}{2}\nu-1)$	Defined in Theorem 4.5.2
$\beta_0'=\beta_0'(k) = \frac{1}{2}-(k\rho)^{-1}$	Defined in Theorem 4.7.2
$q_1(x)$: Chapter 4	$q_1(x) = \exp(-x)\ I\ (x > 0)$
$g_1(y)$: Chapter 4	$g_1(y)=\int_0^{\infty}\{1-\exp(-z-y)\}^{k-1}\exp(-z)dz$, $y>0$
$a=a(k,\alpha)$: Section 5.2.1	$\Phi^k(-a) - \Phi^k(a) = 1 - \alpha$
$\tau_1=\tau_1(m_1,k,\alpha)$	Defined in (5.2.7)
$\psi_1(x)$: Chapter 5	$\psi_1(x)=\{\Phi(x^{\frac{1}{2}})\}^k-\{\Phi(-x^{\frac{1}{2}})\}^k$, $x>0$
$a=a(k,\alpha)$: Section 5.2.2	$a = -\ln\{1-(1-\alpha)^{1/k}\}$
$\psi_2(x)$: Chapter 5	$\psi_2(x)=\{1-\exp(-x)\}^k$, $x>0$; defined in (5.2.40)
$\tau^*=\tau^*(m_1,k,\alpha)$	Defined in (5.2.41)

$f^0(x)$	P.d.f of $\chi^2_{m_1-1}$ in (6.2.4)
$a_1 = a_1(m_1,k,P^*)$	Defined in (6.2.5) and (6.2.8)
$h_1 = h_1(m_1,k,P^*)$	Defined in (6.2.6)
$\psi_3(x)$: Chapter 6	$\psi_3(x) = \Phi(x^{\frac{1}{2}})$, $x > 0$
$h_2 = h_2(m_1,P^*)$	Refer to (6.4.3)
$g_2(x)$, $g_2^*(x)$	Refer to (7.2.1)
$F_1(x)$, $f_1(x)$: Chapter 7	The d.f. and p.d.f. of χ^2_r
$g_3(x)$, $g_3^*(x)$	Refer to (7.3.1)
a_λ^*, Z_λ	Refer to (8.2.1)
ν^*, d_1, d_2	Defined in (8.2.4)
$h_3 = h_3(k,P^*)$	Defined in (8.4.4) and (8.4.6)
$h_4 = h_4(m_1,k,P^*)$	Defined in (8.4.16)
$h_5 = h_5(m_1,P^*)$	Defined in (8.4.17)
$G_4(x)$,$g_4(x)$: Chapter 8	The d.f. and p.d.f. of Student's t with (m_1-1) degrees of freedom
$h_6 = h_6(m_1,k,P^*)$	Defined in (8.4.37)

BIBLIOGRAPHY

Abramowitz, M. and Stegun, I.A.

Handbook of Mathematical Functions. Dover Publications, Inc., New York, 1965.

Anscombe, F.J.

Large sample theory of sequential estimation. *Proc. Camb. Phil. Soc.* (1952), *48*, 600–607.

Sequential estimation. *J. Roy Statist. Soc., B* (1953), *15*, 1–21.

Apostol, T.M.

Mathematical Analysis. Addison–Wesley Publishing Co., Inc., Reading, 1974.

Armitage, P.

Sequential analysis with more than two alternative hypotheses, and its relation to discriminant function analysis. *J. Roy. Statist. Soc., B* (1950), *12*, 137–144.

Bahadur, R.R.

On a problem in the theory of k populations. *Ann. Math. Statist.* (1950), *21*, 362–375.

Bahadur, R.R. and Robbins, H.

The problem of greater mean. *Ann. Math. Statist.* (1950), *21*, 469–487; Corrections in *Ann. Math. Statist.* (1951), *22*, 310.

Bain, L.J.

Statistical Analysis of Reliability and Life–Testing Models. Marcel Dekker, Inc., New York, 1978.

Baker, A.G.

Properties of some tests in sequential analysis. *Biometrika* (1950), *37*, 334–346.

Barlow, R.E., Bartholomew, D.J., Bremner, J.M. and Brunk, H.D.

Statistical Inference Under Order Restrictions. John Wiley & Sons, Inc., New York, 1972.

Basu, A.P.

On a sequential rule for estimating the location parameter of an exponential distribution. *Naval. Res. Logist. Qr.* (1971), *18*, 329–337.

Sequential methods in reliability and life testing. *Handbook of Sequential Analysis* (ed. B.K. Ghosh and P.K. Sen), Marcel Dekker, Inc., New York, 1991, 581–592.

Basu, D.

On statistics independent of a complete sufficient statistic. *Sankhya* (1955), *15*, 377–380.

Bechhofer, R.E.

A single–sample multiple decision procedure for ranking means of normal populations with known variances. *Ann. Math. Statist.* (1954), *25*, 16–39.

Bechhofer, R.E., Dunnett, C.W. and Sobel, M.

A two–sample multiple decision procedure for ranking means of normal populations with a common unknown variance. *Biometrika* (1954), *41*, 170–176.

Bechhofer, R.E. and Goldsman, D.M.

Truncation of the Bechhofer–Kiefer–Sobel sequential procedure for selecting the normal population which has the largest mean. *Commun. Statist. – Simulations* (1987), *16*, 1067–1092.

Truncation of the Bechhofer–Kiefer–Sobel sequential procedure for selecting the normal population which has the largest mean (III): Supplementary truncation numbers and resulting performance characteristics. *Commun. Statist. – Simulations* (1989), *18*, 63–81.

Bechhofer, R.E., Kiefer, J. and Sobel, M.

Sequential Identification and Ranking Procedures. Univ. of Chicago Press, Chicago, 1968.

Berger, J.O.

Statistical Decision Theory and Bayesian Analysis, 2nd Edition, Springer Verlag, Inc., New York, 1985.

Bhattacharyya, P.K. and Mallik, A.

Asymptotic normality of the stopping times of some sequential procedures. *Ann. Statist.* (1973), *1*, 1203–1211.

Birnbaum, A. and Healy, W.C., Jr.

Estimates with prescribed variance based on two–stage sampling. *Ann. Math. Statist.* (1960), *31*, 662–676.

Bristol, D.R., Chen, H.. and Mithongtae, J.S.

Comparing exponential guarantee times with a control. *Private Communication* (1987), 1–64.

Bristol, D.R. and Desu, M.M.

Selection procedures for comparing exponential guarantee times with a standard. *The Frontiers of Modern Statistical Inference Procedures* (ed. E.J. Dudewicz), American Sciences Press, Columbus, 1985, 307–324.

Carroll, R.J.

Some Contributions to the Theory of Parametric and Nonparametric Sequential Ranking and Selection. Ph.D. Thesis (1974), Department of Statistics, Purdue University, West Lafayette.

Carroll, R.J., Gupta, S.S. and Huang, D.Y.

On selection procedures for the t best populations and some related problems. *Commun. Statist.* (1975), *4*, 987–1008.

Chang, I.S. and Hsiung, C.A.

Approximations to the expected sample size of certain sequential procedures. *Proc. Conf. Recent Developments Statist. Methods and Applications* (1979), Institute of Math., Acad. Sinica, Taipei, 71–82.

Chatterjee, S.K.

Two–stage and multistage procedures. *Hankbook of Sequential Analysis* (ed. B.K. Ghosh and P.K. Sen), Marcel Dekker, Inc., New York, 1991, 21–46.

Chen, H.J. and Dudewicz, E.J.

Interval estimation of a ranked mean of k normal populations with unequal sample sizes. *Dept. of Math. Sciences Tech. Report* (1973), No. 73–21, Memphis State Univ., Memphis.

Chow, Y.S., Hsiung, C. and Lai, T.

Extended renewal theory and moment convergence in Anscombe's theorem. *Ann. Prob.* (1979), *7*, 304–318.

Chow, Y.S. and Robbins, H.

On the asymptotic theory of fixed width sequential confidence intervals for the mean. *Ann. Math. Statist.* (1965), *36*, 457–462.

Chow, Y.S., Robbins, H. and Teicher, H.

Moments of randomly stopped sums. *Ann. Math. Statist.* (1965), *36*, 789–799.

Chow, Y.S. and Teicher, H.

Probability Theory. Springer–Verlag, Inc., New York, 1978.

Cohen, A. and Sackrowitz, H.

Estimating the mean of the selected population. *Statist. Decision Theory and Related Topics III* (ed. S.S. Gupta and J. Berger), Academic Press, New York, 1982, 243–270.

Cox, D.R.

Estimation by double sampling. *Biometrika* (1952), *39*, 217–227.

Csenki, A.

On the convergence rate of fixed–width sequential confidence intervals. *Scand. Actuarial J.* (1980), 107–111.

Dantzig, R.B.

On the non–existence of tests of Student's hypothesis having power functions independent of σ. *Ann. Math. Statist.* (1940), *11*, 186–192.

DasGupta, S. and Perlman, M.

Power of the noncentral F–test: Effect of additional variates on Hotelling's T^2–test. *J. Amer. Statist. Assoc.* (1974), *69*, 174–180.

Desu, M.M.

A selection problem. *Ann. Math. Statist.*, (1970), *41*, 1596–1603.

Desu, M.M., Narula, S.C. and Villarreal, B.

A two–stage procedure for selecting the best of k exponential distributions. *Commun. Statist., A* (1977), *6*, 1231–1243.

Desu, M.M. and Sobel, M.

A fixed subset–size approach to the selection problem. *Biometrika* (1968), *55*, 401–410.

Dudewicz, E.J.

Non–existence of a single–sample selection procedure whose P(CS) is independent of the variance. *S. Afr. Statist. J.* (1971), *5*, 37–39.

Dudewicz, E.J. and Dalal, S.R.

Allocation of observations in ranking and selection with unequal variances. *Sankhya, B* (1975), *37*, 28–78.

Dudewicz, E.J. and Koo, J.O.

The Complete Categorized Guide to Statistical Selection and Ranking Procedures. American Sciences Press, Columbus, 1981.

Dudewicz, E.J. and Taneja, V.S.

A multivariate solution of the multivariate ranking and selection problem. *Commun. Statist., A* (1981), *10*, 1849–1868.

Dunnett, C.W. and Sobel, M.

Approximations to the probability integral and certain percentage points of a multivariate analogue of Student's t–distribution. *Biometrika* (1955), *42*, 258–260.

Dykstra, R., Robertson, T. and Wright, F.T.

Advances in Order Restricted Statistical Inference. Springer–Verlag, Inc., New York, 1986.

Fabian, V.

Note on Anderson's sequential procedures with triangular boundary. *Ann. Statist.* (1974a), *2*, 170–176.

Acknowledgement of priority to "Note on Anderson's sequential procedures with triangular boundary." *Ann. Statist.* (1974b), *2*, 1063.

Ferguson, T.S.

Mathematical Statistics. Academic Press, Inc., New York, 1967.

Finney, D.J.

The distribution of the ratio of estimates of the two variances in a sample from a normal bivariate population. *Biometrika* (1941), *30,* 190–192.

Frischtak, R.M.

Statistical multiple–decision procedures for some multivariate selection problems. *Ph.D. dissertation* (1973), *Cornell Univ.*, Ithaca.

Geertsema, J.C.

Nonparametric sequential procedure for selecting the best of k populations. *J. Amer. Statist. Assoc.* (1972), *67,* 614–616.

Ghosh, B.K.

Sequential Tests of Statistical Hypotheses. Addison–Wesley Publishing Co., Inc., Reading, 1970.

Some monotonicity theorems for χ^2, F and t distributions with applications. *J. Roy. Statist. Soc., B* (1973), *35,* 480–492.

Ghosh, B.K. and Sen, P.K.

Handbook of Sequential Analysis. Marcel Deker, Inc., New York, 1991.

Ghosh, M.

Rate of convergence to normality for random means: applications to sequential estimation. *Sankhya, A* (1980), *42,* 231–240.

Hierarchical and empirical Bayes sequential estimation. *Handbook of Sequential Analysis* (ed. B.K. Ghosh and P.K. Sen), Marcel Dekker, Inc., New York, 1991, 441–458.

Ghosh, M. and DasGupta, R.

Berry–Esseen theorems for U–statistics in the non iid case. *Colloq. Math. Soc. János Bolyai*, Budapest (1980), 293–313.

Ghosh, M. and Hoekstra, R.M.

Asymptotically pointwise optimal stopping rules. *Handbook of Sequential Analysis* (ed. B.K. Ghosh and P.K. Sen), Marcel Dekker, Inc., New York, 1991, 429–438.

Ghosh, M. and Mukhopadhyay, N.

Asymptotic normality of stopping times in sequential analysis. *Unpublished manuscript* (1975).

On two fundamental problems of sequential estimation. *Sankhya, B* (1976), *38*, 203–218.

Consistency and asymptotic efficiency of two–stage and sequential estimation procedures. *Sankhya, A* (1981), *43*, 220–227.

Ghosh, M., Sinha, B.K. and Mukhopadhyay, N.

Multivariate sequential point estimation. *J. Mult. Anal.* (1976), *6*, 281–294.

Ghurye, S.G.

Note on sufficient statistics and two–stage procedures. *Ann. Math. Statist.* (1958), *29*, 155–166.

Gibbons, J.D., Olkin, I. and Sobel, M.

Selecting and Ordering Populations: A New Statistical Methodology. John Wiley & Sons, Inc., New York, 1977.

Govindarajulu, Z.

The Sequential Statistical Analysis. American Sciences Press, Inc., Columbus, 1981.

Grubbs, F.E.

Approximate fiducial bounds on reliability for the two parameter negative exponential distribution. *Technometrics* (1971), *13*, 873–876.

Gupta, S.S.

On a decision rule for a problem in ranking means. *Ph.D. dissertation* (1956), *Univ. of North Carolina*, Chapel Hill.

Probability integrals of multivariate normal and multinormal t. *Ann. Math. Statist.* (1963), *34*, 792–828.

Gupta, S.S. and Huang, D.Y.

Multiple Statistical Decision Theory: Recent Developments. Springer Verlag, Inc., New York, 1981.

Gupta, S.S. and Miescke, K.

Sequential selection procedures. *Ann. Statist.* (1984), *12*, 336–350.

On finding the largest normal mean and estimating the selected mean. *Sankhya, B* (1990), *52*, 144–157.

Single–sample Bayes and empirical Bayes rules for ranking and estimating multinomial probabilities. *J. Statist. Plan. Inf.* (1993), *33*, 367–382.

Gupta, S.S., Nagel, K. and Panchapakesan, S.

On the order statistics from equally correlated normal random variables. *Biometrika* (1973), *60*, 403–413.

Gupta, S.S. and Panchapakesan, S.

Multiple Decision Procedures. John Wiley & Sons, Inc., New York, 1979.

On sequential ranking and selection procedures. *Dept. of Statist. Tech. Report* (1988), No. 88–32C, *Purdue University*, West Lafayette.

Sequential ranking and selection procedures. *Handbook of Sequential Analysis* (ed. B.K. Ghosh and P.K. Sen), Marcel Dekker, Inc., New York, 1991, 363–380.

Gupta, S.S. and Sobel, M.

On a statistic which arises in selection and ranking problems. *Ann. Math. Statist.* (1957), *28*, 957–967.

On the smallest of several correlated F–statistics. *Biometrika* (1962), *49*, 509–523.

Hall, P.

Asymptotic theory of triple sampling for estimation of a mean. *Ann. Statist.* (1981), *9*, 1229–1238.

Sequential estimation saving sampling operations. *J. Roy. Statist. Soc., B* (1983), *45*, 219–223.

Hall, W.J.

Some sequential analogs of Stein's two–stage test. *Biometrika* (1962), *49*, 367–378.

Hall, W.J., Wijsman, R.A. and Ghosh, J.K.

The relationship between sufficiency and invariance with applications in sequential analysis. *Ann. Math. Statist.* (1965), *36*, 575–614.

Hamdy, H.I.

Remarks on the asymptotic theory of triple stage estimation of the mean. *Scand. J. Statist.* (1988), *15*, 303–310.

Hamdy, H.I., Mukhopadhyay, N., Costanza, M.C. and Son, M.S.

Triple stage point estimation for the exponential location parameter. *Ann. Inst. Statist. Math.* (1988), *40*, 785–797.

Hartmann, M.

An improvement on Paulson's sequential ranking procedure. *Sequential Analysis* (1988), *7*, 363–372.

An improvement on Paulson's procedure for selecting the population with the largest mean from k normal populations with a common unknown variance. *Sequential Analysis* (1991), *10*, 1–16.

Multi–factor extensions of Paulson's procedures for selecting the best normal population. *Multiple Comparisons, Selection, and Applications in Biometry, C.W. Dunnett Vol.* (ed. F.M. Hoppe), Marcel Dekker, Inc., New York, 1993, 225–245.

Hochberg, Y. and Tamhane, A.C.

Multiple Comparison Procedures. John Wiley & Sons, Inc., New York, 1987.

Hoeffding, W.

A class of statistics with asymptotically normal distribution. *Ann. Math. Statist.* (1948), *19*, 293–325.

Hoel, D.G.

A method for construction of sequential selection procedures. *Ann. Math. Statist.* (1971), *42*, 630–642.

Hoel, D.G. and Mazumder, M.

An extension of Paulson's selection procedure. *Ann. Math. Statist.* (1969), *39*, 2067–2074.

Hogg, R.V. and Craig, A.T.

Introduction to Mathematical Statistics. 4th Edition. Macmillan Publishing Co., Inc., New York, 1978.

Holewijn, P.J.

Note on Weyl's criterion and the uniform distribution of independent random variables. *Ann. Math. Statist.* (1969), *40*, 1124–1125.

Hsieh, H.K.

On estimating the mean of the selected population with unknown variance. *Commun. Statist., Theor. Meth.* (1981), *10*, 1869–1878.

Johnson, N.L. and Kotz, S.

Continuous Univariate Distributions 2. John Wiley & Sons, Inc., New York, 1970.

Distributions in Statistics: Continuous Multivariate Distributions. John Wiley & Sons, Inc., New York, 1972.

Khan, R.A.

On sequential distinguishability. *Ann. Statist.* (1973), *1*, 838–850.

Kimbal, A.W.

On dependent tests of significance in the analysis of variance. *Ann. Math. Statist.* (1951), *22*, 600–602.

Krishnaiah, P.R. and Armitage, P.V.

Distribution of the studentized smallest chi–square with tables and applications. *Report* (1964), No. ARL–64–218. *Aerospace Research Lab*, Wright–Patterson AFB, Dayton.

Percentage points of the multivariate t–distribution. *Report* (1965), No. ARL–65–199. *Aerospace Research Lab*, Wright–Patterson AFB, Dayton.

On a multivariate t distribution. *Essays in Prob. Statist., S.N. Roy Memorial Vol.* (1970), 439–468. Univ. of North Carolina Press, Chapel Hill.

Kuo, L. and Mukhopadhyay, N.

Multi–stage point and interval estimation of the largest mean of k normal populations and the associated second–order properties. *Metrika* (1990a), *37*, 291–300.

Point estimation of the largest location of k negative exponential populations. *Sequential Analysis* (1990b), *9*, 297–304.

Laan, P. Van Der and Verdooren, L. Rob

Selection of populations: An overview and some recent results. *Biom. J.* (1989), *31*, 383–420.

Lai, T.L. and Siegmund, D.

A non–linear renewal theory with applications to sequential analysis I. *Ann. Statist.* (1977), *5*, 946–954.

A non–linear renewal theory with applications to sequential analysis II. *Ann. Statist.* (1979), *7*, 60–76.

Lam, K. and Ng, C.K.

Two–stage procedures for comparing several exponential populations with a control when the scale parameters are unknown and unequal. *Sequential Analysis* (1990), *9*, 151–164.

Landers, D. and Rogge, L.

The exact approximations order in the central–limit–theorem for random summation. *Z. Wahr. Ver. Geb.* (1976), *36*, 269–283.

Lawless, J.F. and Singhal, K.

Analysis of data from life–test experiments under an exponential model. *Naval Res. Logist. Qr.* (1980), *27*, 323–334.

Lehmann, E.L.

Some Model I problems of selection. *Ann. Math. Statist.* (1961), *32*, 990–1012.

A class of selection procedures based on ranks. *Math. Ann.* (1963), *150*, 268–275.

Theory of Point Estimation. John Wiley & Sons, Inc., New York, 1983.

Testing Statistical Hypothesis, 2nd Edition. John Wiley & Sons, Inc., New York, 1986.

Liang, T. and Panchapakesan, S.

A two–stage procedure for selecting δ^*–optimal guaranteed lifetimes in the two–parameter exponential model. *Multiple Comparisons, Selection, and Applications in Biometry, C.W. Dunnett Vol.* (ed. F.M. Hoppe), Marcel Dekker, Inc., New York, 1993, 353–365.

Lohr, S.

Accurate multivariate estimation using triple sampling. *Ann. Statist.* (1990), *18*, 1615–1633.

Lombard, F. and Swanepoel, J.W.H.

On finite and infinite confidence sequences. *S. Afr. Statist. J.* (1978), *12*, 1–24.

Mahalanobis, P.C.

A sample survey of acreage under jute in Bengal, with discussion on planning of experiments. *Proc. 2nd Indian Statist. Conf.* (1940), Statistical Publishing Society, Calcutta.

Martinsek, A.T.

Second order approximation to the risk of a sequential procedure. *Ann. Statist.* (1983), *11*, 827–836.

Miller, D.S. and Mukhopadhyay, N.

Likelihood based sequential procedures for selecting the best exponential population. *Metron* (1990), *48*, 255–282.

Mukhopadhyay, N.

Sequential estimation of location parameter in exponential distributions. *Cal. Statist. Assoc. Bul.* (1974), *23*, 85–95.

Fixed–width confidence intervals for the mean using a three–stage procedure. *Unpublished manuscript* (1976).

Some comments on two–stage selection procedures. *Commun. Statist., A* (1979), *8*, 671–683.

A consistent and asymptotically efficient two–stage procedure to construct fixed–width confidence intervals for the mean. *Metrika* (1980), *27*, 281–284.

Convergence rates of sequential confidence intervals and tests for the mean of a U–statistic. *Commun. Statist., A* (1981), *10*, 2231–2244.

Stein's two–stage procedure and exact consistency. *Skandinavisk Aktuarietdskrift* (1982), 110–122.

Selecting the largest mean through likelihoods. *Statistics and Decisions* (1983a), *1*, 171–182.

Theoretical investigations of some sequential and two–stage procedures to select the larger mean. *Sankhya, A* (1983b), *45*, 346–356.

Sequential and two–stage procedures for selecting the better exponential population covering the case of unknown and unequal scale parameters. *J. Statist. Plan. Inf.* (1984), *9*, 33–43.

A note on three–stage and sequential point estimation procedures for a normal mean. *Sequential Analysis* (1985a), *4*, 311–320.

Asymptotic considerations for selecting the best component of a multivariate normal population. *Metrika* (1985b), *32*, 363–374.

On selecting the best exponential population. *J. Ind. Statist. Assoc.* (1986), *24*, 31–41.

Three–stage procedures for selecting the best exponential population. *J. Statist. Plan. Inf.* (1987), *16*, 345–352.

Sequential estimation problems for negative exponential populations. *Commun. Statist. (Reviews Section), Theory and Methods* (1988), *17*, 2471–2506.

Some properties of a three–stage procedure with applications in sequential analysis. *Sankhya, A* (1990), *52*, 218–231.

Parametric sequential point estimation. *Handbook of Sequential Analysis* (ed. B.K. Ghosh and P.K. Sen), Marcel Dekker, Inc., New York, 1991, 245–267.

Nonlinear renewal theory and beyond: Reviewing the roles in selection and ranking. *Multiple Comparisons, Selection, and Applications in Biometry, C.W. Dunnett Vol.* (ed. F.M. Hoppe), Marcel Dekker, Inc., New York, 1993a, 293–313.

Two–stage and multi–stage estimation. *The Exponential Distribution: Theory, Methods and Applications* (ed. N. Balakrishnan and A.P. Basu), Marcel Dekker, Inc., New York, 1993b.

Mukhopadhyay, N. and Al–Mousawi, J.S.

Fixed–size confidence regions for the mean vector of a multinormal distribution. *Sequential Analysis* (1986), *5*, 139–168.

Mukhopadhyay, N., Chattopadhyay, S. and Sahu, S.K.

Further developments in estimation of the largest mean of k normal populations. *Metrika* (1993), *40*, 173–183.

Mukhopadhyay, N. and Chou, W.S.

Sequential procedures for selecting the largest normal mean when the common variance is unknown. *Unpublished manuscript* (1982a).

Selecting the smallest normal variance through the comparisons of several likelihoods. *Commun. Statist.–Sequential Analysis* (1982b), *1*, 121–144.

On selecting the best component of a multivariate normal population. *Sequential Analysis* (1984), *3*, 1–22.

A further study on selecting the smaller variance: The correlated case. *J. Ind. Statist. Assoc.* (1985), *23*, 45–58.

Mukhopadhyay, N. and Ekwo, M.E.

A note on minimum risk point estimation of the shape parameter of a Pareto distribution. *Cal. Statist. Assoc. Bul.* (1987), *36*, 69–78.

Mukhopadhyay, N. and Hamdy, H.I.

Two–stage procedures for selecting the best exponential population when the scale parameters are unknown and unequal. *Sequential Analysis* (1984a), *3*, 51–74.

On estimating the difference of location parameters of two negative exponential distributions. *Canad. J. Statist.* (1984b), *12*, 67–76.

Mukhopadhyay, N., Hamdy, H.I., Al–Mahmeed, M. and Costanza, M.C.
Three–stage point estimation procedures for a normal mean. *Sequential Analysis* (1987), *6*, 21–36.

Mukhopadhyay, N., Hamdy, H.I. and Darmanto, S.
Simultaneous estimation after selection and ranking and other procedures: The negative exponential case. *Metrika* (1988), *35*, 275–286.

Mukhopadhyay, N. and Hilton, G.F.
Two–stage and sequential procedures for estimating the location parameter of a negative exponential distribution. *S. Afr. Statist. J.* (1986), *20*, 117–136.

Mukhopadhyay, N. and Judge, J.
Second–order expansions for a sequential selection procedure. *Sankhya, A* (1989), *51*, 318–327.

Three–stage procedures for selecting the largest normal mean. *Modern Sequential Statistical Analysis, H. Robbins Vol.* (ed. Z. Govindarajulu), American Sciences Press, Inc., Columbus, 1991, 241–262.

Mukhopadhyay, N. and Kuo, L.
Fixed–width interval estimations of the largest location of k negative exponential populations. *Sequential Analysis* (1988), *7*, 321–332.

Mukhopadhyay, N. and Mauromoustakos, A.
Three–stage estimation for the negative exponential distributions. *Metrika* (1987), *34*, 83–93.

Mukhopadhyay, N. and Miller, D.S.

Sequential likelihood rules for selecting the negative exponential population having the smallest scale parameter. *Cal. Statist. Assoc. Bul.* (1990), *39*, 31–44.

Mukhopadhyay, N. and Panchapakesan, S.

Multistage subset selection procedures and associated second–order asymptotics. *Dept. of Statist. Tech. Report* (1992), No. 92–18, *Univ. of Connecticut*, Storrs.

Mukhopadhyay, N. and Solanky, T.K.S.

Second order properties of accelerated stopping times with applications in sequential estimation. *Sequential Analysis* (1991), *10*, 99–123.

Accelerated sequential procedure for selecting the largest mean. *Sequential Analysis* (1992a), *11*, 137–148.

Accelerated sequential procedure for selecting the best exponential population. *J. Statist. Plan. Inf.* (1992b), *32*, 347–361.

A nonparametric accelerated sequential procedure for selecting the largest center of symmetry. *J. Nonparam. Statist.* (1993), *2*.

Mukhopadhyay, N. and Vik, G.

Asymptotic results for stopping times based on U–statistics. *Sequential Analysis* (1985), *4*, 83–110.

Nagaev, S.V. and Mukhin, A.B.

On a case of convergence to a uniform distribution on an interval. *Limit Theorems and Statist. Inf.* (ed. S. H. Sirazdinov), Tashkent: FAN, 1966, 113–116.

Neyman, J.

Outline of the theory of statistical estimation based on the classical theory of probability. *Phil. Trans. Roy. Soc., London* (1937), 236, 333–380.

Paulson, E.

A sequential procedure for selecting the population with the largest mean from k normal populations. *Ann. Math Statist.* (1964), *35*, 174–180.

Perng, S.K.

A comparison of the asymptotic expected sample sizes of two sequential procedures for ranking problem. *Ann. Math. Statist.* (1969), *40*, 2198–2202.

Pratt, J.W.

Length of confidence intervals. *J. Amer. Statist. Assoc.* (1961), *56*, 549–558.

Raghavachari, M. and Starr, N.

Selection problems for some terminal distributions. *Metron* (1970), *28*, 185–197.

Ramberg, J.S.

A Comparison of the Performance Characteristics of Two Sequential Procedures for Ranking Means of Normal Populations. M.S. Thesis (1966), Dept. of Operations Research, Cornell Univ., Ithaca.

Ramberg, J.S. and Bechhofer, R.E.

A comparison of two multiple decision procedures for ranking means of normal populations with common known variance. *Abstract, IMS Bul.* (1973), *2*, 108.

Rao, C.R.

Linear Statistical Inference and Its Applications, 2nd Edition. John Wiley & Sons, Inc., New York, 1973.

Ray Chaudhuri, A.

A sequential rule for selecting the normal population with the largest mean. *Stat–Math Div. Tech. Report* (1986), No. 27/86, *Indian Statist. Inst.*, Calcutta.

Rinott, Y.

On two–stage selection procedures and related probability–inequalities. *Commun. Statist., A* (1978), *8*, 799–811.

Robbins, H.

Sequential estimation of the mean of a normal population. *Prob. and Statist.* (ed. U. Grenander), Almquist and Wiksell, Uppsala, 1959, 235–245.

Sequential estimation of an integer mean. *Scientists at Work, Festschrift in Honor of H. Wold*, Almquist and Wiksell, Uppsala, 1970.

Robbins, H., Simons, G. and Starr, N.

A sequential analogue of the Behrens–Fisher problem. *Ann. Math. Statist.* (1967), *39*, 88–92.

Robbins, H., Sobel, M. and Starr, N.

A sequential procedure for selecting the best of k populations. *Ann. Math. Statist.* (1968), *39*, 88–92.

Santner, T.J.

A two stage procedure for selecting δ^*–optimal means in the normal model. *Commun. Statist., A* (1976), *5*, 283–292.

Saxena, K.M.L. and Tong, Y.L.

Interval estimation of the largest mean of k normal populations with known variances. *J. Amer. Statist. Assoc.* (1969), *64*, 296–299.

Optimum interval estimation for the largest scale parameter (abstract). *Optimizing Methods in Statistics* (ed. J.S. Rustagi), Academic Press, Inc., New York, 1971, 477.

Sen, P.K.

Sequential Nonparametrics. John Wiley & Sons, Inc., New York, 1981.

Theory and Applications of Sequential Nonparametrics. CBMS–NSF, No. 49. SIAM Publications, Philadelphia, 1985.

Sen, P.K. and Ghosh, M.

Sequential point estimation of estimable parameters based on U–statistics. *Sankhya, A* (1981), *43*, 331–344.

Serfling, R.J.

Approximation Theorems of Mathematical Statistics. John Wiley & Sons, Inc., New York, 1980.

Siegmund, D.

On the asymptotic normality of one–sided stopping times. *Ann. Math. Statist.* (1968), *39*, 1493–1497.

Sequential Analysis. Springer Verlag, Inc., New York, 1985.

Simons, G.

On the cost of not knowing the variance when making a fixed–width confidence interval for the mean. *Ann. Math. Statist.* (1968), *39*, 1946–1952.

Sinha, B.K.

Multivariate problems. *Handbook of Sequential Analysis* (ed. B.K. Ghosh and P.K. Sen), Marcel Dekker, Inc., New York, 1991, 199–228.

Slepian, D.

The one–sided barrier problem for Gaussian noise. *Bell System Tech. J.* (1962), *41*, 463–501.

Smirnov, N.V.

Tables for the Distribution and Density Functions of t–Distribution. Pergamon Press, Oxford, 1961.

Sproule, R.N.

A sequential fixed–width confidence interval for the mean of a U–statistic. *Ph.D. dissertation* (1969), *Univ. of North Carolina,* Chapel Hill.

Asymptotic properties of U–statistics. *Trans. Amer. Math. Soc.* (1974), *199,* 55–64.

Srivastava, M.S.

Some asymptotically efficient sequential procedures for ranking and slippage problems. *J. Roy. Statist. Soc., B* (1966), *28,* 370–380.

Starr, N. and Woodroofe, M.

Further remarks on sequential estimation: The exponential case. *Ann. Math. Statist.* (1972), *43,* 1147–1154.

Stein, C.

A two sample test for a linear hypothesis whose power is independent of the variance. *Ann. Math. Statist.* (1945), *16,* 243–258.

Some problems in sequential estimation (abstract). *Econometrica* (1949), *17,* 77–78.

Swanepoel, J.W.H.

Nonparametric elimination selection procedures based on robust estimators. *S. Afr. Statist. J.* (1977), *11,* 27–41.

Swanepoel, J.W.H. and Geertsema, J.C.

Selection sequences for selecting the best of k normal populations. *S. Afr. Statist. J.* (1973), *7,* 11–21.

Sequential procedures with elimination for selecting the best of k normal populations. *S. Afr. Statist. J.* (1976), *10*, 9–36.

Swanepoel, J.W.H. and vanWyk, J.W.J.

Fixed–width confidence intervals for the location parameter of an exponential distribution. *Commun. Statist., A* (1982), *11*, 1279–1289.

Swanepoel, J.W.H. and Venter, J.H.

A class of elimination selection procedures based on ranks. *S. Afr. Statist. J.* (1975), *9*, 119–128.

Tong, Y.L.

Multi–stage interval estimation of the largest mean of k normal populations. *J. Roy. Statist. Soc., B* (1970), *32*, 272–277.

An asymptotically optimal sequential procedure for the estimation of largest mean. *Ann. Statist.* (1973), *1*, 175–179.

The Multivariate Normal Distribution. Springer–Verlag, Inc., New York, 1990.

Tukey, John W.

The philosophy of multiple comparisons. *Statistical Science* (1991), *6*, 100–116.

Where should multiple comparisons go next? *Multiple Comparisons, Selection, and Applications in Biometry, C.W. Dunnett Vol.* (ed. F.M. Hoppe), Marcel Dekker, Inc., New York, 1993, 187–207.

Vellaisamy, P., Kumar, S. and Sharma, D.

Estimating the mean of the selected uniform population. *Commun. Statist., Theor. Meth.* (1988), *17*, 3447–3476.

Venter, J.H.

Estimation of the mean of the selected population. *Commun. Statist., Theor. Meth.* (1988), *17*, 791–805.

Vik, G. and Mukhopadhyay, N.

Triple sampling to construct fixed–width confidence intervals for estimable parameters based on U–statistics. *Metron* (1988), *46*, 165–174.

Wald, A.

Sequential Analysis. John Wiley & Sons, Inc., New York, 1947.

Wetherill, G.B.

Sequential Methods in Statistics, 2nd Edition. Chapman and Hall, Ltd., London, 1975.

Wiener, N.

The ergodic theorem. *Duke Math.J.* (1939), *5*, 1–18.

Wijsman, R.A.

Stopping times: Termination, moments, distribution. *Handbook of Sequential Analysis* (ed. B.K. Ghosh and P.K. Sen), Marcel Dekker, Inc., New York, 1991, 67–120.

Woodroofe, M.

Second–order approximations for sequential point and interval estimation. *Ann. Statist.* (1977), *5*, 984–995.

Nonlinear Renewal Theory in Sequential Analysis. CBMS–NSF, No. 39. SIAM Publications, Philadelphia, 1982.

Asymptotically optimal sequential point estimation in three stages. *New Perspectives in Theoretical and Applied Statistics* (ed. M.L. Puri), John Wiley & Sons, Inc., New York, 1987, 397–411.

The role of renewal theory in sequential analysis. *Handbook of Sequential Analysis* (ed. B.K. Ghosh and P.K. Sen), Marcel Dekker, Inc., New York, 1991, 145–168.

Zacks, S.

The Theory of Statistical Inference. John Wiley & Sons, Inc., New York, 1971.

Zelen, M.

Applications of exponential models to problems in cancer research. *J. Roy. Statist. Soc.* (1966), *129*, 368–398.

INDEX